OXFORD SERIES ON MATERIALS MODELLING

Series Editors

Oxford Series on Materials Modelling

Materials modelling is one of the fastest growing areas in the science and engineering of materials, both in academe and in industry. It is a very wide field covering materials phenomena and processes that span ten orders of magnitude in length and more than twenty in time. A broad range of models and computational techniques has been developed to model separately atomistic, microstructural and continuum processes. A new field of multi scale modelling has also emerged in which two or more length scales are modelled sequentially or concurrently. The aim of this series is to provide a pedagogical set of texts spanning the atomistic and microstructural scales of materials modelling, written by acknowledged experts. Each book will assume at most a rudimentary knowledge of the field it covers and it will bring the reader to the frontiers of current research. It is hoped that the series will be useful for teaching materials modelling at the postgraduate level.

APS, London
RER, Livermore, California

1. M.W. Finnis: *Interatomic forces in condensed matter*
2. K. Bhattacharya: *Microstructure of martensite - Why it forms and how it gives rise to the shape-memory effects*
3. V.V. Bulatov, W. Cai: *Computer simulations of dislocations*
4. A.S. Argon: *Strengthening mechanisms in crystal plasticity*
5. L.P. Kubin: *Dislocations, mesoscale simulations and plastic flow*
6. A.P. Sutton: *Physics of elasticity and crystal defects*

Forthcoming:
D.N. Theodorou, V. Mavrantzas: *Multiscale modelling of polymers*

From reviews of *Physics of elasticity and crystal defects*

'Sutton is a giant in the field... I am certain this book will be a classic.' *Craig Carter, MIT*

'Superb... and written in an excellent, engaging style... Sutton is an internationally respected expert in structural materials science and condensed matter physics, one of very few people to have such status in these two domains simultaneously.' *T.D. Swinburne, CNRS, Aix-Marseille Université*

'Sutton emphasizes the physical meaning behind the mathematical models he clearly introduces. The style is simple, didactic, and effective. The coverage of some of the Open Questions in Chapter 10 (e.g. electroplasticity) is entirely unique to this book.' *Beñat Gurrutxaga-Lerma, University of Cambridge*

'Although there are other relevant texts in this field, this book includes connections to atomic treatments of defects. These are timely additions, and provide new physical insights. Although the book contains much mathematics, it is essentially readable, and stimulating.' *Sir Peter Hirsch, University of Oxford*

'This is an outstanding book. Students will appreciate the clarity of the arguments, including careful derivations of some important formulas for elasticity.' *Robert Rudd, Series Editor, Oxford Series on Materials Modelling*

'The book is highly accessible, and provides the level of insight into the subject that you would rarely find in academic literature... It is particularly significant that the author has made a clear connection between Physics and Elasticity and Defects in this book. There is an established element of tradition here, where L.D. Landau and E.M. Lifshitz included Theory of Elasticity in their famous Course in Theoretical Physics. This new book by Adrian Sutton matches the Landau-Lifshitz book extremely well, providing new, modern insights into the phenomena, and matching the needs of contemporary generations of students and researchers.' *Sergei Dudarev, UK Atomic Energy Authority*

'It is quite obvious that the majority of the content is material that the author has worked through from scratch, much of it original, and this is especially reflected in the problems, which are detailed and novel.' *Tony Paxton, King's College London*

Also by Adrian P Sutton

Electronic structure of materials, Oxford University Press (1993), 260 pages + xv. Published in German translation as *Elektronische Struktur in Materialen*, by VCH, Weinheim (1996).

Interfaces in crystalline materials, with R W Balluffi, Oxford University Press (1995), 819 pages + xxxi. Reissued in 2006 in the series *Oxford Classic Texts in the Physical Sciences*. Published in Chinese by Higher Education Press Limited, Beijing (2015).

Rethinking the PhD, independently published (2020), 94 pages.

Physics of elasticity and crystal defects

Adrian P. Sutton

OXFORD
UNIVERSITY PRESS

OXFORD
UNIVERSITY PRESS

Great Clarendon Street, Oxford, OX2 6DP,
United Kingdom

Oxford University Press is a department of the University of Oxford. It furthers the University's objective of excellence in research, scholarship, and education by publishing worldwide. Oxford is a registered trade mark of Oxford University Press in the UK and in certain other countries

First Edition published in 2020

Impression: 1

Published in the United States of America by Oxford University Press
198 Madison Avenue, New York, NY 10016, United States of America

British Library Cataloguing in Publication Data
Data available

Library of Congress Control Number: 2019957252

ISBN 978–0–19–886078–5

DOI: 10.1093/oso/9780198860785.001.0001

Printed and bound by
Bell and Bain Ltd, Glasgow

Foreword

All crystalline materials are imperfect. Common defects are vacancies, dislocations and cracks. These affect mechanical properties in important ways—some beneficial, others causing problems. Dislocations confer ductility to crystalline materials, cracks facilitate fracture. Dislocations enable metals and alloys to be mechanically worked, rolled, pressed, drawn etc., to change their shape without cracking, thus increasing immensely their range of use, and enriching our everyday lives. The behaviour of dislocations and cracks is greatly affected by stress, and in order to optimise their effect on mechanical properties, it is important to understand their interaction with stress. This is an important objective of this book. The first part deals with the fundamentals of strain and stress, and discusses the important role of Green's Function in linear elasticity, and introduces Hooke's Law. This is followed by discussion of properties of defects in crystals and their interaction with stress and with each other.

Dislocations are particularly complicated defects, firstly because they are flexible line defects characterised by the atomic displacements (Burgers vector) these cause when moving through the crystal lattice, and whose modes of motion are affected by their directions in the lattice. Secondly, the displacement at the centre of a dislocation results in a long-range elastic stress field which interacts with other dislocations over long distances, while at the centre of the atomic line defect linear elasticity does not apply, but large forces exist which lead to strong interactions with point defects, e.g. solute atoms, causing solution hardening. These short-range interactions require atomic scale quantum mechanical treatments. While such treatments are outside the scope of this book, the author identifies the situations where such treatments are necessary. A further complication is that depending on the crystal lattice and the Burgers vector, such dislocations can dissociate into partial dislocations, with smaller Burgers vector, bounding a stacking fault and lowering their energy.

The mobility of dislocations is important, as it affects the strain-rate at which materials can be deformed. As the dislocation moves from one line in the atomic lattice to the next, bonds are broken and reformed. The mobility is therefore determined by the interatomic forces at the centre (the core) of the dislocation. The fact that diamond is hard and aluminium relatively soft is due to the different interatomic bond strengths in these materials. The classical treatments are discussed carefully by the author. But just as the macroscopic shear of a crystal occurs by the motion of individual dislocations which cause one atomic step displacement at a time, so the individual dislocations do not advance simultaneously along their entire line, but the motion occurs by the creation and movement of atomic scale lengths, called kinks, on the dislocations. So it is these kinks, which are point-type defects, which control mobility. The author includes an example in his book. Sometimes the mobility of a dislocation causing a particular atomic displacement can differ greatly depending on the direction of the dislocation line in the lattice. An extreme example occurs in the body-centred cubic lattice, where transmission electron microscopy observations showed that edge dislocations (Burgers vector normal to the dislocation line) move much more rapidly than screw dislocations (same Burgers vector but parallel to the dislocation line, and to the triad symmetry axis in the crystal). Such screw dislocations can spread out their cores on three planes intersecting along the dislocation. They have to be constricted before movement can take place on a particular atomic slip plane. Such a process is assisted by thermal activation, and results in a sharp rise in yield stress at low temperature, which can cause structural integrity problems under such conditions.

The book includes an excellent discussion of the interaction of dislocations with sharp cracks in the presence of stress. The stress field of a stressed crack is conveniently modelled in terms of arrays of virtual dislocations with infinitesimal Burgers vectors. In the absence of crack tip plasticity, cracks cause brittle fracture at critical stresses, as described in the famous Griffith formula. Such behaviour is found in ceramic crystals in which dislocation mobility is low. But in metals and alloys the plastic deformation generated by the stress concentrations ahead of a stressed sharp crack can shield and blunt the cracks, resulting in stable crack extension and large increases in failure stress. This property is of great importance in the structural integrity of stressed components such as nuclear reactor pressure vessels, or bridges, etc. The variation with temperature of dislocation mobility, and correspondingly the plastic yield stress, gives rise to a brittle–ductile transition with temperature, which affects critically the structural integrity of pressure vessels in nuclear reactors where the yield stress can be impaired by radiation damage.

Much research has been carried out on problems which can be treated by the interactions of individual dislocations with each other, and with other defects. This approach has provided much understanding of the processes occurring in the interaction of dislocations with different Burgers vectors intersecting each other (leading to 'forest hardening'), for which there is much corroborating experimental evidence by transmission electron microscopy. The interaction of dislocations with non-deformable obstacles (e.g. a dispersion of oxide particles in copper) can be solved in a similar way and leads to a simple formula (the Orowan equation) relating yield stress with inter-particle separation (as explained in the text), which can be used to predict the macroscopic yield stress in terms of particle separation.

But other macroscopic mechanical properties, such as workhardening, involve the mutual interactions of many dislocations through their long-range elastic stress field, a many-bodied problem which cannot be treated in this way. Early transmission electron microscope observations showed that in heavily deformed polycrystals (such as aluminium, copper, nickel, etc.) the dislocations tend to form cell structures; in relatively soft materials such as aluminium, the cells are subgrains separated by low-energy low-angle boundaries on the scale of microns. It appears therefore that a mechanism of self-organisation into low-energy structures is operating, very likely facilitated by the ability of screw dislocations to cross-slip from one plane to another and annihilate. The scale at which these substructures occur is not properly understood.

The last chapter of the book on unsolved problems, includes an excellent account by the author of recent theories of workhardening of single crystals. Instead of treating the deformation in terms of the movement of individual dislocations in this many-bodied problem, the unit of deformation is a slip band formed by avalanches of dislocations generated by dislocation sources on individual glide planes. These slip bands are modelled as ellipsoids in which the stress inside the ellipsoid does not affect the primary dislocations forming the slip bands, and producing the shear, but generates dislocations on intersecting slip systems, which stabilise the slip bands and cause 'forest hardening'. It has been shown that such a model gives rise to a linear increase in hardening with increasing strain, as observed experimentally. But, as the author points out, there remain some unanswered questions.

This book provides materials scientists, physicists, and engineers with interest in mechanical properties of crystalline materials with an excellent account of the basic armoury needed to address the many unsolved problems in the mechanical properties of materials. It is beautifully written and the author emphasises the physical basis underlying the mathematical treatment of the various topics, providing insight and understanding. This book should be essential reading for graduate students, and in fact for anyone working in this field.

P.B. Hirsch
22 September 2019

Contents

Preface

My aim in writing this book is to provide mathematically inclined physicists and engineers with an introduction to elasticity, and to take them relatively quickly to advanced concepts and ideas about its application to defects in crystals. The first six chapters are based on a course I gave for ten years in the EPSRC Centre for Doctoral Training (CDT) on Theory and Simulation of Materials at Imperial College London. The last four chapters cover material I would have included had there been more time. The CDT was created in 2009 to attract first class physicists and engineers with a taste for theory into materials science. Some of their research appears in this book.

Some justification is needed for another book in an area that already benefits from excellent texts. By discussing the connections to treatments of defects at the atomic scale I have tried to make the text more appealing to physicists. This approach is also quite novel for engineers, and I suspect that even the treatment of stress in the second chapter contains sections that are new to them. Having given this course to physics graduates from the UK and continental Europe I know that most of the material covered in this course, in many cases all of it, was new to them too. I hope this book treats the subject in a way that will appeal to students with backgrounds in physics and engineering. I hope it will also appeal to materials scientists who would like to see a more mathematical approach to the subject.

Until the late 1960s, in some of the strongest departments of physics around the world, 'metal physics' was focused on the physics of crystal defects and their interactions. But today such metal physics has become unfashionable and it has vanished almost completely from physics departments. The study of metals and alloys by physicists has morphed into functional properties such as superconductivity, plasmonics and magnetism. Undergraduate courses of physics in the UK rarely include anything on defects in crystals, other than dopants in semiconductors. Physics students could be forgiven for thinking the world is made of perfect single crystals. However, the industrial need for an understanding of defect-related mechanisms of deformation and mechanical failures has never been greater. Some major manufacturing companies have recognised they cannot make reliable assessments of the lifetimes of certain components critical for safety unless they understand the defect-related mechanisms that limit their service life. On the other hand, university engineering departments do not have a tradition of thinking about materials at the atomic scale. For example, hydrogen embrittlement is a problem that cuts across swathes of current and proposed technologies, but it has fallen between physics and engineering, and very little progress has been made into the fundamental mechanisms after a century and a half of research. The physics of defects will remain important as long as metals and alloys are used to make things like jet engines, cars, ships, rail track, bridges, skyscrapers, wind turbines and nuclear power stations.

Such a short book as this about such a huge subject cannot be self-contained. I have included a list of recommended books which should be consulted. I have also included references to research papers in footnotes, and a list of them all at the end of the book, with web addresses when I have been able to find them online. Clicking on the URL will take the reader directly to where the paper is located on the web, although most of them are behind paywalls. Exercises and problem sets are included in each chapter, except the last. Some of the problems extend the material covered in the text into more advanced areas, and the reader is guided through them. Solutions to the exercises and problem sets are available free of charge to course instructors at https://global.oup.com/academic/category/science-and-mathematics/physics/solutions/.

The theory of elasticity has a long history going back to Euler in the eighteenth century. Its evolution through the nineteenth and twentieth centuries involved some of the most familiar names in the development of mathematical physics. Brief information is included about these people where it is available. I regret that for some prominent contributors to the subject I could find no information. It will be seen that many of them were Fellows of the Royal Society (FRS), with one President (PRS) and six Foreign Members (ForMemRS).[1] Four recipients of Nobel Prizes in physical sciences also appear in this book. I hope this information will persuade readers that the subject has attracted some of the finest minds in theoretical and mathematical physics, and that it will help to inspire them to make their own contributions to advancing the subject. The subject is still evolving and the final chapter introduces four areas for further research with suggestions for challenging PhD projects.

Since the Second World War the global number of published science papers has doubled approximately every 9 years.[2] In 1665 The Royal Society published the first scientific journal—*The Philosophical Transactions of the Royal Society*. In 2009 it was estimated there were more than 50 million published scholarly papers.[3] Today in 2019 that figure is likely to be more than 100 million, published in tens of thousands of journals. Modern science has become so specialised and fragmented it has become very difficult to raise our eyes above our own narrow furrows of research. I hope this book will provide some satisfaction to readers who feel the need to broaden their horizons.

I am very grateful to Professor Sir Peter Hirsch FRS for writing the Foreword. I thank Luca Cimbaro, Luca Reali, Tchavdar Todorov, Michele Valsecchi and Kang Wang for spotting errors in earlier drafts of the manuscript, and Professor Bob Balluffi for encouragement throughout the writing of this book. Professor Stan Lynch provided helpful comments on an earlier draft of section 10.5. Professors Mick Brown FRS, Tony Paxton and Vasek Vitek read the whole manuscript and provided many helpful suggestions and comments. I am also grateful for useful suggestions from an anonymous reviewer. Remaining errors are entirely my responsibility.

I am grateful to my colleagues in the Department of Physics at Imperial College London who in 2004 took the bold decision to hire a materials scientist, thereby allowing

[1] The Royal Society is the National Academy of Sciences for the UK and the Commonwealth. Founded in 1660 it is the oldest of all the national academies. https://royalsociety.org/

[2] http://blogs.nature.com/news/2014/05/global-scientific-output-doubles-every-nine-years.html

[3] Jinha, AE, Learned Publishing **23**, 258–63 (2010). https://doi.org/10.1087/20100308

a trojan horse into their midst to recruit some of their best students into materials science. Working with these students has been the most fulfilling period of my academic career. I dedicate this book to them.

Imperial College London
August 2019

1 Strain

1.1 The continuum approximation

Condensed matter is lumpy. It comprises very dense atomic nuclei packed a few angstroms apart, with electrons in much less dense clouds between them. The continuum approximation smears out this lumpiness into a uniform, structureless jelly with the same density as a macroscopic lump of the matter it approximates. In addition to its density the continuum is given elastic properties, which characterise how easy it is to deform it in a reversible manner. The elastic properties of the continuum are equated to those of the material it approximates.

What physics do we leave out by approximating the discrete atomic structure of a material with a continuum model? Whenever the discrete atomic structure of the material becomes essential to the physics we can expect the continuum model to be a poor approximation. For example, when we consider structural defects inside the material we can expect the continuum model to become increasingly unreliable as we get closer to the centre of the defect because there the discrete atomic structure of the defect can no longer be ignored. But once we get beyond a few nanometres, in many cases just one nanometre, from the centre of a defect the continuum approximation becomes an accurate description of the distortion the defect generates.

Whereas the smallest separation of atoms in a material provides a natural length scale there is no natural length scale associated with the continuum. This has significant consequences for dynamical properties, such as the propagation of atomic vibrations. There are just three waves that can propagate in the continuum with speeds that vary in general with the direction of propagation. In contrast to vibrations in a crystal the elastic waves are dispersionless, that is, their speed of propagation does not depend on their wavelength. The dispersion of vibrations in a crystal is a direct consequence of its discrete atomic structure. Waves in the crystal and in its continuum representation coincide only in the limit of long wavelengths compared with the spacing of atoms, where the discreteness of the atomic structure no longer plays a significant role in wave propagation.

Physics of elasticity and crystal defects. Adrian P. Sutton, Oxford University Press (2020). © Adrian P. Sutton.
DOI: 10.1093/oso/9780198860785.001.0001

1.2 What is deformation?

When a body changes shape or volume in response to external or internal forces it is *deformed*. In contrast to a rigid rotation or translation, deformation alters distances between points within the body. Materials may undergo changes of volume and shape in response to changes of temperature, applied electric and magnetic fields and other fields including gravitational. In this book we will be concerned principally with deformation created by mechanical forces applied to bodies and by defects within them. The focus of this chapter is the mathematical description of deformation and strain.

The simplest deformation is a homogeneous expansion or contraction where the distance between any two points changes by an amount proportional to their separation in the undeformed state. The word 'homogeneous' here means 'the same everywhere'. When a body is deformed inhomogeneously the deformation depends on position in the undeformed state. The deformation is then a *field*.

Let $\mathbf{X}$ be the position vector of a point in the body before any deformation occurs. Let $\mathbf{X}+\mathrm{d}\mathbf{X}$ be the position vector of a point in the undeformed body infinitesimally close to $\mathbf{X}$. These two points are separated by $|\mathrm{d}\mathbf{X}| = \sqrt{\mathrm{d}X_i \mathrm{d}X_i}$, where X_i is Cartesian component i of $\mathbf{X}$, and summation is implied here and throughout this book whenever subscripts are repeated. Thus, $\mathrm{d}X_i\mathrm{d}X_i$ is shorthand for $\mathrm{d}X_1^2+\mathrm{d}X_2^2+\mathrm{d}X_3^2$.

Suppose $\mathbf{X}$ and $\mathbf{X}+\mathrm{d}\mathbf{X}$ become $\mathbf{x}$ and $\mathbf{x}+\mathrm{d}\mathbf{x}$ in the deformed state. In general $\mathbf{x}$ is a function of $\mathbf{X}$. The chain rule enables us to write down the components of $\mathrm{d}\mathbf{x}$ in terms of the components of $\mathrm{d}\mathbf{X}$:

$$\mathrm{d}x_i = \frac{\partial x_i}{\partial X_j}\mathrm{d}X_j \tag{1.1}$$

The 3×3 matrix $F_{ij} = \partial x_i/\partial X_j$ is called the *deformation tensor*. It follows that the change in the squared separation of the points is given by

$$\mathrm{d}x_i\mathrm{d}x_i - \mathrm{d}X_j\mathrm{d}X_j = \left(\frac{\partial x_i}{\partial X_j}\frac{\partial x_i}{\partial X_k} - \delta_{jk}\right)\mathrm{d}X_j\mathrm{d}X_k, \tag{1.2}$$

where δ_{jk} is the Kronecker delta: $\delta_{jk}=1$ if $j=k$, and $\delta_{jk}=0$ if $j\neq k$.

Exercise 1.1

(i) Prove that the deformation tensor F_{ij} satisfies the tensor transformation law under a rotation of the coordinate system. This is what defines F_{ij} as a tensor.

(ii) Show that in matrix notation eqn. 1.2 becomes

$$(\mathrm{d}\mathbf{x}^T\cdot \mathrm{d}\mathbf{x}) - (\mathrm{d}\mathbf{X}^T\cdot \mathrm{d}\mathbf{X}) = \mathrm{d}\mathbf{X}^T\cdot\left(\mathbf{F}^T\mathbf{F}-\mathbf{I}\right)\cdot \mathrm{d}\mathbf{X},$$

where the T superscript denotes transpose and $\mathbf{I}$ is the identity matrix.

(iii) Hence prove that if the deformation tensor is a rotation the change in the separation of points is zero.

You may recognise $\mathbf{F}^T\mathbf{F}$ as the metric tensor $\mathbf{g}$ which measures the distance between two points in the deformed state. Consider first the distance between two points in the *undeformed* state. If $\mathbf{X} = \mathbf{X}(\lambda)$ is a parametric equation of a path in the undeformed state between two points A and B, where $\lambda = \lambda_A$ and $\lambda = \lambda_B$ respectively, then the distance between these points along the path in the undeformed state is given by

$$
\begin{aligned}
s &= \int_{\lambda=\lambda_A}^{\lambda=\lambda_B} \sqrt{\left(\frac{dX_1}{d\lambda}\right)^2 + \left(\frac{dX_2}{d\lambda}\right)^2 + \left(\frac{dX_3}{d\lambda}\right)^2}\, d\lambda \\
&= \int_{\lambda=\lambda_A}^{\lambda=\lambda_B} \sqrt{\frac{dX_i}{d\lambda} g_{ij} \frac{dX_j}{d\lambda}}\, d\lambda
\end{aligned}
$$

and we see the metric tensor in the undeformed state is the identity matrix.

In the *deformed* state the path becomes $\mathbf{x} = \mathbf{x}(\mathbf{X}(\lambda))$ and points A and B are moved to new positions. The distance between points A and B along the deformed path is given by

$$
s = \int_{\lambda=\lambda_A}^{\lambda_B} \sqrt{dx_k(\lambda) dx_k(\lambda)}\, d\lambda = \int_{\lambda=\lambda_A}^{\lambda_B} \sqrt{\frac{dX_i}{d\lambda} \frac{\partial x_k}{\partial X_i} \frac{\partial x_k}{\partial X_j} \frac{dX_j}{d\lambda}}\, d\lambda \tag{1.3}
$$

and the metric tensor is $g_{ij} = (\partial x_k/\partial X_i)(\partial x_k/\partial X_j)$, or $\mathbf{g} = \mathbf{F}^T\mathbf{F}$ in matrix notation.

1.3 The displacement vector and the strain tensor

We may always express $\mathbf{x}(\mathbf{X})$ as $\mathbf{X} + \mathbf{u}(\mathbf{X})$, where $\mathbf{u}(\mathbf{X})$ is the displacement undergone by a point at $\mathbf{X}$ in the undeformed body when the body is deformed. The gradient of the displacement vector is related to the deformation tensor as follows:

$$
F_{ki} = \frac{\partial x_k}{\partial X_i} = \delta_{ki} + \frac{\partial u_k}{\partial X_i}. \tag{1.4}
$$

Exercise 1.2

Show that the squared separation of points that were at $\mathbf{X}$ and $\mathbf{X} + d\mathbf{X}$ in the undeformed state becomes the following in the deformed state:

$$
(ds)^2 = \left(\delta_{kj} + \frac{\partial u_k}{\partial X_j} + \frac{\partial u_j}{\partial X_k} + \frac{\partial u_i}{\partial X_k}\frac{\partial u_i}{\partial X_j}\right) dX_k dX_j. \tag{1.5}
$$

Equation 1.5 is exact provided $|d\mathbf{X}|$ is infinitesimal, and it leads to nonlinear theories of elasticity. To obtain a linear approximation the assumption is made that the *gradients* of the displacement vector are small in comparison to unity. It is important to recognise that this assumption does not require the displacements themselves to be small. Taking the square root of both sides of eqn. 1.5, and dividing both sides by $|\mathrm{d}\mathbf{X}|$, we obtain

$$\begin{aligned}\frac{\mathrm{d}s}{|\mathrm{d}\mathbf{X}|} &= \sqrt{1 + \hat{l}_k \left(\frac{\partial u_k}{\partial X_j} + \frac{\partial u_j}{\partial X_k} + \frac{\partial u_i}{\partial X_k} \frac{\partial u_i}{\partial X_j} \right) \hat{l}_j} \\ &\approx 1 + \hat{l}_k e_{kj} \hat{l}_j, \end{aligned} \tag{1.6}$$

where the unit vector $\hat{\mathbf{l}}$ is parallel to $\mathrm{d}\mathbf{X}$ and e_{kj} is the *strain tensor*:

$$e_{kj} = \frac{1}{2} \left(\frac{\partial u_k}{\partial X_j} + \frac{\partial u_j}{\partial X_k} + \frac{\partial u_i}{\partial X_k} \frac{\partial u_i}{\partial X_j} \right). \tag{1.7}$$

Since the displacement gradients are assumed to be small the last term in eqn. 1.7 is neglected and thus we obtain the strain tensor used in linear elasticity, and in the rest of this book:

$$e_{kj} = \frac{1}{2} \left(\frac{\partial u_k}{\partial X_j} + \frac{\partial u_j}{\partial X_k} \right). \tag{1.8}$$

We see the strain tensor is symmetric. The displacement gradient may also contain an asymmetric part:

$$\frac{\partial u_k}{\partial X_j} = e_{kj} + \omega_{kj},$$

where $\omega_{kj} = \frac{1}{2} \left(\partial u_k / \partial X_j - \partial u_j / \partial X_k \right)$ is asymmetric because $\omega_{kj} = -\omega_{jk}$.

Exercise 1.3

Given the following general expression for the rotation matrix describing a rotation by θ about an axis $\hat{\rho}$:

$$R_{ij} = \cos\theta \delta_{ij} + \hat{\rho}_i \hat{\rho}_j (1 - \cos\theta) - \varepsilon_{ijk} \hat{\rho}_k \sin\theta,$$

where ε_{ijk} is the permutation tensor, show that in the limit $\theta \to 0$ the rotation matrix becomes

$$R_{ij} = \delta_{ij} - \varepsilon_{ijk} \hat{\rho}_k \theta.$$

Hence show that the axis of the rotation described by the three independent components of ω_{kj} is parallel to $\mathrm{curl}\,\mathbf{u}$.

1.3.1 Normal strain and shear strain

The diagonal and off-diagonal components of the strain tensor of eqn. 1.8 are called *normal* and *shear* strains respectively. An example of a normal strain in two dimensions is illustrated in Fig. 1.1 (a), for which the deformation tensor is

$$\mathbf{F} = \begin{bmatrix} 1+\varepsilon & 0 \\ 0 & 1 \end{bmatrix}$$

and the corresponding strain tensor is

$$\mathbf{e} = \begin{bmatrix} \varepsilon & 0 \\ 0 & 0 \end{bmatrix}.$$

A shear strain arises when the displacement in a particular direction varies with a perpendicular distance. These two types of strain are illustrated in two dimensions in Fig. 1.1.

Two common types of shear strain are a *pure shear* and a *simple shear*, both of which are illustrated in two dimensions in Fig. 1.1. Figure 1.1 (b) illustrates the pure shear described

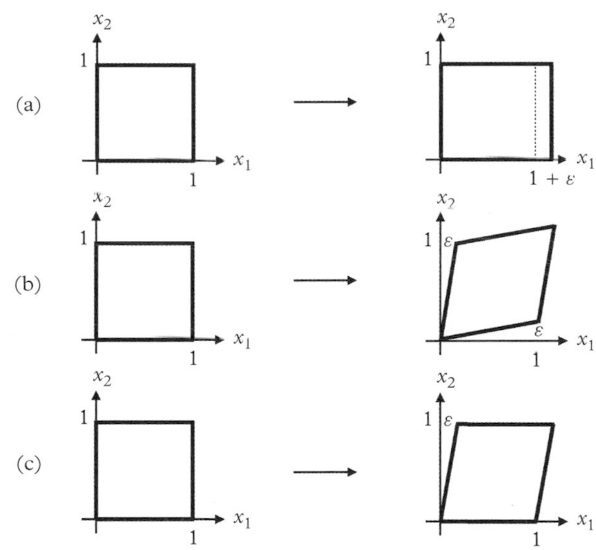

Figure 1.1 *Illustrations in two dimensions of (a) normal strain along x_1 of magnitude ε, (b) pure shear strain of magnitude ε and (c) simple shear strain of magnitude ε. In each case the unit square on the left is deformed into the shape on the right by the corresponding strain.*

by the following deformation tensor:

$$\mathbf{F} = \begin{bmatrix} 1 & \varepsilon \\ \varepsilon & 1 \end{bmatrix}$$

for which the corresponding strain tensor is

$$\mathbf{e} = \begin{bmatrix} 0 & \varepsilon \\ \varepsilon & 0 \end{bmatrix}.$$

The symmetric nature of the deformation tensor ensures it is a pure shear because no rotations are involved.

A simple shear involves a pure shear and a rotation. An example of a simple shear is illustrated in Fig. 1.1(c), for which the deformation tensor is

$$\mathbf{F} = \begin{bmatrix} 1 & \varepsilon \\ 0 & 1 \end{bmatrix}.$$

This may be decomposed into a pure shear strain of $\varepsilon/2$ and a rotation by $\theta = \varepsilon/2$ about an axis normal to the page:

$$\mathbf{F} = \begin{bmatrix} 1 & 0 \\ 0 & 1 \end{bmatrix} + \begin{bmatrix} 0 & \varepsilon/2 \\ \varepsilon/2 & 0 \end{bmatrix} + \begin{bmatrix} 0 & \varepsilon/2 \\ -\varepsilon/2 & 0 \end{bmatrix}.$$

Simple shears occur in *mechanical twinning*, which is a mechanism of deformation of many crystalline materials.

1.4 Closing remarks

The normal strain illustrated in Fig. 1.1(a) changes the volume of the material. The shear strains illustrated in Fig. 1.1(b) and (c) do not change the volume, but unless the material is a liquid[1] there will be an elastic resistance to these deformations. Young[2] was the first to consider shear as an elastic strain in 1807. He noticed that the elastic resistance of a body to shear is different from its resistance to extension or compression, but he did not introduce a separate elastic modulus to characterise the rigidity to shear. The famous 'Young's modulus' refers only to the rigidity of the material to

[1] *Real* liquids do resist shear deformations in a time-dependent manner through their viscosities. We are thinking here of the response of the material after a long period of time when any time-dependent relaxation processes have finished. An *ideal* liquid has no viscosity and displays no resistance to shear stresses.

[2] Thomas Young FRS, 1773–1829, British physicist, physician, optician, linguist, Egyptologist, musician. For an engrossing biography of this exceptional polymath see Robinson, A, *The last man who knew everything*, One World: Oxford (2006). ISBN 978-0452288058

elastic extension or compression. Love[3] commenting[4] on Young's introduction of his modulus wrote

> This introduction of a definite physical concept, associated with the coefficient of elasticity which descends, as it were from a clear sky, on the reader of mathematical memoirs, marks an epoch in the history of the science.

We will come back to the moduli of elasticity in Chapter 3.

It should be noted that the concepts of the deformation tensor and the strain tensor rest on the existence of a reference state of the material where the deformation is sensibly regarded as zero. In a crystalline material the perfect crystal itself is a natural choice for the reference state. But in a glass or amorphous material there is no obvious choice of a reference state, and the concept of strain is much less useful. However, as we shall see in the next chapter the concept of stress is just as applicable in an amorphous material as it is in a crystal.

1.5 Problem set 1

1. Prove that the strain tensor of eqn. 1.8 satisfies the tensor transformation law under a rotation of the coordinate system.
2. Under a homogeneous strain the displacement of a point at $\mathbf{X}$ in the undeformed body is given by $\mathbf{u}(\mathbf{X}) = \mathbf{e} \cdot \mathbf{X}$, where $\mathbf{e}$ is a constant symmetric matrix, in which the elements are small compared to unity. Consider two points $\mathbf{X}^{(1)}$ and $\mathbf{X}^{(2)}$ in the undeformed body. Show that the change in the separation of the two points in the homogeneously deformed state to first order in the strain is given by $|\mathbf{X}^{(2)} - \mathbf{X}^{(1)}|\, l_i e_{ij} l_j$, where l_i is the unit vector $(X_i^{(2)} - X_i^{(1)})/|\mathbf{X}^{(2)} - \mathbf{X}^{(1)}|$.
3. Prove that the trace of the strain tensor, e_{kk}, is invariant with respect to rotations of the coordinate system.
4. Recall the eigenvectors of a symmetric matrix are orthogonal and the eigenvalues are real numbers. Consider a unit sphere $\mathbf{X}^T \cdot \mathbf{X} = 1$ embedded in a body before it is deformed. The body is subjected to a *homogeneous* strain $\mathbf{e}$. Show that the sphere becomes an ellipsoid with its axes aligned along the eigenvectors of the strain tensor. This ellipsoid is called the *strain ellipsoid*. If the eigenvalues of the strain tensor are λ_1, λ_2 and λ_3 determine the equation of the ellipsoid. These eigenvalues are called principal strains. To first order in the strain show that the ratio of the change in the volume of the sphere to its original volume is equal to the trace of the strain tensor. The trace of the strain tensor is called the dilation. Sketch the sphere and the ellipsoid, and for an arbitrarily chosen $\mathbf{X}$, show $\mathbf{x}$ and $\mathbf{u}$ in your sketch.

[3] Augustus Edward Hough Love FRS, 1863–1940, British physicist.

[4] Love, AEH, *A treatise on the mathematical theory of elasticity*, Dover: New York (1944), p.4. ISBN 0-486-60174-9

2

Stress

2.1 What is stress?

Imagine you are stretching an elastic band between your fingers by applying an equal and opposite tensile force F at either end. The elastic band has reached a new stable length and you keep your hands in a fixed position. There is no net force acting on any element of the elastic band. If that were not true the elastic band would not be in equilibrium and some further displacement would take place. But if you release the elastic band from one end it immediately shrinks back to its natural length.

We say 'the elastic band is under tension' to describe its state when it is stretched. If the stretched elastic band were cut anywhere between your fingers the tension would be released. The tension is transmitted across every transverse plane within the elastic band. The atoms on the left side of every transverse plane in the elastic band are exerting forces on the atoms of the right hand side. Conversely the atoms on the right hand side of every transverse plane are exerting forces on the left hand side. The resultant forces exerted by atoms on each side on the other side are equal and opposite when the elastic band is in equilibrium.

Suppose we stretch a thicker elastic band of the same material with the same force F. Obviously it will not stretch as much as the first elastic band. The task of transmitting the tension F across every transverse plane is shared by more atoms on either side of the plane. It follows that the area of the transverse plane is just as significant as the tension F in characterising the internal mechanical state of the elastic band. The concept of stress brings together the force F and the area of the plane on which it acts:

> The stress acting on an element of area of a plane within a body is defined as the resultant force exerted by atoms on one side of the plane on atoms on the other side of the plane, where lines connecting those atoms pass through the element of area. The resultant force is divided by the area of the element through which it acts to yield the stress.

This is illustrated in Fig. 2.1. This atomic-level definition of stress was developed by Cauchy[1] and Saint-Venant[2] in the nineteenth century before the existence of atoms

[1] Baron Augustin-Louis Cauchy ForMemRS 1789–1857, French mathematician and physicist.
[2] Adhémar Jean Claude Barré de Saint-Venant 1797–1886, French engineer and mathematician.

Physics of elasticity and crystal defects. Adrian P. Sutton, Oxford University Press (2020). © Adrian P. Sutton.
DOI: 10.1093/oso/9780198860785.001.0001

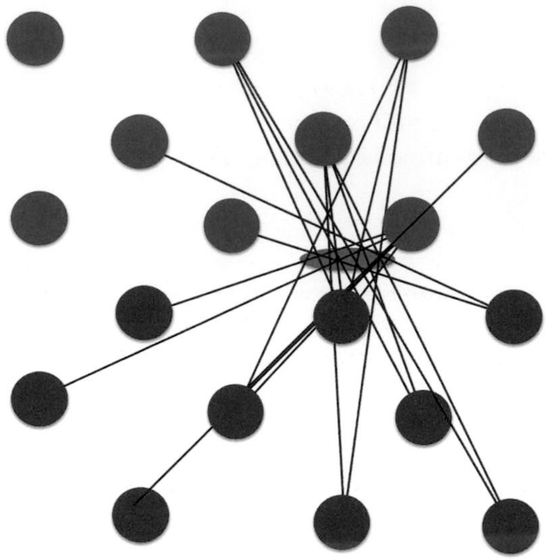

Figure 2.1 *To illustrate the Cauchy–Saint-Venant definition of stress at the atomic scale. The green shaded area is a circle viewed in perspective. The stress acting on it is defined by those forces exerted by blue atoms on red atoms, where the line between the centres of each pair of blue and red atoms passes through the green shaded area. The total force exerted through the shaded area becomes proportional to its size as its size increases. This is the continuum limit.*

was universally accepted, and long before any real understanding of interatomic forces was developed.[3] Given the lack of knowledge about interatomic forces in the 1820s Cauchy developed a pragmatic continuum definition of stress, which is the way most engineers think about stress today. However, physicists have continued to develop atomic-level descriptions of stress up to the present day. In this chapter we will discuss both approaches.

The concept of pressure is closely related to stress: they both have dimensions of force per unit area, with units newtons per square metre, which are called pascals. The key difference is that with pressure the force acting on the plane is always normal to the plane, whereas with stress the force can be inclined to the plane normal and it can even be in the plane. The tensorial nature of stress then becomes apparent because it depends on the direction of the resultant force *and* the direction of the normal to the plane on which it acts.

2.2 Cauchy's stress tensor in a continuum

Cauchy's definition of stress is simple but brilliant. He considered the force $\mathbf{f}$ per unit area acting on a plane in a continuum with unit normal $\hat{\mathbf{n}}$. Not knowing about the spatial range

[3] See Timoshenko, SP, *History of strength of materials*, Dover Publications: New York (1983), p.108. ISBN 0-486-61187-6. Stephen Prokofievitch Timoshenko ForMemRS 1878–1972, Ukrainian born US engineer.

and nature of atomic interactions this force was assumed to have a negligible range and to exist only when the regions of the continuum on either side of the plane were in direct contact. If the plane is not flat its normal will vary with position. Apart from corners, where the normal changes discontinuously, it is always possible to define an infinitesimal element of area where the normal is constant. The concept of the continuum makes all this possible.

Consider a body in mechanical equilibrium but loaded in some arbitrary way, for example by forces applied to its surfaces. How do we calculate the infinitesimal force $\mathbf{df}$ per unit area acting on an infinitesimal area $\mathrm{d}S$ with normal $\hat{\mathbf{n}}$ at any point inside the body? At first sight this might appear to be a hopeless task because there are an infinite number of directions of the plane normals. Cauchy showed it can be done by defining a tensor field of just six independent components, the stress field, inside the body.

Define a right-handed global Cartesian coordinate system in the body, with axes x_1, x_2, x_3 and an arbitrary origin. Consider the plane through the point $\mathbf{x}$ with normal along the positive x_j direction. Let $\mathrm{d}S$ be an infinitesimal element of area of this plane at $\mathbf{x}$. Then component i of the force per unit area acting on $\mathrm{d}S$ is the component $\sigma_{ij}(\mathbf{x})$ of the Cauchy stress tensor $\boldsymbol{\sigma}$.

We will show that the force per unit area acting on an infinitesimal element of area with an *arbitrary* outward unit normal $\hat{\mathbf{n}}$ at $\mathbf{x}$ can be expressed in terms of the stress tensor components $\sigma_{ij}(\mathbf{x})$ and $\hat{\mathbf{n}}$. Let $\hat{\mathbf{n}} = [n_1, n_2, n_3]$ in this coordinate system. A plane with this normal has intercepts $1/n_1, 1/n_2, 1/n_3$ along the x_1, x_2, x_3 axes respectively, which is the plane ABC in Fig. 2.2. The area of the triangle ABC is $1/(2n_1n_2n_3)$.

The plane ABC is parallel to the infinitesimal area element at $\mathbf{x}$. At equilibrium the net force acting on the body OABC must be zero. Component i of the force acting on the face OBC is $-\sigma_{i1} \times 1/(2n_2n_3)$, where the negative sign is because the outward normal to the face OBC is along $-x_1$, and $1/(2n_2n_3)$ is the area of OBC. Similarly, component i of

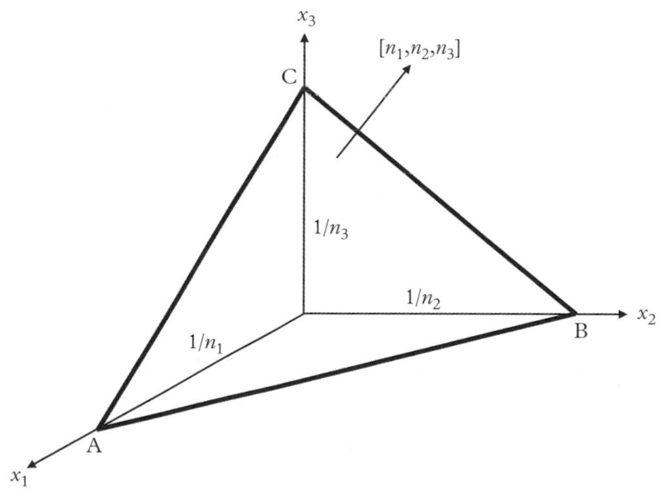

Figure 2.2 *The triangle ABC has unit normal* $\hat{\mathbf{n}} = [n_1, n_2, n_3]$ *and area* $1/(2n_1n_2n_3)$.

the forces acting on the faces OAC and OAB are $-\sigma_{i2} \times 1/(2n_1n_3)$ and $-\sigma_{i3} \times 1/(2n_1n_2)$ respectively. If $\mathbf{f}$ is the force per unit area acting on the plane ABC then equilibrium of OABC requires

$$\frac{f_i}{2n_1n_2n_3} - \frac{\sigma_{i1}}{2n_2n_3} - \frac{\sigma_{i2}}{2n_1n_3} - \frac{\sigma_{i3}}{2n_1n_2} = 0,$$

from which it follows that

$$f_i = \sigma_{i1}n_1 + \sigma_{i2}n_2 + \sigma_{i3}n_3,$$

or

$$f_i = \sigma_{ij}n_j. \tag{2.1}$$

This equation gives the force per unit area acting on a plane with normal $\hat{\mathbf{n}}$ in terms of the components of the stress tensor. In Exercise 2.2 the requirement that there is no torque acting on a volume element in the body leads to the condition $\sigma_{ij} = \sigma_{ji}$, that is, the stress tensor is symmetric. Thus, there are only six independent components of the stress tensor. Equation 2.1 also shows that at equilibrium the forces per unit area acting on either side of a plane are equal and opposite because the sense of the plane normal reverses on either side. Finally, if σ_{ij} varies with position it becomes a stress field and the force per unit area acting on planes with a given normal also varies with position.

Exercise 2.1

Using eqn. 2.1 prove that σ_{ij} satisfies the tensor transformation law under a rotation of the coordinate system.

Exercise 2.2

(a) Sketch a cube of side length equal to unity with edges parallel to the axes Ox_1, Ox_2, Ox_3. Each face has unit area. As a result of a stress field each face experiences a force. On your sketch show the force acting on each face as an arrow and label it in terms of the components of the stress tensor. For example, on the face with normal parallel to the positive x_1 direction the three components of the force are $(\sigma_{11}, \sigma_{21}, \sigma_{31})$ with respect to the x_1, x_2 and x_3 axes respectively. But on the face with normal parallel to the negative x_1-axis the three components of the force are $(-\sigma_{11}, -\sigma_{21}, -\sigma_{31})$, so their arrows point in the opposite directions to those on the first face.

(b) By taking moments of the forces about the centre of the cube show that the condition for there to be no torque on a volume element is $\sigma_{ij} = \sigma_{ji}$.

2.3 Normal stresses and shear stresses

We have seen that the force per unit area acting on an element of area of a plane may have components in the plane as well as normal to it. The components parallel to the plane give rise to *shear* stresses, and they correspond to off-diagonal components of the stress tensor. The components normal to the plane are called *normal* stresses, and they correspond to diagonal components of the stress tensor. However, the designation of normal or shear stress may change under a rotation of the coordinate system, as shown in Exercise 2.3. Tensile and compressive stresses are normal stresses. Shear stresses arise in frictional sliding and they play a central role in plastic deformation of crystalline materials.

Exercise 2.3

A stress σ has the following representation in a Cartesian coordinate system $Ox_1x_2x_3$:

$$\sigma = \begin{bmatrix} -s & 0 & 0 \\ 0 & s & 0 \\ 0 & 0 & 0 \end{bmatrix}.$$

Although this matrix has only diagonal elements it must correspond to a pure shear because the trace of the matrix is zero. When σ is represented in a rotated coordinate system $Ox_1'x_2'x_3'$, obtained by rotating Ox_1 and Ox_2 in a positive sense by $\pi/4$ about Ox_3, show that its matrix representation becomes

$$\sigma = \begin{bmatrix} 0 & s & 0 \\ s & 0 & 0 \\ 0 & 0 & 0 \end{bmatrix}.$$

2.4 Stress at the atomic scale

After studying Cauchy's analysis of stress in terms of forces between elastic continua in contact it may be surprising that the concept of stress can be developed at the atomic scale involving interactions between discrete atoms. An early treatment of stress at the atomic scale may be found in Note B, p.616 of Love's treatise of 1927, albeit for a perfect crystal. The treatment we shall give here is more general, but it does assume the total force on an atom may be expressed as a sum of contributions from surrounding atoms.

Consider a cluster of atoms which may be subject to forces exerted by atoms outside the cluster. The cluster may be any size, from molecules to macroscopic components. The potential energy of the cluster is defined by the energy of interaction between all atoms in the cluster, and between atoms in the cluster and those outside it. Atoms in the cluster may be in a perfect crystal configuration, or a defective crystal or an amorphous state. Let $\mathbf{f}^{(a)}$ be the total force acting on atom a inside the cluster. We assume this force comprises a sum of forces exerted by other atoms in the cluster, and by any atoms outside the cluster. Let $\mathbf{f}^{(m/a)}$ be the force exerted by atom m upon atom a, where atom m may be inside or outside the cluster. Then $\mathbf{f}^{(a)} = \sum_{m \neq a} \mathbf{f}^{(m/a)}$.

Let the position of atom a be $\mathbf{X}^{(a)}$. Imagine all atoms inside the cluster are displaced by infinitesimal amounts $\delta \mathbf{X}^{(a)}$. Different atoms may be displaced in different directions. The change in the total potential energy of the cluster is

$$\begin{aligned} \delta \mathcal{V} &= -\sum_a \mathbf{f}^{(a)} \cdot \delta \mathbf{X}^{(a)} \\ &= -\frac{1}{2} \sum_a \sum_{m \neq a} \mathbf{f}^{(m/a)} \cdot \delta\left(\mathbf{X}^{(a)} - \mathbf{X}^{(m)}\right), \end{aligned} \tag{2.2}$$

where the sum over a is taken over all atoms inside the cluster and the sum over m is taken over all atoms inside and outside the cluster. The factor of $\frac{1}{2}$ takes into account the sharing of the interaction between atoms a and m. Thus, eqn. 2.2 includes all interactions between atoms inside the cluster and between atoms inside and outside the cluster. The latter determine the forces exerted on the cluster by the surrounding medium.

Suppose the infinitesimal displacements of atoms are a result of the application of a homogeneous, infinitesimal strain δe_{ij} applied to all atoms inside and outside the cluster.[4] Since the applied strain is homogeneous and infinitesimal the change of $(X_i^{(a)} - X_i^{(m)})$ is $\delta e_{ij}(X_j^{(a)} - X_j^{(m)})$. Therefore the change in potential energy of the cluster in eqn. 2.2 becomes the following:

$$\delta \mathcal{V} = -\frac{1}{2} \sum_a \sum_{m \neq a} f_i^{(m/a)} \delta e_{ij} \left(X_j^{(a)} - X_j^{(m)}\right). \tag{2.3}$$

We may use eqn. 2.3 to *define* atomic level stresses:

$$\delta \mathcal{V} = \sum_a \Omega^{(a)} \sigma_{ij}^{(a)} \delta e_{ij}, \tag{2.4}$$

where $\Omega^{(a)}$ and $\sigma_{ij}^{(a)}$ are the volume associated with atomic site a and the stress tensor associated with atomic site a inside the cluster. Comparing with eqn. 2.3 we obtain

$$\sigma_{ij}^{(a)} = \frac{1}{2\Omega^{(a)}} \sum_{m \neq a} f_i^{(m/a)} \left(X_j^{(m)} - X_j^{(a)}\right), \tag{2.5}$$

where the sum over m is taken over atomic sites inside and outside the cluster. The atomic volume $\Omega^{(a)}$ may be defined by a Voronoi construction.

[4] When a homogeneous strain is applied to the cluster it also has to be applied to atoms outside the cluster, otherwise the changes of separation between atoms inside and outside the cluster will be incorrect because the cluster will have changed shape and/or volume but the surrounding medium will not have changed.

In eqn. 2.5 it is clear that even if the net force on an atom is zero it may still be in a state of stress. For example, the atoms inside the cluster may be in a perfect crystal environment where each atom experiences zero net force, but the cluster is subjected to compression or tension through forces exerted by atoms outside the cluster that appear in eqn. 2.3.

Provided no heat is exchanged between the cluster and its surroundings as a result of the application of the strain the change in the potential energy of the cluster is equal to the change in its internal energy, δE. Then in a continuum approximation eqn. 2.4 becomes the following volume integral over the cluster:

$$\delta E = \sum_n \Omega^{(n)} \sigma_{ij}^{(n)} \delta e_{ij} \approx \int \mathrm{d}^3 X \sigma_{ij}(\mathbf{X}) \delta e_{ij}(\mathbf{X}),$$

which leads to a new definition of stress in a continuum as the following functional derivative:

$$\sigma_{ij}(\mathbf{X}) = \frac{\delta E}{\delta e_{ij}(\mathbf{X})}, \tag{2.6}$$

where the variation is carried out adiabatically, that is, at constant entropy. The definition of stress in eqn. 2.6 is based on the existence of a strain energy function describing the potential energy of the body as a function of a homogeneous elastic strain applied to it. The notion of a strain energy function was introduced by Green[5] in 1837 and put on a rigorous thermodynamic foundation by Thomson[6] in 1855. This definition of stress appears to be quite different from Cauchy's definition, but they are equivalent.

Not all models of atomic interactions enable the total force on an atom to be expressed as a sum of contributions from surrounding atoms. In quantum mechanics the Ehrenfest–Hellmann–Feynman force on an atomic nucleus depends on the self-consistent electronic charge density at the nucleus. The self-consistent charge density at the nucleus depends on the positions of surrounding atoms in a way that cannot be broken down into a sum of separate contributions from each surrounding atom. However, the definition of stress in eqn. 2.6 may be applied to all models of atomic interactions.

2.5 Invariants of the stress tensor

Let $\hat{\mathbf{e}}_1, \hat{\mathbf{e}}_2, \hat{\mathbf{e}}_3$ be unit vectors along the right-handed Cartesian coordinate system x_1, x_2, x_3. Let $\hat{\mathbf{e}}'_1, \hat{\mathbf{e}}'_2, \hat{\mathbf{e}}'_3$ be unit vectors along the right-handed Cartesian coordinate system x'_1, x'_2, x'_3. The rotation matrix which rotates $\hat{\mathbf{e}}_1, \hat{\mathbf{e}}_2, \hat{\mathbf{e}}_3$ into $\hat{\mathbf{e}}'_1, \hat{\mathbf{e}}'_2, \hat{\mathbf{e}}'_3$ has components $R_{ij} = (\hat{\mathbf{e}}'_i \cdot \hat{\mathbf{e}}_j)$, that is, $\hat{\mathbf{e}}'_i = R_{ij}\hat{\mathbf{e}}_j$. Since stress is a second rank tensor it satisfies the following transformation law:

$$\sigma'_{jk} = R_{ji} R_{kp} \sigma_{ip}, \tag{2.7}$$

[5] George Green 1793–1841, British mathematical physicist and miller.

[6] Sir William Thomson FRS OM 1824–1907, Scots-Irish mathematical physicist and engineer, who became Lord Kelvin in 1892.

where $\boldsymbol{\sigma}'$ and $\boldsymbol{\sigma}$ are the matrix representations of the stress tensor in the primed and unprimed coordinate systems respectively.

Since the components of the matrix representing the stress tensor change under a coordinate transformation, individual stress tensor components have limited physical significance. However, there are three quantities that remain invariant under a rotation of the coordinate system, and they can be used to construct more physically significant quantities. The key to identifying invariants of the stress tensor is to recall that the eigenvalues do not depend on the choice of coordinate system. It follows that the cubic polynomial that defines the eigenvalues must have the same coefficients in all coordinate systems. Let the eigenvalues be s_1, s_2, s_3. These eigenvalues are called principal stresses. When the coordinate system is aligned with the three eigenvectors of the matrix representing the stress tensor, the matrix becomes diagonal with s_1, s_2, s_3 along the leading diagonal. The cubic polynomial defining these eigenvalues is

$$(s-s_1)(s-s_2)(s-s_3)=0,$$

or

$$s^3-(s_1+s_2+s_3)s^2+(s_1s_2+s_2s_3+s_3s_1)s-s_1s_2s_3=0.$$

Therefore, the following three quantities are invariants:

$$\begin{aligned} I_1 &= s_1+s_2+s_3 \\ I_2 &= s_1s_2+s_2s_3+s_3s_1 \\ I_3 &= s_1s_2s_3. \end{aligned} \tag{2.8}$$

Any quantity that may be expressed in terms of these invariants is also invariant. I_1 is the trace of the stress tensor, $\mathrm{Tr}\boldsymbol{\sigma}$. The hydrostatic stress is defined as the average normal stress, which is $I_1/3$. The hydrostatic pressure, p, is the negative of the hydrostatic stress:

$$p=-\frac{1}{3}\mathrm{Tr}\boldsymbol{\sigma}. \tag{2.9}$$

The second and third invariants may be expressed in any coordinate system as follows:

$$I_2=\frac{1}{2}\left[(\mathrm{Tr}\boldsymbol{\sigma})^2-\mathrm{Tr}\boldsymbol{\sigma}^2\right] \tag{2.10}$$

$$I_3=\frac{1}{6}\left[(\mathrm{Tr}\boldsymbol{\sigma})^3+2\mathrm{Tr}\boldsymbol{\sigma}^3-3(\mathrm{Tr}\boldsymbol{\sigma})(\mathrm{Tr}\boldsymbol{\sigma}^2)\right]. \tag{2.11}$$

We note also that I_3 is the determinant of the stress tensor.

Stress invariants are useful for characterising the stress fields of defects in crystals, for example grain boundaries, because they are independent of the coordinate system.

Strain energy functions are also often expressed in terms of stress invariants for similar reasons. The strain tensor has equivalent invariants.

2.6 Shear stress on a plane and the von Mises stress invariant

In the previous section we saw that the hydrostatic stress is an invariant of the stress tensor. As the average of the three normal stresses in any coordinate system it is a scalar quantity which indicates the degree of compression or tension. It is useful to have another invariant quantity that measures the shear content of the stress tensor. This is somewhat more difficult because shear stresses, unlike hydrostatic stresses, depend on the normal of the plane where they act. Therefore, we begin this section by evaluating the magnitude of the shear stress acting on any plane for an arbitrary stress tensor.

Consider a stress tensor σ_{ij} with eigenvalues s_1, s_2, s_3 and corresponding unit eigenvectors $\hat{\mathbf{e}}_1, \hat{\mathbf{e}}_2, \hat{\mathbf{e}}_3$. The eigenvectors form an orthonormal set, which defines the Cartesian coordinate system we shall use. Consider a plane with unit normal $\hat{\mathbf{n}} = n_i\hat{\mathbf{e}}_i$. The force per unit area acting on this plane is $\mathbf{f} = s_1 n_1 \hat{\mathbf{e}}_1 + s_2 n_2 \hat{\mathbf{e}}_2 + s_3 n_3 \hat{\mathbf{e}}_3$. The magnitude of the component of $\mathbf{f}$ along the normal $\hat{\mathbf{n}}$ is $f_n = \mathbf{f}\cdot\hat{\mathbf{n}} = s_1 n_1^2 + s_2 n_2^2 + s_3 n_3^2$. Therefore, the force per unit area normal to the plane is $\mathbf{f_n} = f_n\hat{\mathbf{n}}$. The force per unit area parallel to the plane is $\mathbf{f_p} = \mathbf{f} - (\mathbf{f}\cdot\hat{\mathbf{n}})\hat{\mathbf{n}}$. Thus,

$$\mathbf{f_p} = [s_1 n_1 - f_n n_1, s_2 n_2 - f_n n_2, s_3 n_3 - f_n n_3].$$

We may obtain a useful expression for the square of the magnitude of $\mathbf{f_p}$ as follows (the summation convention is temporarily suspended to derive eqn. 2.12):

$$\begin{aligned}
f_p^2 &= \sum_i (s_i n_i - f_n n_i)^2 \\
&= \sum_i n_i^2 s_i^2 - \sum_i s_i n_i^2 \sum_j s_j n_j^2 \\
&= \sum_i n_i^2 s_i^2 \sum_j n_j^2 - \sum_i s_i n_i^2 \sum_j s_j n_j^2 \\
&= \sum_i \sum_j n_i^2 n_j^2 s_i^2 - n_i^2 n_j^2 s_i s_j \\
&= \frac{1}{2} \sum_i \sum_j n_i^2 n_j^2 (s_i^2 + s_j^2 - 2 s_i s_j) \\
&= \sum_{i<j} \sum_j n_i^2 n_j^2 (s_i - s_j)^2 \\
&= (s_1 - s_2)^2 n_1^2 n_2^2 + (s_2 - s_3)^2 n_2^2 n_3^2 + (s_3 - s_1)^2 n_3^2 n_1^2. \qquad (2.12)
\end{aligned}$$

If the stress is purely hydrostatic then, as expected, eqn. 2.12 shows that the shear stress on all planes is zero. Consider the plane $n_3 = 0$ where $f_p^2 = (s_1 - s_2)^2 n_1^2 n_2^2$. This function

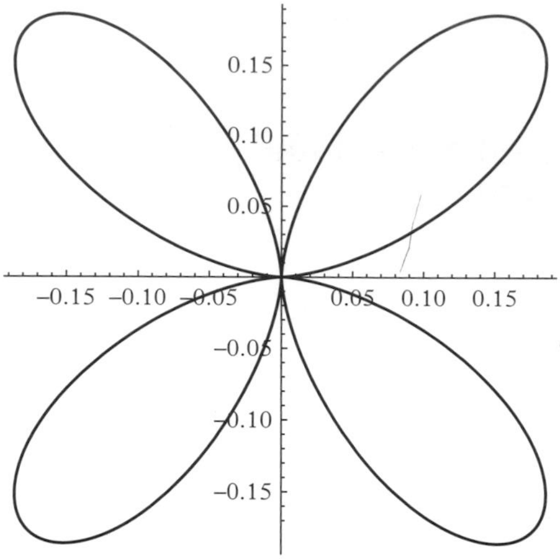

Figure 2.3 *Plot of $f_p^2 = n_1^2 n_2^2$, in the plane $n_3 = 0$. The axes are aligned with the principal stress directions $\hat{\mathbf{e}}_1$ and $\hat{\mathbf{e}}_2$. The maximum value of f_p^2 is located at 45° to the principal stress directions, where it is 0.25.*

is plotted in Fig. 2.3 for $s_1 - s_2 = 1$, where it is seen that the maximum shear stress acts on planes at 45° to the principal stress directions $\hat{\mathbf{e}}_1$ and $\hat{\mathbf{e}}_2$, where $f_p^2 = 0.25$.

Exercise 2.4

(a) Show that $f_p^2 = \hat{\mathbf{n}}^T\sigma^2\hat{\mathbf{n}} - \left(\hat{\mathbf{n}}^T\sigma\hat{\mathbf{n}}\right)^2$, where $\hat{\mathbf{n}}^T$ is the transpose of $\hat{\mathbf{n}}$.
(b) Show that $\mathbf{f_p}$ is unaffected if σ_{ij} is replaced by $\sigma_{ij}^{(D)} = \sigma_{ij} - \frac{1}{3}(\mathrm{Tr}\sigma)\delta_{ij}$, where $\sigma^{(D)}$ is called the *deviatoric* stress.
(c) Using eqn. 2.12 show that the average value of f_p^2, where the averaging is over all orientations of $\hat{\mathbf{n}}$ on the surface of a unit sphere, is

$$\begin{aligned} < f_p^2 > &= \frac{1}{15}\left((s_1 - s_2)^2 + (s_2 - s_3)^2 + (s_3 - s_1)^2\right) \\ &= \frac{1}{15}\left(3\mathrm{Tr}\sigma^2 - (\mathrm{Tr}\sigma)^2\right). \end{aligned} \tag{2.13}$$

The von Mises[7] stress is an invariant used to characterise the degree of shear in a stress tensor. It is used for example as a yield criterion to decide whether the stress in a body enables plastic deformation to take place. It is defined by

[7] Richard Edler von Mises 1883–1953, US mathematician, engineer and philosopher, born in what is now Ukraine.

$$\sigma^{vM} = \frac{1}{\sqrt{2}}\sqrt{(s_1 - s_2)^2 + (s_2 - s_3)^2 + (s_3 - s_1)^2}. \tag{2.14}$$

The von Mises shear stress is $\sigma^{vM} = \sqrt{15/2} \times \sqrt{< f_p^2 >}$, and is therefore directly related to the average shear stress on a plane.

2.7 Mechanical equilibrium

We come now to one of the most important ideas in the continuum theory of elasticity, that of mechanical equilibrium. We have seen already that the stress tensor has to be symmetric if there are no torques acting. We consider now the equilibrium of a region $\mathcal{R}$ that experiences a force per unit volume $\mathbf{f}(\mathbf{X})$ within a body. This force may be gravitational for example, and it is called a *body force*. The continuum surrounding $\mathcal{R}$ distorts generating stresses that balance the net body force acting on $\mathcal{R}$. These stresses are transmitted to $\mathcal{R}$ through the surface $\mathcal{S}$ surrounding it.

Mechanical equilibrium requires that the total force acting on $\mathcal{R}$ is zero:

$$\int_{\mathcal{R}} f_i(\mathbf{X}) \mathrm{d}V - \int_{\mathcal{S}} \sigma_{ij} n_j \mathrm{d}S = 0,$$

where n_j is component j of the *inward* pointing normal $\hat{\mathbf{n}}$ at the surface $\mathcal{S}$ of the region $\mathcal{R}$. We have chosen the inward pointing normal because we are considering the force that the surrounding medium exerts on the region $\mathcal{R}$. But we normally use the *outward* pointing normal in which case we must change the minus sign to a plus sign:

$$\int_{\mathcal{R}} f_i(\mathbf{X}) \mathrm{d}V + \int_{\mathcal{S}} \sigma_{ij} n_j \mathrm{d}S = 0,$$

where n_j is now component j of the *outward* pointing normal $\hat{\mathbf{n}}$ at the surface $\mathcal{S}$ of the region $\mathcal{R}$. Using the divergence theorem this may be rewritten as

$$\int_{\mathcal{R}} \left(f_i(\mathbf{X}) + \sigma_{ij,j} \right) \mathrm{d}V = 0,$$

where the comma denotes differentiation ($Z_{,j}$ means $\partial Z / \partial X_j$) and $\sigma_{ij,j}$ is the divergence of the stress tensor. Since this balance of forces must hold for all regions $\mathcal{R}$ within the body we arrive at the following differential equations for mechanical equilibrium:

$$\sigma_{ij,j} + f_i = 0. \tag{2.15}$$

There are three equations here, one for each component i of the body force, and $\sigma_{ij,j}$ consists of three derivatives for each value of i. These equations are analogous to

Poisson's equation in electromagnetism where the divergence of the electric displacement vector is the electric charge density, and electric charges are the sources of electric fields. In eqn. 2.15 body forces are sources of stress fields originating within the body.[8]

Exercise 2.5

Consider the resultant torque T_i acting on a region $\mathcal{R}$ with surface $\mathcal{S}$ within a body. It is acted on by a distribution of body forces $f_i(\mathbf{x})$ within $\mathcal{R}$ and surface tractions[a] $t_i(\mathbf{x})$ on $\mathcal{S}$. If the region $\mathcal{R}$ is in mechanical equilibrium prove that the condition for the torque T_i to be zero is that the stress tensor is symmetric.

Hint
With respect to an arbitrary origin the torque acting on the region $\mathcal{R}$ is given by

$$\mathbf{T} = \int_{\mathcal{R}} \mathbf{x} \times \mathbf{f} \, \mathrm{d}V + \int_{\mathcal{S}} \mathbf{x} \times \mathbf{t} \, \mathrm{d}S.$$

Rewrite this equation in component form using suffix notation and use the divergence theorem to convert the surface integral into a volume integral. Then use the equilibrium condition $\sigma_{ij,j} + f_i = 0$ to simplify the terms and deduce the symmetry of the stress tensor as the condition for $T_i = 0$.

Why is the expression for the torque $\mathbf{T}$ independent of the choice of origin?

[a] A 'surface traction' is a force per unit area acting on a surface.

Exercise 2.6

In this question we use the equilibrium condition $\sigma_{ij,j} + f_i = 0$ to prove $\sigma_{ij} = \delta E / \delta e_{ij}(\mathbf{x})$.

Consider a region $\mathcal{R}$ with surface $\mathcal{S}$ within a body in which there is a distribution of body forces $f_i(\mathbf{x})$ within $\mathcal{R}$ and surface tractions $t_i(\mathbf{x})$ on $\mathcal{S}$. There is no net force acting on $\mathcal{R}$. Suppose an infinitesimal displacement field $\delta u_i(\mathbf{x})$ is applied to points within $\mathcal{R}$ and on $\mathcal{S}$, which does not disturb the equilibrium of the body. The work done δW by the body forces in $\mathcal{R}$ and surface tractions on $\mathcal{S}$ is

$$\delta W = -\left\{ \int_{\mathcal{R}} f_i(\mathbf{x}) \delta u_i(\mathbf{x}) \, \mathrm{d}V + \int_{\mathcal{S}} \sigma_{ij}(\mathbf{x}) \delta u_i(\mathbf{x}) n_j(\mathbf{x}) \, \mathrm{d}S \right\},$$

where the normal vector in the surface integral points outwards. The corresponding change in the internal energy of the region $\mathcal{R}$ is $\delta E = -\delta W$. Thus,

continued

[8] There may also be stress fields generated by forces applied to the surface of the body, but their divergence is zero.

$$\delta E = \int_{\mathcal{R}} f_i(\mathbf{x})\delta u_i(\mathbf{x})\,\mathrm{d}V + \int_{\mathcal{S}} \sigma_{ij}(\mathbf{x})\delta u_i(\mathbf{x})n_j(\mathbf{x})\,\mathrm{d}S.$$

Use the divergence theorem to express the surface integral as a volume integral and simplify the resulting terms using the equilibrium condition to show that

$$\delta E = \int_{\mathcal{R}} \sigma_{ij}\delta u_{i,j}\mathrm{d}V.$$

Using the symmetry of the stress tensor show that this expression is equivalent to

$$\delta E = \int_{\mathcal{R}} \sigma_{ij}\delta e_{ij}\mathrm{d}V.$$

Hence deduce

$$\sigma_{ij}(\mathbf{x}) = \delta E/\delta e_{ij}(\mathbf{x}). \tag{2.16}$$

2.8 Adiabatic and isothermal stress

We will show in this section that the stress in eqn. 2.16 is the isentropic or adiabatic stress because it ignores any exchange of heat. In a real solid it is not the same as a stress calculated when heat flows to maintain a constant temperature, which is called the isothermal stress. Both adiabatic and isothermal stresses may arise depending on how the strain is applied. If a material is deformed so rapidly that there is insufficient time for heat to flow the immediate stress response will be adiabatic. But with the passage of time heat will flow, and the stress will evolve to the isothermal limit.

In a continuum model there is no distinction between isothermal and adiabatic stresses unless the continuum model also displays thermal strain, which is strain caused by a change of temperature at constant stress. In a continuum model the influence of the rate of elastic deformation is usually limited to distinguishing between adiabatic and isothermal elastic constants, which we shall return to in the next chapter.

In an atomistic model the isothermal and adiabatic stresses are not the same in general. But the definition of an isothermal stress atomistically has been controversial, with some investigators believing that the momenta of atoms contribute to the Cauchy stress, and others claiming that the Cauchy stress arises only from forces acting between atoms. The controversy appears to originate from a failure to distinguish conceptually between the pressure exerted by an ideal gas on the walls of a container and the Cauchy stress.

In an ideal gas there are no forces acting between atoms except when they collide. The gas exerts a pressure on the wall of a container through the exchange of momentum when gas atoms bounce off it. Therefore, the pressure exerted by the gas on the wall of a container is determined by the distribution of atomic momenta in the gas, as in the kinetic theory of gases.

In contrast, we see in Fig. 2.1 that the Cauchy stress does not arise from an exchange of momentum: it arises *exclusively* from interatomic forces acting across a plane. For example, if the plane is within the solid it is only notional in the sense that no atoms are bouncing off it and no impulses are imparted to it. The role of the plane is merely to separate the medium into two distinct parts so that the total force per unit area acting on one part due to the other can be calculated. However, *interatomic forces vary with temperature owing to their anharmonicity.*[9] It is this temperature dependence that gives rise to thermal strain. If atomic interactions were strictly harmonic there would be no thermal strain no matter how much atoms vibrate about their equilibrium positions: the solid would get hot but its shape and volume would not change because no internal stresses would be generated to drive those changes.

The combined first and second laws of thermodynamics for a solid may be expressed as follows:

$$\mathrm{d}E = T\mathrm{d}S + \int \sigma_{ij}^{S}(\mathbf{x})\,\mathrm{d}e_{ij}(\mathbf{x})\,\mathrm{d}V, \tag{2.17}$$

where E is the internal energy of the solid and S is the entropy. The integral $\int \sigma_{ij}^{S}(\mathbf{x})\,\mathrm{d}e_{ij}(\mathbf{x})\,\mathrm{d}V$ is the work done by the solid when an infinitesimal strain field $\mathrm{d}e_{ij}(\mathbf{x})$ is applied to the solid and there is a pre-existing stress field $\sigma_{ij}^{S}(\mathbf{x})$. The position dependence of the stress tensor allows for the possibility that the body is not homogeneous. The local stress in eqn. 2.16 is obtained from eqn. 2.17 by an adiabatic variation of the internal energy with respect to a local strain, that is, at constant entropy. The superscript S on the stress tensor is to remind us that it is obtained by a variation of the internal energy at constant entropy:

$$\sigma_{ij}^{S}(\mathbf{x}) = \left(\frac{\delta E}{\delta e_{ij}(\mathbf{x})}\right)_S. \tag{2.18}$$

The Helmholtz free energy of the solid is defined by $A = E - TS$, so that

$$\mathrm{d}A = -S\mathrm{d}T + \int \sigma_{ij}^{T}(\mathbf{x})\,\mathrm{d}e_{ij}(\mathbf{x})\,\mathrm{d}V. \tag{2.19}$$

It follows that

$$\sigma_{ij}^{T}(\mathbf{x}) = \left(\frac{\delta A}{\delta e_{ij}(\mathbf{x})}\right)_T. \tag{2.20}$$

In contrast to the local adiabatic stress in eqn. 2.18, the local isothermal stress is obtained by a variation of the Helmholtz free energy with respect to the local strain tensor at

[9] There is a related discussion of the temperature dependence of interatomic forces in section 3.9 of Sutton, AP and Balluffi, RW, *Interfaces in crystalline materials*, Oxford classic texts in the physical sciences, Clarendon Press: Oxford (2006). ISBN 978-0-19-921106-7. Robert Weierter Balluffi (1924–), US materials physicist.

constant temperature. That is indicated by the superscript T on the stress tensor. The two stress tensors σ_{ij}^S and σ_{ij}^T are equal at absolute zero.[10] But in general they differ as the temperature becomes finite.

As an illustration of eqn. 2.20 consider a perfect crystal with one atom in each unit cell. The crystal is initially in equilibrium at absolute zero with a volume V, so that the stress tensor is zero throughout the crystal. The temperature of the crystal is then raised to T, less than the melting point, while its shape and volume V are constrained to remain as they were at absolute zero. We will show that a stress is generated within the crystal if and only if atomic interactions are anharmonic. It is this stress which drives the thermal strain of the crystal when the constraints on its shape and volume are relaxed.

The homogeneity of the crystal enables eqn. 2.19 to be rewritten as follows:

$$\mathrm{d}A = -S\mathrm{d}T + V\sigma_{ij}^T \mathrm{d}e_{ij}. \tag{2.21}$$

Since the Helmholtz free energy is a state function the following Maxwell relation must apply:

$$-\left(\frac{\partial \sigma_{ij}^T}{\partial T}\right)_e = \frac{1}{V}\left(\frac{\partial S}{\partial e_{ij}}\right)_{T,e'}. \tag{2.22}$$

$-(\partial \sigma_{ij}^T/\partial T)_e = \beta_{ij}$ is known as the thermal stress tensor. It is evaluated with *all* strain components held constant, which is indicated by the e outside the bracket. Similarly, the thermal strain tensor is defined as $\alpha_{ij} = (\partial e_{ij}/\partial T)_\sigma$, where the σ outside the bracket indicates that *all* stress components are held constant, and usually that constant is zero. The partial derivative on the right is evaluated at constant temperature and all strain components *except* e_{ij} and e_{ji}, which is indicated by the prime on the e outside the bracket.[11] When this Maxwell relation is applied to a crystal it shows that the temperature dependence of a stress component, at a constant crystal configuration, is determined by the dependence of the entropy of the crystal on the same component of the strain tensor when all other strain components and the temperature are held constant. This immediately tells us that anharmonicity is involved in the temperature dependence of the stress, because in a harmonic crystal the elastic stiffness matrix is independent of strain. We will now show this explicitly.

Let the sum of the atomic interaction energies at absolute zero be the potential energy E_P. Then since the crystal is in equilibrium at absolute zero we have $\partial E_P/\partial e_{kl} = 0$ because there are no internal stresses. When the temperature is raised to T there is a free energy associated with the thermal vibrations. In the harmonic approximation the Helmholtz free energy of the crystal becomes

[10] This statement assumes the zero point energy is included in both the internal energy E and the free energy A.

[11] When this Maxwell relation is applied to an ideal gas it becomes the usual $(\partial P/\partial T)_V = (\partial S/\partial V)_T$, which is satisfied by $PV = RT$ with $S - S_o = R\ln(V/V_o)$, where S_o and V_o refer to a reference state of the gas and R is the gas constant.

$$A = E_P + k_B T \sum_n \ln\left[2\sinh\left(\frac{\hbar\omega_n}{2k_BT}\right)\right], \tag{2.23}$$

where k_B is the Boltzmann constant and ω_n is the angular frequency of normal mode n. The normal modes are obtained by solving the equations of motion:

$$m\ddot{u}_{Ai} = -\sum_{Bj} S_{AiBj} u_{Bj}, \tag{2.24}$$

where u_{Ai} is the displacement of atom A along the x_i direction, m is the atomic mass and S is the elastic stiffness matrix consisting of the second derivatives of the potential energy[12] with elements $S_{AiBj} = \partial^2 E_P/\partial u_{Ai}\partial u_{Bj}$. Since we are assuming each atom is performing harmonic vibrations we write $u_{Ai} = U_{Ai}e^{i\omega t}$. Then the angular frequency of normal mode n satisfies the following equation:

$$\omega_n^2 = \frac{1}{m}\sum_{Ai}\sum_{Bj} U_{Ai}^{(n)} U_{Bj}^{(n)} S_{BjAi}, \tag{2.25}$$

where we have used the symmetry of the stiffness matrix to write $S_{AiBj} = S_{BjAi}$, the orthonormality of its eigenvectors $U^{(n)}$, and the eigenvectors of a symmetric matrix can always be expressed as real numbers.

Differentiating the free energy in eqn. 2.23 with respect to a homogeneous strain at constant temperature, and remembering $\partial E_P/\partial e_{kl} = 0$, we obtain

$$\begin{aligned}
\sigma_{kl}^T &= \frac{1}{2}\sum_n \coth\left(\frac{\hbar\omega_n}{2k_BT}\right)\frac{\hbar}{2\omega_n}\frac{\partial\omega_n^2}{\partial e_{kl}} \\
&= \frac{1}{2}\sum_n \coth\left(\frac{\hbar\omega_n}{2k_BT}\right)\frac{\hbar}{2m\omega_n}\sum_{Ai}\sum_{Bj} U_{Ai}^{(n)} U_{Bj}^{(n)}\frac{\partial S_{BjAi}}{\partial e_{kl}} \\
&= \frac{1}{2}\sum_{Ai}\sum_{Bj}\langle U_{Ai}U_{Bj}\rangle\frac{\partial S_{BjAi}}{\partial e_{kl}}.
\end{aligned} \tag{2.26}$$

$\langle U_{Ai}U_{Bj}\rangle$ is the equal time displacement–displacement correlation function. We see that the stress is non-zero if and only if there are non-zero derivatives of the elastic stiffness matrix with respect to strain. If E_P consists of only harmonic interactions each S_{BjAi} is constant and the strain derivatives are all zero. In that case the stress is independent of temperature and $\sigma_{ij}^T = \sigma_{ij}^S$ at all temperatures. But in a real crystal atomic interactions are never purely harmonic and there are higher order derivatives of E_P. The stress is then dependent on temperature and $\sigma_{ij}^T \neq \sigma_{ij}^S$. By taking the limit $T \to 0$ in eqn. 2.26 it is seen

[12] These second derivatives are sometimes called force constants. Since they are constant only if E_P is a sum of harmonic interactions (i.e. a sum of quadratic functions of the atomic separations) we prefer to call them stiffnesses. This terminology allows for their variation when separations between atoms change owing to the existence of higher order derivatives in E_P.

that there is a contribution to σ_{ij}^T even at absolute zero which arises from the zero point motion.

When the constraints on the surface of the crystal to maintain its shape and volume are removed it undergoes a spontaneous strain to relieve the stress σ^T of eqn. 2.26. This is the origin of thermal strain. Once this strain has occurred the time-average separations of atoms in the crystal change slightly, and they have new mean positions, as determined by the anharmonicity of the interatomic forces. The model of atomic interactions is then called quasi-harmonic because the potential energy E_P is still expanded only to second order in the displacements of atoms from their mean positions, but the stiffnesses S_{AiBj}, which are evaluated at the new mean atomic positions, change owing to the existence of higher order derivatives in E_P.

To evaluate the isothermal Cauchy stress of eqn. 2.20 in a molecular dynamics simulation we calculate the time average of the expression in eqn. 2.5, with the vectors defining the periodic supercell and the temperature held constant. When the supercell vectors are allowed to relax, thermal stresses create thermal strains changing the volume and/or shape of the supercell. But it should be remembered that at temperatures below the Debye temperature quantum effects become significant and classical molecular dynamics does not capture them.

To summarise, isothermal and adiabatic stresses differ only because interatomic forces are anharmonic. Increasing the kinetic energies of atoms enables them to experience forces that are increasingly anharmonic, but atomic momenta do not appear explicitly in the Cauchy stress at a finite temperature.

2.9 Problem set 2

1. With respect to Cartesian axes x_1, x_2, x_3 a stress tensor σ is represented by the matrix

$$\sigma = \begin{bmatrix} 2 & 1 & 3 \\ 1 & 0 & -1 \\ 3 & -1 & 1 \end{bmatrix}.$$

   (a) Show that the force per unit area on the plane $2x_1 + x_2 - 2x_3 = 0$ is $\mathbf{f} = 1/3[-1, 4, 3]$.

   (b) Show that the normal stress on this plane is $-4/9$.

   (c) Show that the shear stress on this plane is $\sqrt{1962}/27$ and that it acts along the $[-1, 40, 19]$ direction.

   (d) Show that the stress tensor when referred to a new set of Cartesian axes obtained by rotating the x_1 and x_3 axes by $-45°$ about the positive x_2-axis (i.e. [010]) is given by

$$\sigma = \frac{1}{2}\begin{bmatrix} 9 & 0 & -1 \\ 0 & 0 & -2\sqrt{2} \\ -1 & -2\sqrt{2} & -3 \end{bmatrix}.$$

2. Consider an atomistic model in which the total potential energy is described by a sum of pairwise interactions:

$$E = \frac{1}{2}\sum_{m}\sum_{n\neq m} V(X^{(mn)}),$$

where $X^{(mn)}$ is the separation $\left|\mathbf{X}^{(m)} - \mathbf{X}^{(n)}\right|$ between atoms m and n and $V(X)$ is a function of the separation between pairs of atoms, for example a Lennard-Jones potential. The factor of one half is to correct for the double counting in the sum over m and n. Show that the atomic level stress tensor at atom k is given by

$$\sigma_{ij}^{(k)} = \frac{1}{2\Omega^{(k)}}\sum_{n\neq k}\frac{\left(X_i^{(n)} - X_i^{(k)}\right)\left(X_j^{(n)} - X_j^{(k)}\right)}{X^{(nk)}}\left(\frac{\mathrm{d}V}{\mathrm{d}X}\right)_{X=X^{(nk)}}.$$

3. (a) Using eqns. 2.14 and 2.13 show that

$$\sigma^{vM} = \frac{1}{\sqrt{2}}\sqrt{(\sigma_{11}-\sigma_{22})^2 + (\sigma_{22}-\sigma_{33})^2 + (\sigma_{33}-\sigma_{11})^2 + 6(\sigma_{12}^2+\sigma_{23}^2+\sigma_{31}^2)}. \tag{2.27}$$

(b) Show that the second invariant I_2 (see eqn. 2.10) of the deviatoric stress tensor $(\sigma_{ij} - \delta_{ij}\sigma_{kk}/3)$ is directly proportional to the square of the von Mises stress σ^{vM}, eqn. 2.13.

4. *This question is more advanced. It provides insight into the relationship between the equilibrium condition $\sigma_{ij,j} + f_i$ in a continuum and the Cauchy–Saint-Venant definition of stress in terms of interatomic forces illustrated in Fig. 2.1. It is based on the work of Noll.*[13]

We return to the definition of stress at the beginning of this chapter in terms of atomic interactions. Atoms are discrete objects and this poses a mathematical difficulty in applying the condition for mechanical equilibrium in a continuum embodied in eqn. 2.15. Noll's analysis overcomes this difficulty by replacing the discrete force that one atom exerts on another with a continuous force density $\mathbf{f}(\mathbf{x}',\mathbf{x})$ with units of force per unit volume squared. Then $\mathbf{f}(\mathbf{x}',\mathbf{x})\mathrm{d}V_{\mathbf{x}'}\mathrm{d}V_{\mathbf{x}}$ is the force that a volume element $\mathrm{d}V_{\mathbf{x}'}$ at $\mathbf{x}'$ exerts on a volume element $\mathrm{d}V_{\mathbf{x}}$ at $\mathbf{x}$. Notice that $\mathbf{f}(\mathbf{x}',\mathbf{x}) = -\mathbf{f}(\mathbf{x},\mathbf{x}')$. If the stress tensor at $\mathbf{x}$ is $\sigma_{ij}(\mathbf{x})$ then Noll proves that

[13] Walter Noll 1925–2017, US mathematician, born in Germany, see http://www.math.cmu.edu/~wn0g/. On this website there are many fascinating articles, including a very thought-provoking short essay on *The role of the professor* at http://www.math.cmu.edu/~wn0g/RP.pdf.

$$\sigma_{ij,j}(\mathbf{x}) = \int_V f_i(\mathbf{x}', \mathbf{x}) \mathrm{d}V_{\mathbf{x}'}. \tag{2.28}$$

This is the resultant force acting on a volume element at $\mathbf{x}$ arising from the surrounding medium. It is important to recognise that although $\mathbf{f}(\mathbf{x}', \mathbf{x})$ depends only on $\mathbf{x}'$ and $\mathbf{x}$ this does not amount to an assumption of pairwise interactions that depend only on the separation of $\mathbf{x}'$ and $\mathbf{x}$. In other words $\mathbf{f}(\mathbf{x}', \mathbf{x})$ may depend on the *environments* of $\mathbf{x}'$ and $\mathbf{x}$. It is also *not* necessarily the case that the force $\mathbf{f}(\mathbf{x}', \mathbf{x})$ is parallel to $\mathbf{x} - \mathbf{x}'$.

Following Noll we will show that

$$\sigma_{ij}(\mathbf{x}) = \frac{1}{2} \int_S \mathrm{d}\Omega_m \int_{r=0}^{\infty} \mathrm{d}r r^2 \int_{\alpha=0}^{1} \mathrm{d}\alpha f_i(\mathbf{x} + \alpha r \hat{\mathbf{m}}, \mathbf{x} - (1-\alpha) r \hat{\mathbf{m}}) r m_j, \tag{2.29}$$

where $\mathrm{d}\Omega_m$ is an element of solid angle centred on the direction $\hat{\mathbf{m}}$ and the integral over S is over the unit sphere centred at $\mathbf{x}$. The magnitude of the vector $\mathbf{r}$ is r. The function $f_i(\mathbf{x} + \alpha r \hat{\mathbf{m}}, \mathbf{x} - (1-\alpha) r \hat{\mathbf{m}})$ is the force exerted by a volume element at $\mathbf{x} + \alpha r \hat{\mathbf{m}}$ on the volume element at $\mathbf{x} - (1-\alpha) r \hat{\mathbf{m}}$, where these volume elements are separated by r at all values of $0 \leq \alpha \leq 1$. As α varies between 0 and 1 in the third integral the forces of interaction that pass through $\mathbf{x}$ between all points separated by r along the direction $\hat{\mathbf{m}}$ are included. In the second integral r ranges over all possible separations, and in the first integral all possible directions $\hat{\mathbf{m}}$ are considered. In this way all forces of interaction that pass through $\mathbf{x}$ between points on either side of $\mathbf{x}$ contribute to the stress, in accord with the definition of stress due to Cauchy and Saint-Venant in section 2.1. Each interaction is counted twice, and this is corrected by the factor of one half. The reason for the final factor $r m_j$ will become clear shortly.

To prove eqn. 2.29 we show that it satisfies eqn. 2.28.

Let $\mathbf{u} = \mathbf{x} + \alpha r \hat{\mathbf{m}}$ and $\mathbf{v} = \mathbf{x} - (1-\alpha) r \hat{\mathbf{m}}$. Show that

$$\frac{\partial f_i}{\partial x_j} = \frac{\partial f_i}{\partial u_j} + \frac{\partial f_i}{\partial v_j}.$$

Using the chain rule show that

$$\frac{\partial f_i}{\partial \alpha} = \left(\frac{\partial f_i}{\partial u_j} + \frac{\partial f_i}{\partial v_j} \right) r m_j = \frac{\partial f_i}{\partial x_j} r m_j$$

Using these results obtain eqn (2.28). We observe the following:

- The stress in eqn. 2.29 has the correct units, that is, force per unit area.
- The stress at $\mathbf{x}$ is attributed to forces that act not only on $\mathbf{x}$ but also through $\mathbf{x}$.

- Equation 2.29 is a continuum version of the discrete atomic-level stress tensor of eqn. 2.5.
- In eqn. 2.28 the force flux is inward towards the point $\mathbf{x}$: it is the resultant force the surrounding medium exerts on the point $\mathbf{x}$. If this resultant force is not zero mechanical equilibrium requires there is an equal and opposite force exerted from $\mathbf{x}$ on the surrounding medium. This is the body force at $\mathbf{x}$. The right hand side of eqn. 2.28 is therefore equal and opposite to the body force at $\mathbf{x}$.

3

Hooke's law and elastic constants

3.1 Generalised Hooke's law: elastic constants and compliances

Robert Hooke 1635–1703 was one of the most versatile and accomplished experimentalists of all time.[1] He was appointed 'Curator of Experiments' in the Royal Society in 1662, two years after the Society was formed, a post he held for 40 years until his death in 1703.

The modern form of the law which takes his name is that the stress tensor is proportional to the strain tensor, and conversely the strain tensor is proportional to the stress tensor. Since both stress and strain are second rank tensors the proportionality constants are fourth rank tensors:

$$\sigma_{ij} = c_{ijkl} e_{kl} \tag{3.1}$$

$$e_{ij} = s_{ijkl} \sigma_{kl}, \tag{3.2}$$

where c_{ijkl} is called the elastic stiffness tensor, or elastic constant tensor, and s_{ijkl} is called the elastic compliance tensor. By substituting eqn. 3.2 into eqn. 3.1, and noting that the stress and strain tensors are symmetric, it is seen that the elastic constant and compliance tensors are related as follows:

$$c_{ijkl} s_{klmn} = \frac{1}{2} \left(\delta_{im} \delta_{jn} + \delta_{in} \delta_{jm} \right). \tag{3.3}$$

Exercise 3.1

Verify eqn. 3.3.

The direct proportionality between stress and strain is the basis of *linear* elasticity. Hooke's 'law' is an approximation because nonlinear terms become significant as the magnitude of the strain increases, but are neglected in the linear theory. Physically, stiffer

[1] See Jardine, L. *The curious life of Robert Hooke*, Harper Collins: London (2003). ISBN 978-0007151752

Physics of elasticity and crystal defects. Adrian P. Sutton, Oxford University Press (2020). © Adrian P. Sutton.
DOI: 10.1093/oso/9780198860785.001.0001

bonds between atoms lead to larger elastic constants and smaller elastic compliances. Thus, in diamond, which is a very stiff insulator, c_{1111} is 1079 GPa, while in lead, which is a soft metal, it is 49.66 GPa. In most metals the elastic constants are of order 10^{11} Pa. One GPa (gigapascal) is 10^9 Pa, and 1 Pa = 1 N m^{-2} = 1 J m^{-3}; 1 GPa $\approx 6.24\times10^{-3}$ eV Å^{-3}.

3.2 The maximum number of independent elastic constants in a crystal

Since the elastic constant tensor is a fourth rank tensor it appears at first sight that there are $3^4 = 81$ independent elastic constants. If that were true the theory of elasticity would be much less useful because it would require the measurement of 81 material parameters. Symmetry enables the number of independent elastic constants to be reduced to a much more manageable number. The smallest number of independent elastic constants is just two, and this is the case in an elastically isotropic material like rubber. In cubic crystals there are just three independent elastic constants and in hexagonal crystals five. Since these restrictions are determined by symmetry they apply to the elastic compliance tensor in the same way as they do to the elastic constant tensor.

The largest number of independent elastic constants in any material is 21. This is the case in a triclinic crystal where there are no rotational symmetries in the point group. The first reduction is achieved by enforcing the symmetry of the stress and strain tensors: $\sigma_{ij} = \sigma_{ji}$, $e_{kl} = e_{lk}$. Therefore we must have $c_{ijkl} = c_{jikl} = c_{ijlk} = c_{jilk}$. The second reduction is more subtle and was first shown by George Green when he introduced the strain energy function, or elastic energy density.

3.2.1 The elastic energy density

In linear elasticity the elastic energy density is given by

$$\begin{aligned} E &= \frac{1}{2}\sigma_{ij}e_{ij} \\ &= \frac{1}{2}c_{ijkl}e_{ij}e_{kl}. \end{aligned} \tag{3.4}$$

This expression comes from integrating $dE = \sigma_{ij}de_{ij}$ with respect to strain from $e_{ij} = 0$ to the final strain and using Hooke's law to express stress in terms of strain. The factor of one half is a consequence of the linear relationship between stress and strain. The elastic energy density has units of J m^{-3}, the same as the elastic constants. The total elastic energy is then the integral of the elastic energy density over the volume of the body.

The elastic constants are second derivatives of the elastic energy density with respect to strain. For example, consider the terms involving the product $e_{12}e_{32}$. Since $e_{12} = e_{21}$ and $e_{32} = e_{23}$ we cannot vary e_{12} and e_{32} without also varying e_{21} and e_{23}. Therefore, there are four terms in the elastic energy density to consider: $\frac{1}{2}[c_{1232}e_{12}e_{32} + c_{2132}e_{21}e_{32} + c_{1223}e_{12}e_{23} + c_{2123}e_{21}e_{23}]$. We have already seen that $c_{1232} = c_{2132} = c_{1223} = c_{2123}$. Therefore these four terms amount to $2c_{1232}e_{12}e_{32}$, and

$$c_{1232} = \frac{1}{2}\frac{\partial^2 E}{\partial e_{12}\partial e_{32}}.$$

But there are also four terms in the elastic energy density involving the product $e_{32}e_{12}$. They amount to $2c_{3212}e_{32}e_{12}$, and

$$c_{3212} = \frac{1}{2}\frac{\partial^2 E}{\partial e_{32}\partial e_{12}}.$$

Green argued that the order of differentiation in these two second derivatives cannot matter. It follows that $c_{1232} = c_{3212}$. More generally,

$$c_{ijkl} = c_{klij}. \tag{3.5}$$

Thus the elastic constant tensor displays the following symmetries in all materials:

$$c_{ijkl} = c_{jikl} = c_{ijlk} = c_{jilk} = c_{klij} = c_{lkij} = c_{klji} = c_{lkji}. \tag{3.6}$$

There are six independent $\{ij\}$ and $\{kl\}$ combinations: 11, 22, 33, 23, 13 and 12. The symmetry embodied in eqn. 3.5 reduces the number of independent elastic constants from $6\times 6 = 36$ to $6+5+4+3+2+1 = 21$. This was first demonstrated by Green in 1837.[2] Any further reduction in the number of independent elastic constants depends on the point group symmetry of the material.

3.2.2 Matrix notation

Green's analysis above suggests that Hooke's law can be expressed in a convenient matrix form where each index signifies two indices in the tensor form of the equation:

$$11 \to 1,\ 22 \to 2,\ 33 \to 3,\ 23 \text{ or } 32 \to 4,\ 13 \text{ or } 31 \to 5,\ 12 \text{ or } 21 \to 6.$$

For example, $\sigma_{31} \to \sigma_5$ and $c_{1232} \to c_{64}$. Hooke's law may then be written in the following matrix form:

$$\begin{bmatrix} \sigma_1 \\ \sigma_2 \\ \sigma_3 \\ \sigma_4 \\ \sigma_5 \\ \sigma_6 \end{bmatrix} = \begin{bmatrix} c_{11} & c_{12} & c_{13} & c_{14} & c_{15} & c_{16} \\ c_{21} & c_{22} & c_{23} & c_{24} & c_{25} & c_{26} \\ c_{31} & c_{32} & c_{33} & c_{34} & c_{35} & c_{36} \\ c_{41} & c_{42} & c_{43} & c_{44} & c_{45} & c_{46} \\ c_{51} & c_{52} & c_{53} & c_{54} & c_{55} & c_{56} \\ c_{61} & c_{62} & c_{63} & c_{64} & c_{65} & c_{66} \end{bmatrix} \begin{bmatrix} e_1 \\ e_2 \\ e_3 \\ e_4 \\ e_5 \\ e_6 \end{bmatrix}. \tag{3.7}$$

[2] Green, G, *Trans. Cambridge Philos. Soc.*, 7, 1 (1839), https://archive.org/details/transactionsofca07camb

Now we see explicitly that there are just 21 independent components of the matrix **c**. Let us compare this equation with the tensor form of Hooke's law, eqn. 3.1. For example, consider σ_{11} in eqn. 3.1:

$$\sigma_{11} = c_{1111}e_{11} + c_{1122}e_{22} + c_{1133}e_{33} + 2c_{1123}e_{23} + 2c_{1113}e_{13} + 2c_{1112}e_{12}.$$

This has to be equivalent to σ_1 in eqn. 3.7:

$$\sigma_1 = c_{11}e_1 + c_{12}e_2 + c_{13}e_3 + c_{14}e_4 + c_{15}e_5 + c_{16}e_6.$$

For these two expressions to be equivalent we must have:

$$\begin{bmatrix} e_1 \\ e_2 \\ e_3 \\ e_4 \\ e_5 \\ e_6 \end{bmatrix} = \begin{bmatrix} e_{11} \\ e_{22} \\ e_{33} \\ 2e_{23} \\ 2e_{13} \\ 2e_{12} \end{bmatrix}.$$

Note the factors of 2 for the off-diagonal elements of the strain tensor.

It is important to recognise that $\boldsymbol{\sigma}$, **c** and **e** in eqn. 3.7 are *not* tensors because they do not transform according to the tensor transformation law under a rotation. They are merely a convenient way of writing the tensor relationship in eqn. 3.1 as a matrix equation. This highlights the difference between tensors and their representations as matrices.

An example of a triclinic crystal for which all 21 elastic constants have been determined experimentally is low albite ($NaAlSi_3O_8$), which is a plagioclase feldspar mineral.[3] With the x_2-axis parallel to the crystal *b*-axis, the x_1-axis perpendicular to crystal *b* and *c* axes and the x_3-axis completing a right-handed Cartesian coordinate system, the matrix **c** is as follows:

$$\mathbf{c} = \begin{bmatrix} 69.1 & 34.0 & 30.8 & 5.1 & -2.4 & -0.9 \\ 34.0 & 183.5 & 5.5 & -3.9 & -7.7 & -5.8 \\ 30.8 & 5.5 & 179.5 & -8.7 & 7.1 & -9.8 \\ 5.1 & -3.9 & -8.7 & 24.9 & -2.4 & -7.2 \\ -2.4 & -7.7 & 7.1 & -2.4 & 26.8 & 0.5 \\ -0.9 & -5.8 & -9.8 & -7.2 & 0.5 & 33.5 \end{bmatrix} \text{GPa.}$$

[3] Brown, JM, Abramson, EH and Angel, RJ, *Phys. Chem. Minerals* **33**, 256–65 (2006), https://doi.org/10.1007/s00269-006-0074-1

3.3 Transformation of the elastic constant tensor under a rotation

We have seen that since stress and strain are second rank tensors they are related in Hooke's law by a fourth rank tensor, which is either the elastic constant tensor or the elastic compliance tensor. For a rotation of the Cartesian coordinate system defined as in eqn. 2.7 the elastic constant tensor transforms as follows:

$$c'_{ijkl} = R_{im}R_{jn}R_{kp}R_{lq}c_{mnpq}. \tag{3.8}$$

We shall use this transformation to reduce the number of independent elastic constants to less than 21 when **R** represents a rotational symmetry of the material.

We note a further useful transformation property. If (x'_1, x'_2, x'_3) and (x_1, x_2, x_3) are the coordinates of a point in the rotated and unrotated coordinate systems respectively then

$$x'_i x'_j x'_k x'_l = R_{im}R_{jn}R_{kp}R_{lq}x_m x_n x_p x_q. \tag{3.9}$$

This equation shows that the elastic constant tensor c_{mnpq} transforms under a rotation in exactly the same way as the product of coordinates $x_m x_n x_p x_q$. We shall make use of this observation extensively below.

3.3.1 Neumann's principle

This is a fundamental principle that relates the symmetry displayed by a physical property of a crystal to the point group symmetry of the crystal. It is arguably the most fundamental structure–property relationship in materials science. It was formulated by Neumann[4] and first appeared in print in 1885.[5] Here is how the International Union of Crystallography states the principle:

> The symmetry elements of any physical property of a crystal must include all the symmetry elements of the point group of the crystal.

Mathematically, Neumann's principle means that any physical property is invariant with respect to every symmetry operation of the crystal. This means that when we transform the elastic constant tensor according to eqn. 3.8, with the rotation **R** being one of the symmetry rotations of the crystal, we must obtain an elastic constant tensor that is equivalent to the elastic constant tensor before the rotation was applied.

Note the word 'include' in Neumann's principle: the physical property may display more symmetry than the point group of the crystal. For example, the diffusivity tensor in a cubic crystal is isotropic, so that it displays the symmetry of a sphere in 3D, that is, the rotation group SO(3), which has infinitely more rotational symmetries than a

[4] Franz Ernst Neumann 1798–1895, German mineralogist, physicist and mathematician.

[5] Neumann, FE, *Vorlesungen über die Theorie der Elastizität der festen Körper und des Lichtäthers*, ed. OE Meyer, Leipzig: B G Teubner-Verlag, (1885).

cube or octahedron or tetrahedron. The elastic constant tensor always displays inversion symmetry because if a homogeneous stress and strain were inverted through any centre no change would be apparent in the elastic properties since a state of homogeneous stress or strain is centrosymmetrical.[6] This remains true even in a crystal that does not display inversion symmetry in its point group. All point group operations are either rotations or rotations combined with an inversion (e.g. mirror planes are two fold rotations followed (or preceded) by an inversion). Since the elastic constant tensor already displays inversion symmetry it is necessary to ask how it transforms under only the rotational symmetries of the point group. Eleven of the 32 point groups contain only rotational symmetries, and they are known as the proper groups, or enantiomorphous groups. They are the point groups that determine the numbers of independent elastic constants in all 32 point groups.

3.4 Isotropic materials

An elastically isotropic material is one in which the elastic constants do not depend on direction in the material: they have the symmetry of SO(3). Examples of isotropic materials are rubber, glass and amorphous materials.

If c_{ijkl} is the same in all directions then $c_{ijkl} = \langle c_{ijkl} \rangle$ where $<\cdots>$ means an average taken over all radial directions within a sphere. It follows from eqn. 3.9 that $\langle c_{ijkl} \rangle$ is proportional to $\langle x_i x_j x_k x_l \rangle$, where x_i are the coordinates of a point on the surface of the unit sphere, with respect to an origin at its centre. We find

$$
\begin{aligned}
< x_i x_j x_k x_l > &= < x_1^2 x_2^2 > \delta_{ij}\delta_{kl}(1-\delta_{jk}) + < x_1^2 x_2^2 > \delta_{ik}\delta_{jl}(1-\delta_{kj}) \\
&\quad + < x_1^2 x_2^2 > \delta_{il}\delta_{jk}(1-\delta_{jl}) + < x_1^4 > \delta_{ij}\delta_{jk}\delta_{kl} \\
&= \frac{1}{15}(\delta_{ij}\delta_{kl} + \delta_{ik}\delta_{jl} + \delta_{il}\delta_{jk}) \\
&\quad + \frac{1}{15}(3\delta_{ij}\delta_{jk}\delta_{kl} - \delta_{ij}\delta_{kl}\delta_{jk} - \delta_{ik}\delta_{jl}\delta_{ij} - \delta_{il}\delta_{jk}\delta_{ik}) \\
&= \frac{1}{15}(\delta_{ij}\delta_{kl} + \delta_{ik}\delta_{jl} + \delta_{il}\delta_{jk}), \qquad (3.10)
\end{aligned}
$$

where $< x_1^2 x_2^2 >=< x_2^2 x_3^2 >=< x_3^2 x_1^2 >= \frac{1}{15}$ and $< x_1^4 >=< x_2^4 >=< x_3^4 >= \frac{1}{5}$ have been used. It follows that an isotropic elastic constant tensor has the following form:

$$
c_{ijkl} = \lambda\delta_{ij}\delta_{kl} + \mu\delta_{ik}\delta_{jl} + \mu'\delta_{il}\delta_{jk},
$$

where λ, μ and μ' are constants.

[6] Nye, JF, *Physical properties of crystals*, Oxford University Press: Oxford (1957), p.21. ISBN 0-19-851165-5. John Frederick Nye FRS 1923–2019, British physicist.

Since $c_{ijij} = c_{ijji}$ (no summation) we must have $\mu = \mu'$. Therefore there are just two independent elastic constants in an isotropic material:

$$c_{ijkl} = \lambda\delta_{ij}\delta_{kl} + \mu\left(\delta_{ik}\delta_{jl} + \delta_{il}\delta_{jk}\right). \tag{3.11}$$

λ is called Lamé's first constant,[7] and μ is sometimes called Lamé's second constant but more commonly the shear modulus.

Exercise 3.2

Derive eqn. 3.10 in detail.

Exercise 3.3

Verify that $c_{ijkl} = \lambda\delta_{ij}\delta_{kl} + \mu\left(\delta_{ik}\delta_{jl} + \delta_{il}\delta_{jk}\right)$ is invariant when it substituted into eqn. 3.8 for any rotation **R**.

When we substitute the isotropic elastic constants, eqn. 3.11, into Hooke's law, eqn. 3.1, we obtain the following equations:

$$\begin{aligned}
\sigma_{11} &= 2\mu e_{11} + \lambda(e_{11} + e_{22} + e_{33})\\
\sigma_{22} &= 2\mu e_{22} + \lambda(e_{11} + e_{22} + e_{33})\\
\sigma_{33} &= 2\mu e_{33} + \lambda(e_{11} + e_{22} + e_{33})\\
\sigma_{23} &= 2\mu e_{23}\\
\sigma_{13} &= 2\mu e_{13}\\
\sigma_{12} &= 2\mu e_{12},
\end{aligned} \tag{3.12}$$

where we recognise $e_{kk} = e_{11} + e_{22} + e_{33}$ as the dilation $\Delta V/V$. Thus $c_{11} = 2\mu + \lambda$, $c_{12} = \lambda$ and $c_{44} = \mu$. Therefore in an isotropic material we have

$$A = \frac{2c_{44}}{c_{11} - c_{12}} = 1. \tag{3.13}$$

This is called the anisotropy ratio, about which we will say more in the context of cubic crystals where $A \neq 1$.

To relate λ and μ to Young's modulus Y consider a tensile test where a sample is loaded in tension along the x_3-axis and no constraints or loads are applied along x_1 and x_2. There are no shear strains and eqns. 3.12 become

[7] Named after Gabriel Lamé 1795–1870, French mathematician.

$$
\begin{aligned}
0 &= 2\mu e_{11} + \lambda(e_{11} + e_{22} + e_{33}) \\
0 &= 2\mu e_{22} + \lambda(e_{11} + e_{22} + e_{33}) \\
\sigma_{33} &= 2\mu e_{33} + \lambda(e_{11} + e_{22} + e_{33}).
\end{aligned}
$$

Solving these equations for e_{11}, e_{22} and e_{33} we find $e_{11} = e_{22} = -\lambda\sigma_{33}/\{2\mu(2\mu + 3\lambda)\}$ and $e_{33} = 2(\mu + \lambda)\sigma_{33}/\{2\mu(2\mu + 3\lambda)\}$. From these relations we deduce the following:

$$Y = \frac{\mu(2\mu + 3\lambda)}{\mu + \lambda} \tag{3.14}$$

$$\nu = -\frac{e_{11}}{e_{33}} = \frac{\lambda}{2(\mu + \lambda)} \tag{3.15}$$

$$\mu = \frac{Y}{2(1 + \nu)} \tag{3.16}$$

$$\lambda = \frac{2\mu\nu}{1 - 2\nu}, \tag{3.17}$$

where ν is called Poisson's[8] ratio. Poisson's ratio is the ratio of the contraction in the lateral x_1 and x_2 directions to the tensile strain along x_3. Most materials contract along the lateral directions when they are stretched, and expand along the lateral directions when they are compressed. Materials that do the opposite are called 'auxetic', and they have negative Poisson's ratios.[9] In terms of the Young's modulus and Poisson's ratio the strains may be expressed in terms of the stresses as follows:

$$
\begin{aligned}
e_{11} &= \frac{\sigma_{11}}{Y} - \frac{\nu\sigma_{22}}{Y} - \frac{\nu\sigma_{33}}{Y} \\
e_{22} &= \frac{\sigma_{22}}{Y} - \frac{\nu\sigma_{11}}{Y} - \frac{\nu\sigma_{33}}{Y} \\
e_{33} &= \frac{\sigma_{33}}{Y} - \frac{\nu\sigma_{11}}{Y} - \frac{\nu\sigma_{22}}{Y} \\
e_{23} &= \frac{1 + \nu}{Y}\sigma_{23} = \frac{\sigma_{23}}{2\mu} \\
e_{13} &= \frac{1 + \nu}{Y}\sigma_{13} = \frac{\sigma_{13}}{2\mu} \\
e_{12} &= \frac{1 + \nu}{Y}\sigma_{12} = \frac{\sigma_{12}}{2\mu}.
\end{aligned} \tag{3.18}
$$

Another commonly used elastic constant is the bulk modulus, B. This relates the hydrostatic pressure $p = -\mathrm{Tr}\sigma/3$ to the dilation $\Delta V/V = \mathrm{Tr}e$:

8 Siméon Denis Poisson 1781–1840. French mathematician, engineer and physicist.

9 Most auxetic materials are cellular solids such as honeycombs and foams. But they also occur naturally, for example human artery walls and skin and a form of silica (SiO_2) known as α-cristobalite. The Poisson's ratio of cork is almost zero, which makes it ideal for sealing wine in bottles.

$$p = -B\frac{\Delta V}{V}. \tag{3.19}$$

Using eqns. 3.12 it is deduced that

$$B = \frac{1}{3}(2\mu + 3\lambda) = \frac{2\mu(1+\nu)}{3(1-2\nu)}. \tag{3.20}$$

It is stressed that in isotropic elasticity only two of the Young's modulus Y, the shear modulus μ, Poisson's ratio ν, the bulk modulus B and Lamé's first constant λ are independent.

Exercise 3.4

(a) Why must the value of ν always be between -1 and $\frac{1}{2}$?
(b) What do these two limits correspond to physically?

Exercise 3.5

Show that in an isotropic medium Hooke's law may be expressed in the following equivalent ways:

$$\sigma_{ij} = 2\mu e_{ij}^{(d)} + \delta_{ij}Be_{kk}$$
$$e_{ij} = \frac{1}{2\mu}\sigma_{ij}^{(d)} + \delta_{ij}\frac{\sigma_{kk}}{9B},$$

where $e_{ij}^{(d)}$ and $\sigma_{ij}^{(d)}$ are the deviatoric strain and stress tensors. By introducing the deviatoric stress and strain tensors we see a clear separation between shear and dilational contributions, involving the shear modulus and bulk modulus respectively, to the total stress and strain tensors.

Exercise 3.6

(a) By orienting the axes along the eigenvectors of the stress tensor show that the elastic energy density in an isotropic medium may be expressed as follows:

$$E = \frac{1}{2Y}\left(s_1^2 + s_2^2 + s_3^2 - 2\nu(s_1 s_2 + s_2 s_3 + s_3 s_1)\right),$$

where s_i are the eigenvalues of the stress tensor.

continued

Exercise 3.6 Continued

(b) Show that the elastic energy density may be expressed as

$$E = \frac{I_1^2}{18B} + \frac{I_1^2 - 3I_2}{6\mu} = \frac{p^2}{2B} + \frac{\left(\sigma^{vM}\right)^2}{6\mu},$$

where I_1 and I_2 are the first and second invariants of the stress tensor, p is the hydrostatic pressure and σ^{vM} is the von Mises shear stress given by eqn. 2.14. We see here that the elastic energy density in an isotropic medium also separates into dilational and shear contributions.

3.5 Anisotropic materials

There are no crystalline materials that are exactly elastically isotropic, but tungsten is almost isotropic. In this section we will illustrate how point group symmetry is used to reduce the number of independent elastic constants from 21 in a crystal. As an example we will show there are three independent elastic constants in cubic crystals.

3.5.1 Cubic crystals

In this section we will make use of the observation in eqn. 3.9 that the elastic constant tensor c_{mnpq} transforms under a rotation in exactly the same way as the product of coordinates $x_m x_n x_p x_q$.

Cubic crystals are defined by four three fold rotational symmetry axes along $\langle 111\rangle$ directions. These rotational symmetries generate a further three two fold rotation axes along $\langle 100\rangle$. In this way we obtain the cubic point group '23' in Hermann–Mauguin notation or T in Schönflies notation.

Rotating the coordinate axes by π about [100] results in $x_1' = x_1, x_2' = -x_2, x_3' = -x_3$. Therefore the following eight elastic constants must be zero because they are equal to their own negative under this rotation: $c_{1112} = c_{16}$, $c_{1113} = c_{15}$, $c_{2212} = c_{26}$, $c_{2213} = c_{25}$, $c_{3312} = c_{36}$, $c_{3313} = c_{35}$, $c_{2312} = c_{46}$, $c_{2313} = c_{45}$, where we are specifying the 4-index tensor component and its corresponding element of the 6×6 matrix in eqn. 3.7.

Similarly rotating the coordinate axes by π about [010] results in $x_1' = -x_1$, $x_2' = x_2, x_3' = -x_3$, and four additional elastic constants are found to be zero: $c_{1123} = c_{14}$, $c_{2223} = c_{24}$, $c_{3323} = c_{34}$, $c_{1312} = c_{56}$.

No additional information is obtained by rotating by π about [001]. Rotating by $2\pi/3$ anti-clockwise about [111] results in $x_1' \to x_2, x_2' \to x_3, x_3' \to x_1$. Therefore the following elastic constants must be equal: $c_{1111} = c_{2222} = c_{3333}$; $c_{1122} = c_{2233} = c_{3311}$; $c_{2323} = c_{3131} = c_{1212}$, which in matrix notation are $c_{11} = c_{22} = c_{33}$; $c_{12} = c_{23} = c_{31}$; $c_{44} = c_{55} = c_{66}$. No additional information is obtained by invoking any of the other symmetry operations.

The conclusion is that there are three independent elastic constants in a cubic crystal: c_{11}, c_{12}, c_{44}:

$$\mathbf{c} = \begin{bmatrix} c_{11} & c_{12} & c_{12} & 0 & 0 & 0 \\ c_{12} & c_{11} & c_{12} & 0 & 0 & 0 \\ c_{12} & c_{12} & c_{11} & 0 & 0 & 0 \\ 0 & 0 & 0 & c_{44} & 0 & 0 \\ 0 & 0 & 0 & 0 & c_{44} & 0 \\ 0 & 0 & 0 & 0 & 0 & c_{44} \end{bmatrix}. \tag{3.21}$$

This conclusion remains the same with the cubic point group '432' in Hermann–Mauguin notation or O in Schönflies notation. Therefore, all cubic point groups have an elastic constant matrix of the same form as that shown in eqn. 3.21. This is conveniently summarised in the following formula for the elastic constants in cubic crystals:

$$c_{ijkl} = c_{12}\delta_{ij}\delta_{kl} + c_{44}\left(\delta_{ik}\delta_{jl} + \delta_{il}\delta_{jk}\right) + (c_{11} - c_{12} - 2c_{44})\delta_{ij}\delta_{jk}\delta_{kl}. \tag{3.22}$$

The elements of the elastic constant matrix with value zero in a cubic crystal are the same as those in an isotropic medium. The only difference between the cubic and isotropic cases is that the anisotropy ratio, eqn. 3.13, in a cubic crystal is not unity. Let us look at this more closely. A pure shear strain e_{23} in a cubic crystal is on (010) and (001) planes, and it is created by the shear stress $\sigma_{23} = 2c_{44}e_{23}$. Therefore, c_{44} measure the resistance to shear on $\{100\}$ planes in the cubic crystal. If we rotate the coordinate system by $\pi/4$ about [100] then e'_{23} is a pure shear on (011) and $(01\bar{1})$ planes. After transforming the elastic constant tensor it is found that $c'_{2323} = (c_{11} - c_{12})/2$. Therefore, $(c_{11} - c_{12})/2$ measures the resistance to shear on $\{110\}$ planes in the cubic crystal, and it is called C' (pronounced 'C prime'). It follows that the anisotropy ratio in a cubic crystal is the ratio of the shear resistance on $\{100\}$ planes to the shear resistance on $\{110\}$ planes. In an isotropic crystal the resistances are the same. The anisotropy ratio can have a strong influence on the elastic fields of defects and modes of plastic deformation in cubic crystals.

3.5.2 The directional dependence of the elastic constants in anisotropic media

In an anisotropic medium the elastic constants vary with direction. For a chosen elastic constant this variation can be depicted graphically by plotting a surface $r(\theta,\phi)$ where r is the magnitude of the elastic constant along the direction (θ,ϕ) in spherical coordinates. In an isotropic medium this surface is a sphere.

As a first example consider the variation of c_{11} with direction in a cubic crystal. Orienting the Cartesian axes along the $\langle 100\rangle$ directions the variation of $c_{11} = c_{1111}$ as the coordinate system is rotated is given by eqn. 3.8:

$$c'_{1111} = R_{1i}R_{1j}R_{1k}R_{1l}c_{ijkl}. \tag{3.23}$$

Let $R_{1i} = \hat{\eta}_i$. This vector is parallel to the x'_1-axis. eqn. 3.23 provides the value of c_{11} along the direction $\hat{\eta}$ with respect to the cube axes. We obtain

$$c'_{11} = c'_{1111} = c_{11} + 2(c_{11} - c_{12})\left(\frac{2c_{44}}{c_{11} - c_{12}} - 1\right)\left(\hat{\eta}_1^2\hat{\eta}_2^2 + \hat{\eta}_2^2\hat{\eta}_3^2 + \hat{\eta}_3^2\hat{\eta}_1^2\right), \tag{3.24}$$

where we see that the directional dependence is proportional to the deviation of the anisotropy ratio A (eqn. 3.13) from unity. For $A > 1$ the maximum value of c'_{11} is along $\langle 111\rangle$ directions. A plot of c'_{11} is shown in Fig. 3.1 for copper, where $c_{11} = 168.4$ GPa, $c_{12} = 121.4$ GPa, $c_{44} = 75.4$ GPa and the anisotropy ratio is $A = 3.21$. The average value of c'_{11}, where the averaging is over all directions, is $c_{11} + \frac{2}{5}(A-1)(c_{11} - c_{12})$, and in copper this is 210 GPa.

The variation of the shear elastic constant c_{44} with direction is more complicated because it depends on two directions: the plane normal and the direction of shear. Invoking the transformation law:

$$c'_{1212} = R_{1i}R_{2j}R_{1k}R_{2l}c_{ijkl} = \hat{\eta}_i\hat{\xi}_j\hat{\eta}_k\hat{\xi}_l c_{ijkl}, \tag{3.25}$$

where $\hat{\xi}_j = R_{2j}$ is any unit vector perpendicular to $\hat{\eta}$. In this equation $\hat{\eta}$ may be interpreted as the normal to the plane where c'_{1212} is evaluated and $\hat{\xi}$ as the direction of shear in that plane. Thus, c'_{1212} is a function of three independent variables. After some algebraic manipulations we obtain

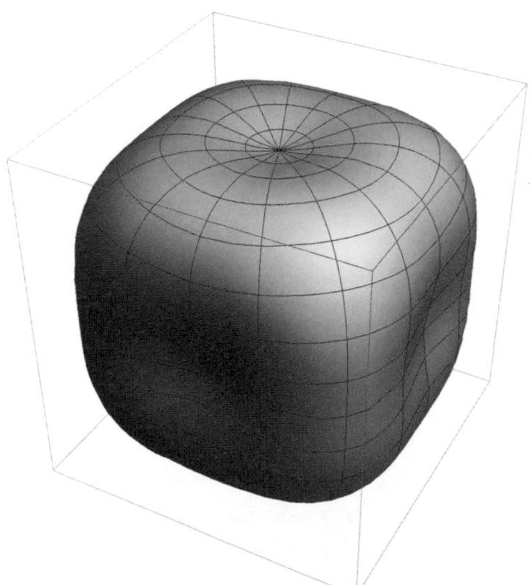

Figure 3.1 *Polar plot of c'_{11} given by eqn. 3.24 for copper inside a bounding cube aligned with the $\langle 100\rangle$ directions of the fcc crystal.*

$$c'_{44} = c'_{66} = c'_{1212} = c_{44} - (c_{11} - c_{12})\left(\frac{2c_{44}}{c_{11} - c_{12}} - 1\right)\left(\hat{\eta}_1^2\hat{\xi}_1^2 + \hat{\eta}_2^2\hat{\xi}_2^2 + \hat{\eta}_3^2\hat{\xi}_3^2\right). \tag{3.26}$$

For example, when $\hat{\eta} = [110]/\sqrt{2}$ and $\hat{\xi} = [1\bar{1}0]/\sqrt{2}$ we obtain $c'_{44} = \frac{1}{2}(c_{11} - c_{12})$, which is C'. For completeness we also state the variation of the elastic constant c_{12} with two orthonormal directions $\hat{\eta}$ and $\hat{\xi}$:

$$c'_{12} = c'_{1122} = c_{12} - (c_{11} - c_{12})\left(\frac{2c_{44}}{c_{11} - c_{12}} - 1\right)\left(\hat{\eta}_1^2\hat{\xi}_1^2 + \hat{\eta}_2^2\hat{\xi}_2^2 + \hat{\eta}_3^2\hat{\xi}_3^2\right). \tag{3.27}$$

3.6 Further restrictions on the elastic constants

For structural stability the elastic energy density must be positive definite. Otherwise the material will spontaneously distort to a lower energy structure. The elastic energy density may be written in matrix notation as $\frac{1}{2}\sigma_i e_i = \frac{1}{2}c_{ij}e_i e_j$. For the quadratic form $c_{ij}e_i e_j$ to be positive definite all six of the leading principal minors of the 6×6 matrix **c** must be positive definite.

In an isotropic medium this condition leads to $\lambda + \frac{2}{3}\mu > 0$ and $\mu > 0$. Since the bulk modulus is given by $B = \lambda + \frac{2}{3}\mu$ (see eqn. 3.20) the first condition is equivalent to requiring the bulk modulus is positive. In Exercise 3.6 it was also shown that the elastic energy density is positive definite provided $B > 0$ and $\mu > 0$.

In a cubic crystal the elastic energy density is positive definite provided $c_{11} - c_{12} > 0$, $c_{44} > 0$ and $c_{11} + 2c_{12} > 0$. Thus, the elastic energy density is positive definite provided the two shear elastic constants and the bulk modulus are all positive. c_{12} has to lie between $-c_{11}/2$ and c_{11}, and therefore Poisson's ratio, which is $c_{12}/(c_{11} + c_{12})$, has to lie between -1 and $\frac{1}{2}$.

3.7 Elastic constants and atomic interactions

Elastic constants may be calculated for a crystal if we have a description of atomic interactions. Consider a crystal with one atom at each lattice site. Since all atoms are at centres of inversion there is no net force acting on any atom. Let $u_i^{(n)}$ be a small arbitrary displacement of atom n. Then the change in the energy of the crystal to second order in the displacements is

$$E = \sum_n \frac{\partial E}{\partial u_i^{(n)}} u_i^{(n)} + \frac{1}{2} \cdot \sum_n \sum_p \frac{\partial^2 E}{\partial u_i^{(n)} \partial u_j^{(p)}} u_i^{(n)} u_j^{(p)}. \tag{3.28}$$

This is the usual harmonic expansion of the energy of the crystal, where the derivatives are evaluated in the perfect crystal configuration. The first term on the right is zero because the forces on all atoms are zero at equilibrium.

If $u_i^{(n)} = t_i$ for all n, where t_i is a small constant vector, then the energy E in eqn. 3.28 should be invariant, because the crystal has undergone a rigid body translation. This is achieved if the following equation is satisfied:

$$\frac{\partial^2 E}{\partial u_i^{(n)} \partial u_j^{(n)}} = -\sum_{p \neq n} \frac{\partial^2 E}{\partial u_i^{(n)} \partial u_j^{(p)}},$$

for each site n. When this is substituted into eqn. 3.28 the second order term becomes

$$\frac{1}{2} \cdot \sum_n \sum_p \frac{\partial^2 E}{\partial u_i^{(n)} \partial u_j^{(p)}} u_i^{(n)} u_j^{(p)} = \frac{1}{2} \cdot \sum_n \sum_{p \neq n} \frac{\partial^2 E}{\partial u_i^{(n)} \partial u_j^{(p)}} u_i^{(n)} \left(u_j^{(p)} - u_j^{(n)} \right)$$

$$= -\frac{1}{2} \cdot \frac{1}{2} \cdot \sum_n \sum_{p \neq n} \frac{\partial^2 E}{\partial u_i^{(n)} \partial u_j^{(p)}} \left(u_i^{(p)} - u_i^{(n)} \right) \left(u_j^{(p)} - u_j^{(n)} \right). \tag{3.29}$$

Let the displacements be created by a small homogeneous strain e_{kl}, such that $u_i^{(p)} - u_i^{(n)} = e_{ik} \left(X_k^{(p)} - X_k^{(n)} \right)$, where $\mathbf{X}^{(p)}$ is the position of atom p in the unstrained crystal. All atoms remain at centres of inversion during this operation and therefore the net force on any atom remains zero. Since the strain is homogeneous and since all atoms in the crystal remain equivalent we need to consider the change in the energy of just atom n, which we call δE_n:

$$\delta E_n = -\frac{1}{4} \sum_{p \neq n} e_{ik} \left(X_k^{(p)} - X_k^{(n)} \right) S_{ij}^{(np)} e_{jl} \left(X_l^{(p)} - X_l^{(n)} \right), \tag{3.30}$$

where $S_{ij}^{(np)} = \partial^2 E / \partial u_i^{(n)} \partial u_j^{(p)}$. We may rewrite this equation in terms of the elastic constants:

$$\delta E_n = \frac{1}{2} \Omega \, c_{ikjl} \, e_{ik} \, e_{jl}$$

where Ω is the volume of a primitive unit cell of the crystal. Comparing this with eqn. 3.30 we obtain

$$c_{ikjl} = -\frac{1}{2\Omega} \sum_{p \neq n} \left(X_k^{(p)} - X_k^{(n)} \right) S_{ij}^{(np)} \left(X_l^{(p)} - X_l^{(n)} \right). \tag{3.31}$$

It is evident that this expression satisfies the symmetry $c_{ikjl} = c_{jlik}$. To satisfy the other symmetries of eqns. 3.6 we set $c_{ikjl} = \frac{1}{4} \left(c_{ikjl} + c_{kijl} + c_{iklj} + c_{kilj} \right)$:

$$c_{ikjl} = -\frac{1}{8\Omega}\left\{\sum_{p\neq n}\left(X_k^{(p)} - X_k^{(n)}\right)S_{ij}^{(np)}\left(X_l^{(p)} - X_l^{(n)}\right) + \sum_{p\neq n}\left(X_i^{(p)} - X_i^{(n)}\right)S_{kj}^{(np)}\left(X_l^{(p)} - X_l^{(n)}\right) + \sum_{p\neq n}\left(X_k^{(p)} - X_k^{(n)}\right)S_{il}^{(np)}\left(X_j^{(p)} - X_j^{(n)}\right) + \sum_{p\neq n}\left(X_i^{(p)} - X_i^{(n)}\right)S_{kl}^{(np)}\left(X_j^{(p)} - X_j^{(n)}\right)\right\}.$$

If there is more than one atom associated with each lattice site those atoms not on lattice sites may undergo small displacements in addition to those prescribed by a homogeneous strain. These additional displacements are sometimes called the 'internal strain'. Although the strain is still imposed by displacing atoms at lattice sites, atoms between lattice sites will experience net forces as a result of the strain if they are not at centres of inversion. Relaxation of those forces reduces the energy of the homogeneously strained crystal, and therefore it affects the calculated elastic constants.

3.8 Isothermal and adiabatic elastic moduli

So far the elastic moduli[10] we have considered are those obtained by an adiabatic variation of the internal energy. In this section we follow the treatment[11] by Wallace[12] to show the relationships between elastic moduli obtained adiabatically and isothermally. Insightful relationships for the temperature dependences of the isothermal elastic constant tensor and the isothermal elastic compliance tensor are also derived. Whether adiabatic or isothermal moduli should be used in any given thermoelastic process depends on the rate of elastic deformation. For example, ultrasonic pulse experiments measure adiabatic elastic moduli, whereas isothermal elastic moduli are measured in tensile and torsion tests at constant temperature.

In a homogeneous crystal the adiabatic and isothermal elastic moduli differ because there is always a finite thermal strain or stress tensor owing to the anharmonicity of atomic interactions. However the difference between them tends to zero as the temperature approaches absolute zero.[13] The elastic moduli normally assigned to a continuum are either adiabatic or isothermal. But if a continuum model is required to display either adiabatic or isothermal moduli over a range of thermoelastic conditions,[14] the continuum must also be assigned a thermal stress tensor and/or a thermal strain tensor and specific

[10] That is, elastic constants and elastic compliances.

[11] Wallace, DC, *Thermodynamics of crystals*, John Wiley & Sons Inc.: New York (1972), section 2. ISBN 9780471918554.

[12] Duane C Wallace 1931–, US materials physicist.

[13] This is a requirement of the third law of thermodynamics.

[14] For example, in a simulation of a shock impact.

heats at constant strain and/or constant stress spanning the range of temperatures required.

The combined first and second laws of thermodynamics for a solid was given in eqn. 2.17. For simplicity we consider a crystal of just one atomic species in which there is one atom associated with each lattice site.[15] The homogeneity of the crystal enables the combined first and second laws to be written as follows:

$$\mathrm{d}E = T\mathrm{d}S + V\sigma_{ij}\mathrm{d}e_{ij}, \tag{3.32}$$

where V is the volume of the crystal in its current state of strain.

The Gibbs free energy is defined by $G = E - TS - V\sigma_{ij}e_{ij}$. Then we have

$$\mathrm{d}G = -S\mathrm{d}T - Ve_{ij}\mathrm{d}\sigma_{ij}. \tag{3.33}$$

Since $\mathrm{d}G$ is an exact differential the following Maxwell relation holds:

$$\frac{1}{V}\left(\frac{\partial S}{\partial \sigma_{ij}}\right)_{T,\sigma'} = \left(\frac{\partial e_{ij}}{\partial T}\right)_{\sigma} = \alpha_{ij}, \tag{3.34}$$

where α_{ij} is the thermal strain tensor.[16] We will use this Maxwell relation below.

Writing the strain component as a function of all the stress tensor components and temperature, $e_{ij} = e_{ij}(\sigma, T)$, we have

$$\begin{aligned} \mathrm{d}e_{ij} &= \left(\frac{\partial e_{ij}}{\partial \sigma_{kl}}\right)_{T} \mathrm{d}\sigma_{kl} + \left(\frac{\partial e_{ij}}{\partial T}\right)_{\sigma} \mathrm{d}T \\ &= s^{T}_{ijkl}\mathrm{d}\sigma_{kl} + \alpha_{ij}\mathrm{d}T, \end{aligned} \tag{3.35}$$

where s^{T}_{ijkl} is a component of the isothermal elastic compliance tensor. Since $\mathrm{d}e_{ij}$ is an exact differential we have the following Maxwell relation:

$$\left(\frac{\partial s^{T}_{ijkl}}{\partial T}\right)_{\sigma} = \left(\frac{\partial \alpha_{ij}}{\partial \sigma_{kl}}\right)_{T,\sigma'}. \tag{3.36}$$

This Maxwell relation shows that the temperature dependence of the isothermal elastic compliance tensor at constant stress is determined by the stress dependence of the thermal strain tensor at constant temperature.

[15] If there is more than one atomic species present and/or more than one atom associated with each lattice site we have to include the relaxations of atoms not carried to their final positions by the strain tensor.

[16] In the partial derivative $(\partial S/\partial \sigma_{ij})_{T,\sigma'}$ all stress components except σ_{ij} and σ_{ji} are held constant as well as the temperature. In $(\partial e_{ij}/\partial T)_{\sigma}$ all stress components are held constant.

Writing the entropy as a function of the stress tensor and temperature, $S = S(\sigma, T)$, we have

$$\begin{aligned} dS &= \left(\frac{\partial S}{\partial \sigma_{kl}}\right)_{T,\sigma'} d\sigma_{kl} + \left(\frac{\partial S}{\partial T}\right)_{\sigma} dT \\ &= V\alpha_{kl} d\sigma_{kl} + \frac{C_\sigma}{T} dT, \end{aligned} \tag{3.37}$$

where we have used the Maxwell relation of eqn. 3.34 and C_σ is the specific heat at constant stress of the crystal. During an adiabatic change $dS = 0$. Using eqn. 3.37 we find that the corresponding change in temperature when the stress is changed adiabatically is as follows:

$$dT = -\frac{VT}{C_\sigma} \alpha_{kl} d\sigma_{kl}. \tag{3.38}$$

Putting this expression for dT into eqn. 3.35 we obtain the following relationship between the adiabatic elastic compliance tensor s^S_{ijkl} and the isothermal elastic compliance tensor s^T_{ijkl}:

$$s^T_{ijkl} = s^S_{ijkl} + \frac{VT}{C_\sigma} \alpha_{ij} \alpha_{kl}. \tag{3.39}$$

This equation shows that the isothermal and adiabatic elastic compliances of a crystal differ owing to the anharmonicity of atomic interactions, without which the thermal strain is zero. A similar conclusion was reached in section 2.8 where the anharmonicity of atomic interactions was shown to be responsible for the difference between isothermal and adiabatic stresses.

Exercise 3.7

By writing the stress as a function of strain and temperature, $\sigma_{ij} = \sigma_{ij}(e, T)$, show that

$$d\sigma_{ij} = c^T_{ijkl} de_{kl} - \beta_{ij} dT, \tag{3.40}$$

where $\beta_{ij} = -(\partial \sigma^T_{ij}/\partial T)_e$ is the thermal stress tensor (see section 2.8). Setting $d\sigma_{ij} = 0$, show that

$$\beta_{ij} = c^T_{ijkl} \alpha_{kl}. \tag{3.41}$$

$d\sigma_{ij}(e, T)$ is an exact differential. Show that the corresponding Maxwell relation is as follows:

$$\left(\frac{\partial c^T_{ijkl}}{\partial T}\right)_e = -\left(\frac{\partial \beta_{ij}}{\partial e_{kl}}\right)_{T,e'}. \tag{3.42}$$

continued

Exercise 3.7 Continued

The temperature dependence of the isothermal elastic constant tensor at constant strain is seen here to be determined by the strain dependence of the thermal stress tensor at constant temperature.

By considering the entropy S as a function of strain and temperature show that at constant entropy:

$$dT = -\frac{VT}{C_e}\beta_{kl}de_{kl}, \tag{3.43}$$

where $C_e = T(\partial S/\partial T)_e$ is the specific heat of the volume V at constant strain. Inserting this expression for dT into eqn. 3.40 show that the isothermal and adiabatic elastic constants are related as follows:

$$c^T_{ijkl} = c^S_{ijkl} - \frac{VT}{C_e}\beta_{ij}\beta_{kl}. \tag{3.44}$$

Substituting $d\sigma_{kl} = c^T_{klmn}de_{mn} - \beta_{kl}dT$ into eqn. 3.37, and using eqn. 3.41, show that

$$C_\sigma - C_e = VT\alpha_{ij}C^T_{ijkl}\alpha_{kl}. \tag{3.45}$$

Since $\alpha_{ij}C^T_{ijkl}\alpha_{kl} > 0$ for mechanical stability this equation demonstrates that C_σ is always larger than C_e.

For copper at 25°C the linear thermal expansion coefficient is 17.1×10^{-6} K^{-1}, the density is 8.96 g cm^{-3} and the specific heat is 0.385 J g^{-1}K^{-1}. The adiabatic elastic constants at 25°C are $c^S_{11} = 1.684 \times 10^{11}$ Pa, $c^S_{12} = 1.214 \times 10^{11}$ Pa and $c^S_{44} = 0.754 \times 10^{11}$ Pa.

Using eqns. 3.41 and 3.44 calculate the isothermal elastic constants c^T_{11}, c^T_{12}, c^T_{44} at 25°C.

Hence calculate $C_P - C_V$ for copper at 25°C, where C_P and C_V are the specific heats in units of J g^{-1} K^{-1} at constant pressure and constant volume respectively.

3.9 Problem set 3

1. In a face-centred cubic crystal with lattice constant a the 12 nearest neighbours of an atom are at $\pm a/2[110], \pm a/2[1\bar{1}0], \pm a/2[101], \pm a/2[10\bar{1}], \pm a/2[011]$, $\pm a/2[01\bar{1}]$. Consider a model of the crystal in which the bonding is represented by linear springs, with spring constant k, between nearest neighbours only. By considering the elastic energy density of the crystal when an arbitrary small elastic strain is applied calculate the elastic constants c_{11}, c_{12} and c_{44}.

   Answers: $c_{11} = 2k/a$ and $c_{12} = c_{44} = k/a$. It is interesting to note that the anisotropy ratio for this simple model is 2 and Poisson's ratio is $\frac{1}{3}$.

2. This question and the next are based on sections 5.3.1 and 5.3.2 of the book[17] by Finnis.[18] Let a homogeneous strain tensor be written as $e_{ij} = \gamma T_{ij}$, where γ is a scalar which scales the magnitude of the strain and the matrix elements T_{ij} are constant. The component X_i of a vector $\mathbf{X}$ becomes $(\delta_{ij} + \gamma T_{ij})X_j$. If $X = |\mathbf{X}|$ show that

$$\frac{dX}{d\gamma} = \frac{X_i T_{ij} X_j}{X} \tag{3.46}$$

$$\frac{d^2X}{d\gamma^2} = \frac{X_i T_{ij} T_{jk} X_k}{X} - \frac{(X_i T_{ij} X_j)^2}{X^3}. \tag{3.47}$$

3. In a pairwise interaction model of a crystal the interaction energy between any two atoms is a function of their separation only. The total energy is then the sum of all such pairwise interactions. Let $V(X)$ be the interaction energy between two atoms separated by X. For example, $V(X)$ might be a Lennard-Jones potential:

$$V(X) = \varepsilon\left[\left(\frac{X_0}{X}\right)^{12} - 2\left(\frac{X_0}{X}\right)^{6}\right],$$

which leads to repulsion between atoms separated by $X < X_0$, attraction when $X > X_0$ and $V(X)$ has a minimum at $X = X_0$ where $V(X) = -\varepsilon$. Consider a *cubic* crystal in which there is just one atom per lattice site. Let the origin of a Cartesian coordinate system be located at an atomic site, and orient the axes along the $\langle 100 \rangle$ directions of the crystal. The energy of the atom at the origin is then

$$E_c = \frac{1}{2}\sum_n V(X^{(n)}),$$

where $X^{(n)}$ is the distance to atom n, and the factor of one half is because each pairwise interaction is shared by two atoms.

In this question you will prove the following:

- The equilibrium volume of the crystal is determined by the equation:

$$\frac{1}{6\Omega}\sum_n X^{(n)} V'(X^{(n)}) = 0,$$

where the prime denotes differentiation.

[17] Finnis, MW, *Interatomic forces in condensed matter*, Oxford University Press: Oxford (2003). ISBN 978-0198509776.

[18] Michael William Finnis 1950–, British materials physicist.

- The elastic constants c_{11}, c_{12} and c_{44} are as follows:

$$c_{11} = \frac{1}{6\Omega}\sum_n V'(X^{(n)})X^{(n)}$$
$$+\left(V''(X^{(n)})\left(X^{(n)}\right)^2 - V'(X^{(n)})X^{(n)}\right)F(X_1^{(n)},X_2^{(n)},X_3^{(n)})$$

$$c_{12} = \frac{1}{6\Omega}\sum_n \left(V''(X^{(n)})\left(X^{(n)}\right)^2 - V'(X^{(n)})X^{(n)}\right)H(X_1^{(n)},X_2^{(n)},X_3^{(n)})$$

$$c_{44} = \frac{1}{12\Omega}\sum_n V'(X^{(n)})X^{(n)}$$
$$+\frac{1}{6\Omega}\sum_n \left(V''(X^{(n)})\left(X^{(n)}\right)^2 - V'(X^{(n)})X^{(n)}\right)H(X_1^{(n)},X_2^{(n)},X_3^{(n)}),$$

where

$$F(X_1^{(n)},X_2^{(n)},X_3^{(n)}) = \frac{\left(X_1^{(n)}\right)^4 + \left(X_2^{(n)}\right)^4 + \left(X_3^{(n)}\right)^4}{\left(X^{(n)}\right)^4}$$

$$H(X_1^{(n)},X_2^{(n)},X_3^{(n)}) = \frac{\left(X_1^{(n)}\right)^2\left(X_2^{(n)}\right)^2 + \left(X_2^{(n)}\right)^2\left(X_3^{(n)}\right)^2 + \left(X_3^{(n)}\right)^2\left(X_1^{(n)}\right)^2}{\left(X^{(n)}\right)^4}$$
$$= \frac{1}{2}\left(1 - F(X_1^{(n)},X_2^{(n)},X_3^{(n)})\right).$$

- At the equilibrium volume of the crystal it follows that $c_{12} = c_{44}$ for all pairwise interaction models of cubic crystals.

The equilibrium volume of the crystal is determined by the condition that the pressure arising from all the pairwise interactions is zero. Mathematically, this amounts to the condition that $\mathrm{d}E_c/\mathrm{d}\Omega = 0$ where Ω is the atomic volume. Setting the tensor T_{ij} of the previous question equal to δ_{ij} show that

$$\frac{\mathrm{d}E_c}{\mathrm{d}\Omega} = \frac{1}{3\Omega}\left(\frac{\mathrm{d}E_c}{\mathrm{d}\gamma}\right)_{\gamma=0}.$$

Hence show that

$$\frac{dE_c}{d\Omega} = \frac{1}{6\Omega}\sum\nolimits_n X^{(n)} V'(X^{(n)}). \tag{3.48}$$

In a pairwise interaction model the volume per atom in cubic crystal with just one per lattice site is determined by the condition that the above sum is zero.

For an arbitrary homogeneous strain $e_{ij} = \gamma T_{ij}$ the elastic energy per atom in the cubic crystal is

$$E_{el} = \frac{1}{2}\gamma^2 \Omega c_{ijkl} T_{ij} T_{kl},$$

so that

$$\frac{d^2 E_{el}}{d\gamma^2} = \Omega c_{ijkl} T_{ij} T_{kl}.$$

Show that

$$\frac{d^2 E_c}{d\gamma^2} = \frac{1}{2}\sum_n V''(X^{(n)})\left(\frac{dX^{(n)}}{d\gamma}\right)^2 + V'(X^{(n)})\frac{d^2 X^{(n)}}{d\gamma^2},$$

where the derivatives with respect to γ are given by eqn. 3.46 and eqn. 3.47.

Equating the change in E_c to the elastic energy we obtain

$$c_{ijkl} T_{ij} T_{kl} = \frac{1}{2\Omega}\sum_n V''(X^{(n)})\left(\frac{dX^{(n)}}{d\gamma}\right)^2 + V'(X^{(n)})\frac{d^2 X^{(n)}}{d\gamma^2}. \tag{3.49}$$

By choosing $T_{ij} = \delta_{ij}$ show that

$$c_{ijkl} T_{ij} T_{kl} = 3(c_{11} + 2c_{12})$$

$$\frac{dX^{(n)}}{d\gamma} = X^{(n)}$$

$$\frac{d^2 X^{(n)}}{d\gamma^2} = 0$$

and hence

$$c_{11} + 2c_{12} = \frac{1}{6\Omega}\sum_n V''(X^{(n)})\left(X^{(n)}\right)^2, \tag{3.50}$$

which is three times the bulk modulus.

To obtain an expression for $C' = \frac{1}{2}(c_{11} - c_{12})$ we may set $T_{ij} = \delta_{i1}\delta_{j1} - \delta_{i2}\delta_{j2}$. Then $c_{ijkl}T_{ij}T_{kl} = 2(c_{11} - c_{12}) = 4C'$. Using cubic symmetry show that

$$C' = \frac{1}{24\Omega}\sum_n V''(X^{(n)})\left(X^{(n)}\right)^2\left(3F(X_1^{(n)}, X_2^{(n)}, -1)\right)$$

$$+ 3V'(X^{(n)})X^{(n)}\left[1 - F(X_1^{(n)}, X_2^{(n)}, X_3^{(n)})\right]. \tag{3.51}$$

Using eqn. 3.50 and eqn. 3.51 derive the formulae for c_{11} and c_{12} stated above.

To obtain c_{44} we may set $T_{ij} = \delta_{i1}\delta_{j2} + \delta_{i2}\delta_{j1} + \delta_{i2}\delta_{j3} + \delta_{i3}\delta_{j2} + \delta_{i3}\delta_{j1} + \delta_{i1}\delta_{j3}$. Show that

$$c_{ijkl}T_{ij}T_{kl} = 12c_{44}$$

$$\frac{\mathrm{d}X^{(n)}}{\mathrm{d}\gamma} = \frac{2\left(X_1^{(n)}X_2^{(n)} + X_2^{(n)}X_3^{(n)} + X_3^{(n)}X_1^{(n)}\right)}{X^{(n)}}$$

$$\frac{\mathrm{d}^2X^{(n)}}{\mathrm{d}\gamma^2} = \frac{2\left[\left(X_1^{(n)}\right)^2 + \left(X_2^{(n)}\right)^2 + \left(X_3^{(n)}\right)^2\right]}{X^{(n)}}$$

$$+ \frac{2\left[\left(X_1^{(n)}\right)\left(X_2^{(n)}\right) + \left(X_2^{(n)}\right)\left(X_3^{(n)}\right) + \left(X_3^{(n)}\right)\left(X_1^{(n)}\right)\right]}{X^{(n)}}$$

$$- \frac{4\left[\left(X_1^{(n)}\right)\left(X_2^{(n)}\right) + \left(X_2^{(n)}\right)\left(X_3^{(n)}\right) + \left(X_3^{(n)}\right)\left(X_1^{(n)}\right)\right]^2}{\left(X^{(n)}\right)^3}.$$

Hence, using cubic symmetry derive the formula stated above for c_{44}.

In the first question of this problem set the harmonic springs between nearest neighbours are described by $V(X) = \frac{1}{2}k\left[X - (a/\sqrt{2})\right]^2$, where $a/\sqrt{2}$ is the equilibrium bond length. Show that the equations derived in this question for c_{11}, c_{12} and c_{44} are consistent with the elastic constants obtained in the first question.

Comments: The difference $c_{12} - c_{44} = -(1/12\Omega)\sum_n V'(X^{(n)})X^{(n)}$ is called the *Cauchy pressure*. In a model where all the cohesion is provided by only pairwise interactions the Cauchy pressure is zero when the crystal is at its equilibrium volume. Therefore, $c_{12} = c_{44}$ for all pairwise interaction models of cubic crystals at their equilibrium volumes. The experimental fact that c_{12} and c_{44} are not equal to each other in any cubic crystal at equilibrium indicates that the pairwise interaction model fails as a description of atomic interactions. Not knowing any better in the nineteenth century Cauchy assumed a pairwise interaction model for all atomic interactions and showed there are six such relations between the elastic constants. They are: $c_{23} = c_{44}$; $c_{14} = c_{56}$; $c_{31} = c_{55}$; $c_{25} = c_{46}$; $c_{12} = c_{66}$; $c_{45} = c_{36}$. In a cubic crystal $c_{23} = c_{31} = c_{12}$ and $c_{66} = c_{55} = c_{44}$ so we expect c_{12} to equal c_{44}. These six equalities are known as the Cauchy relations and Cauchy used them to argue (incorrectly) that the maximum number of independent elastic constants is not 21 but 15. The Cauchy relations are now only of historical significance. But the extent to which they are violated is an indication of how well atomic interactions in a crystal may be described by pair potentials only.

In metals with nearly free electrons, such as aluminium and the alkali metals, the cohesive energy of the crystal is described quite accurately by a term that depends only on the average electron density in the metal together with a sum of pairwise interaction energies which are themselves dependent on the local electron density.[19] The presence of the density-dependent energy and the density dependence of the pairwise interactions ensures that the Cauchy pressure is not zero, and that $c_{12} \neq c_{44}$. Similarly, in transition metals the Finnis–Sinclair model[20] also ensures $c_{12} \neq c_{44}$ through the addition to a sum of pairwise interaction energies a new term that is the square root of a sum of pairwise interactions. The square root ensures the Cauchy pressure is not zero.

4. With respect to arbitrary rotations of the coordinate system prove that c_{ijij} and c_{iijj} are invariant in all crystals. In a cubic crystal show that the invariant $c_{ijij} - c_{iijj}$ is equal to $6(c_{12} - c_{44})$.
5. Show that $c_{ijij} = c_{11} + c_{22} + c_{33} + 2(c_{44} + c_{55} + c_{66})$ and $c_{iijj} = c_{11} + c_{22} + c_{33} + 2(c_{12} + c_{13} + c_{23})$. Show that c_{iijj} is directly related to the bulk modulus in any crystal structure.
6. Consider a polycrystal in which all crystals are elastically anisotropic and identical except for the orientations of their crystal axes. In general, when an arbitrary homogeneous stress is applied to the polycrystal the strain generated within each crystal is different owing to the elastic anisotropy. Consequently to maintain continuity of displacements and tractions at the grain boundaries additional stress fields are generated. These additional stresses are called *compatibility stresses*, and they are often of the same order of magnitude as the applied stress. However, there is a notable exception to this general rule. Show that if a purely hydrostatic stress is applied to a polycrystal, in which all crystals have the same *cubic* structure, there are no compatibility stresses required at the grain boundaries because the strain in all crystals is the same pure dilation.

[19] See Chapter 6 of the book by Finnis.
[20] See section 7.10 of the book by Finnis.

7. In a monoclinic crystal the only rotational symmetry is a two fold rotation axis along x_3. Prove that the following eight elastic constants are zero: c_{14}, c_{15}, c_{24}, c_{25}, c_{34}, c_{35}, c_{46}, c_{56}.
8. In an orthorhombic crystal there are two fold rotation axes along each of the Cartesian axes x_1, x_2, x_3. Starting from the reduced elastic constant matrix of the monoclinic crystal prove that the effect of the two fold rotation axes along x_1 or x_2 is to make the following four additional elastic constants zero: c_{16}, c_{26}, c_{36}, c_{45}. Hence show there are nine independent elastic constants in an orthorhombic crystal.
9. In a tetragonal crystal the x_3-axis is a four fold rotational symmetry axis. Starting from the reduced elastic constant matrix of the orthorhombic crystal prove that the effect of the four-fold rotation axis is to make: $c_{11} = c_{22}$; $c_{13} = c_{23}$; $c_{44} = c_{55}$. Hence show that there are six independent elastic constants in a tetragonal crystal with point group 422 (D_4): c_{11}, c_{12}, c_{13}, c_{33}, c_{44}, c_{66}. The only difference between this elastic constant matrix and the elastic constant matrix for hexagonal crystals is that c_{66} is no longer independent in a hexagonal crystal: $c_{66} = \frac{1}{2}(c_{11} - c_{12})$.
10. Show that in hexagonal crystals the elastic constant matrix is invariant with respect to rotations about the x_3-axis. The elastic constant matrix in a hexagonal crystal is as follows:

$$\mathbf{c} = \begin{bmatrix} c_{11} & c_{12} & c_{13} & 0 & 0 & 0 \\ c_{12} & c_{11} & c_{13} & 0 & 0 & 0 \\ c_{13} & c_{13} & c_{33} & 0 & 0 & 0 \\ 0 & 0 & 0 & c_{44} & 0 & 0 \\ 0 & 0 & 0 & 0 & c_{44} & 0 \\ 0 & 0 & 0 & 0 & 0 & \frac{1}{2}(c_{11} - c_{12}) \end{bmatrix}. \tag{3.52}$$

    This is why hexagonal crystals are described as 'transversely isotropic'.
11. Show that the following restrictions must apply to the elastic constants in a hexagonal crystal: $c_{11} > 0$; $c_{11} > c_{12}$; $c_{44} > 0$; $c_{33}(c_{11} + c_{12}) > 2c_{13}^2$. Hence deduce that $c_{33} > 0$; $c_{11} + c_{12} > 0$.
12. If the elastic constants of a crystal are averaged over all possible orientations with respect to a fixed coordinate system the elastic constants obtained are those of an isotropic medium. To obtain the elastic constants λ and μ of this isotropic medium we can use the two invariants of the elastic constant tensor c_{ijij} and c_{iijj} mentioned in problem 4.

    Show that in an isotropic medium $c_{ijij} = 3\lambda + 12\mu$ and $c_{iijj} = 9\lambda + 6\mu$.

    In a cubic crystal show that $c_{ijij} = 3c_{11} + 6c_{44}$ and $c_{iijj} = 3c_{11} + 6c_{12}$.

Hence show that

$$\lambda = c_{12} + \frac{1}{5}(c_{11} - c_{12} - 2c_{44})$$
$$\mu = c_{44} + \frac{1}{5}(c_{11} - c_{12} - 2c_{44}).$$

Hence show that the average values of c_{11}, c_{12} and c_{44} over all orientations of a cubic crystal are given by

$$\begin{aligned} 5\langle c_{11}\rangle &= 3c_{11} + 2c_{12} + 4c_{44} \\ 5\langle c_{12}\rangle &= c_{11} + 4c_{12} - 2c_{44} \\ 5\langle c_{44}\rangle &= c_{11} - c_{12} + 3c_{44} \end{aligned} \tag{3.53}$$

where $\langle c_{11}\rangle = \langle c_{12}\rangle + 2\langle c_{44}\rangle$. These averages may be obtained directly from eqns. 3.24, 3.27 and 3.26 respectively.

4

The Green's function in linear elasticity

4.1 Differential equation for the displacement field

Imagine you are applying pressure with your finger against a piece of supported rubber. You see the rubber distorts until you can't push any further. Equilibrium has been achieved between the force you are applying through your finger and the stresses created within the rubber by the distortion. This is the physics of eqn. 2.15: $\sigma_{ij,j}(\mathbf{x}) + f_i(\mathbf{x}) = 0$. The force f_i acting at $\mathbf{x}$ creates a stress field with a divergence that exactly balances the force and keeps the body in equilibrium. We can make this more explicit by using the divergence theorem. Let the body force be a point force located at $\mathbf{x}_0$, such that $f_i(\mathbf{x}) = \mathcal{F}_i \delta(\mathbf{x} - \mathbf{x}_0)$. Then we can write

$$\int_{\mathcal{R}} f_i \, \mathrm{d}V = \int_{\mathcal{R}} \mathcal{F}_i \delta(\mathbf{x} - \mathbf{x}_0) \, \mathrm{d}V = \mathcal{F}_i = -\int_{\mathcal{R}} \sigma_{ij,j} \, \mathrm{d}V = -\int_{\mathcal{S}} \sigma_{ij} n_j \, \mathrm{d}S. \tag{4.1}$$

Here $\mathcal{R}$ is any region in the continuum containing the body force at $\mathbf{x}_0$. The surface of $\mathcal{R}$ is $\mathcal{S}$. The stresses created by the distortion of the medium in response to the body force at $\mathbf{x}_0$ give rise to tractions, $\sigma_{ij} n_j$, on the surface $\mathcal{S}$, which when integrated over the whole surface surrounding $\mathcal{R}$ exactly cancel the force $\mathcal{F}_i$.

If we substitute Hooke's law into the equilibrium condition $\sigma_{ij,j}(\mathbf{x}) + f_i(\mathbf{x}) = 0$ we obtain a differential equation for the strain field created by the body force: $c_{ijkl} e_{kl,j}(\mathbf{x}) + f_i(\mathbf{x}) = 0$. If we then use the relationship between the strain and the displacement field, $e_{kl}(\mathbf{x}) = \frac{1}{2}(u_{k,l}(\mathbf{x}) + u_{l,k}(\mathbf{x}))$, we obtain a second order differential equation for the displacement field:

$$c_{ijkl} u_{k,lj}(\mathbf{x}) + f_i(\mathbf{x}) = 0, \tag{4.2}$$

where we have used the symmetry of the elastic constant tensor to write $c_{ijkl} u_{l,k} = c_{ijlk} u_{l,k}$. The assertion that the strain field is the symmetrised gradient of the displacement

Physics of elasticity and crystal defects. Adrian P. Sutton, Oxford University Press (2020). © Adrian P. Sutton.
DOI: 10.1093/oso/9780198860785.001.0001

field, $e_{kl}(\mathbf{x}) = \frac{1}{2}(u_{k,l}(\mathbf{x}) + u_{l,k}(\mathbf{x}))$, ensures that the material surrounding the region of the body force fits together compatibly, that is with no holes or overlapping material.[1] This assertion cannot be made where the body forces act.

Equation 4.2 plays the same role in elasticity as Poisson's equation in electrostatics. The equivalent in elasticity of Laplace's equation in electrostatics is $c_{ijkl}u_{k,lj} = 0$. Whereas electrostatics relies on invisible electric fields and 'action at a distance' to transmit forces between charges, in elasticity the forces acting between defects in the material are conveyed by the elastic displacement field which is real and can sometimes be seen in detail in the electron microscope using modern imaging techniques.

The apparent simplicity of eqn. 4.2 is deceptive. It comprises three equations, one for each component of the body force. Each of these equations involves second derivatives of all three components of the displacement field. Below we consider the simplest case, which arises when the isotropic elastic approximation is made.

Boundary value problems in elasticity amount to finding solutions of eqn. 4.2 subject to boundary conditions of three principal types. The first is where displacements are prescribed on the surface of the body. The second is where tractions on the surface of the body are prescribed. The third is a mixture of the first two, where displacements are prescribed on parts of the surface of the body and tractions on the surface of the body where displacements are not prescribed. There are uniqueness theorems for the solutions of eqn. 4.2 for both simply connected and multiply connected bodies.[2]

4.1.1 Navier's equation

The elastic constant tensor in isotropic elasticity is conveniently expressed in eqn. 3.11. When this is inserted into eqn. 4.2 we obtain the following differential equation:

$$\mu u_{i,jj} + (\lambda + \mu)u_{k,ki} + f_i = 0, \tag{4.3}$$

which can be expressed in vector form as follows:

$$\mu\nabla^2\mathbf{u} + (\lambda + \mu)\nabla(\nabla\cdot\mathbf{u}) + \mathbf{f} = 0. \tag{4.4}$$

This is known as Navier's equation.[3]

[1] For an illuminating discussion of compatibility and incompatibility see section 10, p.65-73 of Mura, T, *Micromechanics of defects in solids*, 2nd edn., Kluwer: Dordrecht (1991). ISBN 90-247-3256-5. Toshio Mura 1925–2009. Japanese born US scientist and engineer.

[2] See section 6.2 of Teodosiu, C, *Elastic models of crystal defects*, Springer Verlag: Berlin (1982). ISBN 0-387-11226-X. Cristian Victor Teodosiu 1937–, Romanian mathematician, physicist and engineer.

[3] Claude-Louis Navier 1785–1836, French engineer and physicist.

Exercise 4.1

Derive eqn. 4.3 from eqn. 3.11 and eqn. 4.2.

4.2 The physical meaning of the elastic Green's function

4.2.1 Definition of the Green's function in linear elasticity

We consider an infinite homogeneous elastic medium in which there is a point force $\mathbf{f}$ acting at $\mathbf{x} = \mathbf{x}_0$. In linear elasticity the Green's function gives the elastic displacement field created by this point force:

$$u_i(\mathbf{x}) = G_{ij}(\mathbf{x} - \mathbf{x}_0) f_j(\mathbf{x}_0). \tag{4.5}$$

This may be taken as the definition of the Green's function in linear elasticity, although it does not by itself enable the Green's function to be evaluated. Since eqn. 4.2 is linear the displacement due to a distribution of body forces $\mathbf{f}(\mathbf{x})$ is found by linear superposition:

$$u_i(\mathbf{x}) = \int G_{ij}(\mathbf{x} - \mathbf{x}') f_j(\mathbf{x}') \mathrm{d}^3 x'. \tag{4.6}$$

The elastic fields of many structural defects in crystals arise from forces between atoms in the centre of the defect, otherwise known as the core of the defect. Atoms move from their perfect crystal positions until the forces on them are counteracted by forces from their neighbours, including those further from the defect. The neighbours of the neighbours move from their ideal crystal positions until the net forces on them return to zero, and so the displacement field spreads from the defect. The displacements each neighbour shell undergoes decay with distance from the defect because the forces are distributed among more atoms. For example, consider a missing atom in a crystal, which is called a vacancy. The atoms neighbouring the vacancy experience a net force as a result of the missing atom. They move in response to the net forces acting on them, which sets up forces on their neighbours, and so their neighbours also move but by generally smaller amounts. When equilibrium is re-established the net force acting on any atom is zero and an elastic displacement field is set up in the crystal which decays, in an infinite crystal, as the inverse square of the distance from the vacancy.

The Green's function is unlikely to estimate accurately the displacements of atoms in the core of the defect where the forces may be very large. In this region it is best to use an atomistic model with a sound description of atomic interactions. But it is generally found that at distances of no more than a nanometre or so from the defect core the description of the relaxation displacements is quite accurately described by linear anisotropic elasticity. Conversely, atomistic models are inappropriate for long-range interactions between defects partly because the numbers of atoms involved may be far greater than can be treated computationally, but also because elasticity theory is likely to be more accurate and certainly a lot more informative for long-range interactions.

Before we derive the equation that determines the Green's function in linear elasticity we note some of its properties. First, since the medium is assumed to be homogeneous it follows that G_{ij} depends only on the relative position of the field point $\mathbf{x}$ and the location $\mathbf{x}_0$ of the point force. If the material were inhomogeneous then G_{ij} would become a function of both $\mathbf{x}$ and $\mathbf{x}_0$. Secondly, $G_{ij}(\mathbf{x}) = G_{ij}(-\mathbf{x})$ because the infinite, homogeneous elastic continuum is everywhere centrosymmetric.

Thirdly, $G_{ij}(\mathbf{x}) = G_{ji}(\mathbf{x})$. This follows from a result in mechanics, known as Maxwell's reciprocity theorem[4], which may be stated as follows: the work done by a force $\mathbf{F}^{(1)}$ when its point of application is displaced by $\mathbf{u}^{(2)}$ due to another force $\mathbf{F}^{(2)}$ is equal to the work done by the force $\mathbf{F}^{(2)}$ when its point of application is displaced by $\mathbf{u}^{(1)}$ due to the force $\mathbf{F}^{(1)}$.

To prove this theorem we apply a point force $\mathbf{F}^{(1)}$ gradually at $\mathbf{x}^{(1)}$. It produces a displacement $\mathbf{u}^{(1)}$ at $\mathbf{x}^{(1)}$, and the work done is $W^{(1)} = \frac{1}{2}\mathbf{F}^{(1)} \cdot \mathbf{u}^{(1)}(\mathbf{x}^{(1)})$. The factor of a half is because the force is gradually built up from zero to $\mathbf{F}^{(1)}$, during which the displacement of $\mathbf{x}^{(1)}$ increases linearly from zero to $\mathbf{u}^{(1)}$.

Now we introduce the force $\mathbf{F}^{(2)}$ gradually at $\mathbf{x}^{(2)}$. The additional work done is $W^{(2)} + W^{(12)}$ where $W^{(2)} = \frac{1}{2}\mathbf{F}^{(2)} \cdot \mathbf{u}^{(2)}(\mathbf{x}^{(2)})$ and $W^{(12)} = \mathbf{F}^{(1)} \cdot \mathbf{u}^{(2)}(\mathbf{x}^{(1)})$. Note the absence of the factor of a half in $W^{(12)}$ because the force $\mathbf{F}^{(1)}$ at $\mathbf{x}^{(1)}$ already exists in full. Thus the total work done is $\frac{1}{2}\mathbf{F}^{(1)} \cdot \mathbf{u}^{(1)}(\mathbf{x}^{(1)}) + \frac{1}{2}\mathbf{F}^{(2)} \cdot \mathbf{u}^{(2)}(\mathbf{x}^{(2)}) + \mathbf{F}^{(1)} \cdot \mathbf{u}^{(2)}(\mathbf{x}^{(1)})$.

Repeat the process, but introduce the force $\mathbf{F}^{(2)}$ gradually at $\mathbf{x}^{(2)}$ and then introduce the force $\mathbf{F}^{(1)}$ gradually at $\mathbf{x}^{(1)}$. The total work done must be same, but now it is $\frac{1}{2}\mathbf{F}^{(1)} \cdot \mathbf{u}^{(1)}(\mathbf{x}^{(1)}) + \frac{1}{2}\mathbf{F}^{(2)} \cdot \mathbf{u}^{(2)}(\mathbf{x}^{(2)}) + \mathbf{F}^{(2)} \cdot \mathbf{u}^{(1)}(\mathbf{x}^{(2)})$. Thus, we arrive at Maxwell's reciprocity relation:

$$\mathbf{F}^{(1)} \cdot \mathbf{u}^{(2)}(\mathbf{x}^{(1)}) = \mathbf{F}^{(2)} \cdot \mathbf{u}^{(1)}(\mathbf{x}^{(2)}). \tag{4.7}$$

Since $u_i^{(1)}(\mathbf{x}^{(2)}) = G_{ij}(\mathbf{x}^{(2)} - \mathbf{x}^{(1)})F_j^{(1)}$ and $u_i^{(2)}(\mathbf{x}^{(1)}) = G_{ij}(\mathbf{x}^{(1)} - \mathbf{x}^{(2)})F_j^{(2)}$ then the left hand side of eqn. 4.7 becomes

$$F_i^{(1)} u_i^{(2)}(\mathbf{x}^{(1)}) = F_i^{(1)} G_{ij}(\mathbf{x}^{(1)} - \mathbf{x}^{(2)})F_j^{(2)}$$

and the right hand side becomes

$$F_i^{(2)} u_i^{(1)}(\mathbf{x}^{(2)}) = F_i^{(2)} G_{ij}(\mathbf{x}^{(2)} - \mathbf{x}^{(1)})F_j^{(1)}.$$

[4] James Clerk Maxwell FRS 1831–79. The theorem is in this paper: Maxwell, JC, L. On the calculation of the equilibrium and stiffness of frames, *Phil. Mag.* **27**, 294–9 (1864). http://www.tandfonline.com/doi/abs/10.1080/14786446408643668

Swopping the dummy indices i and j in the last line and using $G_{ij}(\mathbf{x}) = G_{ij}(-\mathbf{x})$ the right hand side becomes $F_j^{(2)} G_{ji}(\mathbf{x}^{(1)} - \mathbf{x}^{(2)}) F_i^{(1)}$. Equating this to the left hand side we obtain the final result, $G_{ij}(\mathbf{x}) = G_{ji}(\mathbf{x})$.

Green's functions can be defined for other *linear* differential equations. For example, for Poisson's equation in a vacuum, $\nabla^2 V(\mathbf{x}) = -\rho(\mathbf{x})/\varepsilon_0$, the Green's function is the familiar potential at $\mathbf{x}$ of a unit point charge at $\mathbf{x}'$: $G(\mathbf{x}-\mathbf{x}') = 1/(4\pi\varepsilon_0 |\mathbf{x}-\mathbf{x}'|)$. The linearity of Poisson's equation enables the potential at $\mathbf{x}$ of a charge density $\rho(\mathbf{x}')$ to be calculated as a linear superposition of the potentials from the charges $\rho(\mathbf{x}')\mathrm{d}^3x'$ in the distribution:

$$V(\mathbf{x}) = \int G(\mathbf{x}-\mathbf{x}')\rho(\mathbf{x}')\mathrm{d}^3x'$$
$$= \int \frac{\rho(\mathbf{x}')}{4\pi\varepsilon_0 |\mathbf{x}-\mathbf{x}'|}\mathrm{d}^3x'.$$

4.2.2 The equation for the Green's function in an infinite medium

In this section we derive a partial differential equation for the Green's function in linear elasticity. Consider a point force $\mathbf{F}$ applied at $\mathbf{x}_0$ in an infinite, homogeneous elastic continuum. From the definition of the Green's function, eqn. 4.5, this force sets up an elastic displacement field given by $u_i(\mathbf{x}) = G_{ij}(\mathbf{x}-\mathbf{x}_0)F_j(\mathbf{x}_0)$. Differentiating this displacement field and using Hooke's law we find the stress field associated with this displacement field is as follows:

$$\sigma_{kp}(\mathbf{x}) = c_{kpim} G_{ij,m}(\mathbf{x}-\mathbf{x}_0)F_j(\mathbf{x}_0),$$

where the derivative of the Green's function is with respect to x_m. Consider any region $\mathcal{R}$, with surface $\mathcal{S}$, containing $\mathbf{x}_0$. For mechanical equilibrium we must have

$$F_k(\mathbf{x}_0) + \int_{\mathcal{S}} \sigma_{kp}(\mathbf{x})\, n_p\, \mathrm{d}S(\mathbf{x}) = 0,$$

where n_p is the outward normal at $\mathbf{x}$ to the surface $\mathcal{S}$. Therefore,

$$F_k(\mathbf{x}_0) + \int_{\mathcal{S}} c_{kpim} G_{ij,m}(\mathbf{x}-\mathbf{x}_0)F_j(\mathbf{x}_0)\, n_p\, \mathrm{d}S(\mathbf{x}) = 0.$$

Applying the divergence theorem to the surface integral we transform it into a volume integral over $\mathcal{R}$:

$$F_k(\mathbf{x}_0) + \int_{\mathcal{R}} c_{kpim} G_{ij,mp}(\mathbf{x} - \mathbf{x}_0) F_j(\mathbf{x}_0)\,\mathrm{d}V = 0.$$

We bring $F_k(\mathbf{x}_0)$ inside the volume integral by writing it as $\int_{\mathcal{R}} \delta_{jk}\,\delta(\mathbf{x} - \mathbf{x}_0) F_j(\mathbf{x}_0)\mathrm{d}V$, where the integration is again over $\mathbf{x}$ inside the region $\mathcal{R}$. We then obtain

$$\int_{\mathcal{R}} \left[c_{kpim} G_{ij,mp}(\mathbf{x} - \mathbf{x}_0) + \delta_{jk}\,\delta(\mathbf{x} - \mathbf{x}_0) \right] F_j(\mathbf{x}_0)\,\mathrm{d}V = 0.$$

Since this must hold for all point forces at $\mathbf{x}_0$, and for all regions $\mathcal{R}$ containing $\mathbf{x}_0$, the expression in square brackets must be zero:

$$c_{kpim} G_{ij,mp}(\mathbf{x} - \mathbf{x}_0) + \delta_{jk}\,\delta(\mathbf{x} - \mathbf{x}_0) = 0. \tag{4.8}$$

This is the partial differential equation that enables us to calculate the Green's function.

Exercise 4.2

Show that $\delta(h\mathbf{x}) = (1/|h|^3)\delta(\mathbf{x})$, where h is any scaling factor. Therefore the delta function in eqn. 4.8 behaves as a homogeneous function of degree -3. Hence show that

$$G_{ij}(\mathbf{x} - \mathbf{x}_0) = \frac{1}{|\mathbf{x} - \mathbf{x}_0|} g_{ij}, \tag{4.9}$$

where g_{ij} depends only on the orientation of $\mathbf{x} - \mathbf{x}_0$. This separation of the Green's function into radial and orientational dependencies applies in all cases regardless of the degree of anisotropy.

4.2.3 Solving elastic boundary value problems with the Green's function

This section illustrates the usefulness of the Green's function for solving boundary value problems. To do this we do not need explicit functional forms for the Green's function, only its defining differential equation, eqn. 4.8. We will see how and why we can use the Green's function for an infinite medium even when we are dealing with a finite medium, which of course has a surface. We shall use two spatial variables $\mathbf{x}$ and $\mathbf{x}'$. Differentiation with respect to the primed variable will be indicated by a prime on the subscript, thus $\partial f/\partial x'_m \equiv f_{,m'}$. If there is no prime on a subscript after a comma it signifies differentiation with respect to the corresponding component of $\mathbf{x}$.

All linear elastic fields must satisfy the equation of mechanical equilibrium, eqn. 4.2:

$$c_{kpim}\, u_{i,m'p'}(\mathbf{x}') + f_k(\mathbf{x}') = 0. \tag{4.10}$$

Replacing $\mathbf{x}_0$ in eqn. 4.8 with $\mathbf{x}'$ the Green's function $G_{ij}(\mathbf{x}-\mathbf{x}')$ satisfies

$$c_{kpim}\, G_{ij,mp}(\mathbf{x}-\mathbf{x}') + \delta_{jk}\,\delta(\mathbf{x}-\mathbf{x}') = 0. \tag{4.11}$$

Multiplying eqn. 4.11 by $u_k(\mathbf{x}')$ and eqn. 4.10 by $G_{kj}(\mathbf{x}-\mathbf{x}')$ and subtracting we obtain

$$\begin{aligned} &c_{kpim}\, G_{ij,mp}(\mathbf{x}-\mathbf{x}')u_k(\mathbf{x}') + \delta_{jk}\,\delta(\mathbf{x}-\mathbf{x}')u_k(\mathbf{x}') \\ &\quad - c_{kpim}\, u_{i,m'p'}(\mathbf{x}')G_{kj}(\mathbf{x}-\mathbf{x}') - f_k(\mathbf{x}')G_{kj}(\mathbf{x}-\mathbf{x}') = 0. \end{aligned}$$

Integrating this equation with respect to $\mathbf{x}'$ over a region $\mathcal{R}$ containing $\mathbf{x}$ we get

$$\begin{aligned} u_j(\mathbf{x}) = &\int_{\mathcal{R}} G_{jk}(\mathbf{x}-\mathbf{x}')f_k(\mathbf{x}')\mathrm{d}^3x' \\ &+ \int_{\mathcal{R}} c_{kpim}\, u_{i,m'p'}(\mathbf{x}')\, G_{kj}(\mathbf{x}-\mathbf{x}')\,\mathrm{d}^3x' \\ &- \int_{\mathcal{R}} c_{kpim}\, G_{ij,m'p'}(\mathbf{x}-\mathbf{x}')\, u_k(\mathbf{x}')\mathrm{d}^3x', \end{aligned}$$

where we have used $G_{ij,mp}(\mathbf{x}-\mathbf{x}') = G_{ij,m'p'}(\mathbf{x}-\mathbf{x}')$. We can combine the last two volume integrals as follows:

$$\begin{aligned} &\int_{\mathcal{R}} c_{kpim}\, u_{i,m'p'}(\mathbf{x}')\, G_{kj}(\mathbf{x}-\mathbf{x}')\,\mathrm{d}^3x' - \int_{\mathcal{R}} c_{kpim}\, G_{ij,m'p'}(\mathbf{x}-\mathbf{x}')\, u_k(\mathbf{x}')\mathrm{d}^3x' \\ &\quad = \int_{\mathcal{R}} c_{kpim}\left[u_{i,m'}(\mathbf{x}')\, G_{kj}(\mathbf{x}-\mathbf{x}') - u_k(\mathbf{x}')G_{ij,m'}(\mathbf{x}-\mathbf{x}')\right]_{,p'}\mathrm{d}^3x', \end{aligned}$$

where we have used the symmetry of the elastic constant tensor $c_{kpim} = c_{imkp}$ to achieve a cancellation of two additional terms that arise in the differentiation of the expression in square brackets. Applying the divergence theorem to the resulting volume integral we obtain the following surface integrals over the surface $\mathcal{S}$ of the region $\mathcal{R}$:

$$\int_{\mathcal{R}} c_{kpim}\left[u_{i,m'}(\mathbf{x}')\,G_{kj}(\mathbf{x}-\mathbf{x}') - u_k(\mathbf{x}')G_{ij,m'}(\mathbf{x}-\mathbf{x}')\right]_{,p'} \mathrm{d}^3x'$$
$$= \int_{\mathcal{S}} G_{jk}(\mathbf{x}-\mathbf{x}')\,c_{kpmi}\,u_{i,m'}(\mathbf{x}')\,n_{p'}\,\mathrm{d}^2x' - \int_{\mathcal{S}} G_{ji,m'}(\mathbf{x}-\mathbf{x}')\,c_{mikp}\,u_k(\mathbf{x}')\,n_{p'}\,\mathrm{d}^2x',$$

where we observe that $c_{kpmi}\,u_{i,m'}\,n_{p'} = \sigma_{kp}(\mathbf{x}')\,n_{p'} = t_k(\mathbf{x}')$ is the traction acting on $\mathcal{S}$ at $\mathbf{x}'$. Replacing the last two volume integrals in the above equation for $u_j(\mathbf{x})$ with these two surface integrals we obtain finally:

$$\begin{aligned} u_j(\mathbf{x}) = &\int_{\mathcal{R}} G_{jk}(\mathbf{x}-\mathbf{x}')f_k(\mathbf{x}')\,\mathrm{d}^3x' \\ &+ \int_{\mathcal{S}} G_{jk}(\mathbf{x}-\mathbf{x}')\,t_k(\mathbf{x}')\mathrm{d}^2x' \\ &- \int_{\mathcal{S}} G_{ji,m'}(\mathbf{x}-\mathbf{x}')\,c_{mikp}\,u_k(\mathbf{x}')\,n_{p'}\,\mathrm{d}^2x'. \end{aligned} \tag{4.12}$$

Here are some observations to illustrate the usefulness of this result. First, consider a distribution of body force, such as that created by a structural defect, inside a finite body with free surfaces. If we use only the first line of eqn. 4.12 to evaluate the displacement field in the finite body the answer will be wrong because the Green's function is constructed for an infinite body. In particular, this displacement field will predict tractions at the surface of the finite body, when the surface should be free of such tractions. To correct for this we may calculate the displacement field caused by an equal and opposite distribution of surface tractions using the second line of eqn. 4.12. We may then use the superposition principle and add this to the displacements caused by the distribution of body forces in the first line to obtain the correct solution with surfaces free of tractions. In this way we are able to use the Green's function for an infinite body to solve a problem in a finite body, which is very convenient because all bodies are finite in practice. Secondly, if there are additional tractions applied to the surface of the body they may be included in the surface integral on the second line. The surface integral on the third line is used when displacements are specified on the surface $\mathcal{S}$. Here the surface may include a cut made from the external surface of the body into some point inside the body, where there is a discontinuity in the displacement across the cut. This will be very useful when we discuss dislocations in Chapter 6.

It may be puzzling that the derivative of the Green's function appears in the third line of eqn. 4.12 to satisfy the boundary condition where a displacement is specified. To get some insight into this we offer a slightly modified version of an argument from the book[5]

[5] Landau, LD and Lifshitz, EM, *Theory of elasticity*, 3rd edn., Pergamon Press: Oxford (1986), section 27, p.111. ISBN 0-08-033916-6.

by Landau[6] and Lifshitz.[7] Consider an infinite planar fault with unit normal $\hat{\mathbf{n}}$ in an infinite continuum. The half space on the negative side of the fault has the displacement vector $\mathbf{u} = \tau/2$, and $\mathbf{u} = -\tau/2$ in the positive half space. To deal with the singularity in the displacement gradient in the plane of the fault we assign a finite thickness Δ to the fault, and subsequently take the limit $\Delta \to 0$. The displacement gradient is then $u_{i,k} = -\tau_i n_k/\Delta$ inside the fault and zero outside. The strain tensor is the constant value $e_{ik} = -\frac{1}{2}(\tau_i n_k + n_i \tau_k)/\Delta$ inside the fault and zero outside. Thus the stress changes discontinuously at the planes where the faulted region begins and ends. Discontinuous changes in stress are unphysical because they correspond to infinite forces. Therefore we use the equilibrium condition $\sigma_{im,m} + f_i = 0$ to introduce sheets of body force to cancel these discontinuities. The displacement field is then given by

$$
\begin{aligned}
u_i(\mathbf{x}) &= \int_{\text{all space}} G_{ij}(\mathbf{x}-\mathbf{x}')f_j(\mathbf{x}')\mathrm{d}^3x' \\
&= -\int_{\text{all space}} G_{ij}(\mathbf{x}-\mathbf{x}')\sigma_{jm,m'}(\mathbf{x}')\mathrm{d}^3x' \\
&= -\int_{\text{all space}} G_{ij}(\mathbf{x}-\mathbf{x}')c_{jmkl}e_{kl,m'}(\mathbf{x}')\mathrm{d}^3x' \\
&= -\int_{\text{all space}} \left\{\left(G_{ij}(\mathbf{x}-\mathbf{x}')c_{jmkl}e_{kl}(\mathbf{x}')\right)_{,m'} - G_{ij,m'}(\mathbf{x}-\mathbf{x}')c_{jmkl}e_{kl}(\mathbf{x}')\right\}\mathrm{d}^3x' \\
&= -\int_{\text{fault surfaces}} \left(G_{ij}(\mathbf{x}-\mathbf{x}')c_{jmkl}e_{kl}(\mathbf{x}')\right) n_{m'}\mathrm{d}^2x' \\
&\quad + \int_{\text{all space}} G_{ij,m'}(\mathbf{x}-\mathbf{x}')c_{jmkl}e_{kl}(\mathbf{x}')\mathrm{d}^3x'.
\end{aligned}
$$

The surface integral is over the surfaces of the fault region where the displacement gradient begins and ends. In the limit $\Delta \to 0$ these surface integrals cancel because the surface normal changes sign on either side of the fault. The region of integration in the volume integral in the last line reduces to the volume occupied by the fault, which is Δ per unit area; this cancels the Δ in the denominator of e_{kl}. As $\Delta \to 0$ we obtain

[6] Lev Davidovich Landau 1908–68, Nobel Prize-winning Soviet physicist.
[7] Evgeny Mikhailovich Lifshitz ForMemRS 1915–85, Soviet physicist.

$$u_i(\mathbf{x}) = -\int_{\text{fault plane}} G_{ij,m'}(\mathbf{x}-\mathbf{x}')c_{jmkl}\frac{1}{2}(\tau_k n_l + n_k \tau_l)\mathrm{d}^2x'$$
$$= -\int_{\text{fault plane}} G_{ij,m'}(\mathbf{x}-\mathbf{x}')c_{jmkl}\tau_k n_l \mathrm{d}^2x', \tag{4.13}$$

where we have used the symmetry of the elastic constant tensor $c_{jmkl} = c_{jmlk}$ in the final integral.

In summary, to create a fault with normal $\hat{\mathbf{n}}$ and a constant relative translation of $\boldsymbol{\tau}$ we introduce dipolar sheets of forces on either side of the fault. It is because the constant displacement is created by force dipoles that the derivative of the Green's function is involved. There is an analogy here with a jump in the electrostatic potential on either side of an interface at which there are dipolar sheets of charges.

4.3 A general formula for the Green's function in anisotropic elastic media

We turn now to deriving a general formula for the Green's function in an anisotropic elastic continuum. We shall follow the treatment given in the review[8] by Bacon, Barnett and Scattergood. There are only two cases where the Green's function can be obtained in closed form analytically for any direction within a crystal, and they are the cases of elastic isotropy and hexagonal symmetry. However, in the third problem of the set at the end of this chapter perturbation theory is used to derive an approximate analytic form of the elastic Green's function in a cubic crystal when the anisotropy ratio differs only slightly from unity.

We start from eqn. 4.8:

$$c_{jlms} G_{mp,ls}(\mathbf{x}) + \delta_{jp}\,\delta(\mathbf{x}) = 0,$$

where we have used the translational invariance of the infinite, homogeneous continuum to replace $\mathbf{x}-\mathbf{x}_0$ with just $\mathbf{x}$. Taking the Fourier transform of this equation we obtain

$$c_{jlms} k_l k_s \tilde{G}_{mp}(\mathbf{k}) = \delta_{jp},$$

[8] Bacon, DJ, Barnett, DM and Scattergood RO, Anisotropic continuum theory of lattice defects, *Prog. Mater. Sci.* **23**, 51–262 (1979). ISBN 0080242472. https://doi.org/10.1016/0079-6425(80)90007-9. David J Bacon, British materials physicist and engineer. David M Barnett, US materials physicist and engineer. Ronald O Scattergood, US materials engineer.

where

$$\tilde{G}_{mp}(\mathbf{k}) = \int G_{mp}(\mathbf{x}) e^{i\mathbf{k}\cdot\mathbf{x}} \mathrm{d}^3 x$$

$$G_{mp}(\mathbf{x}) = \frac{1}{(2\pi)^3} \int \tilde{G}_{mp}(\mathbf{k}) e^{-i\mathbf{k}\cdot\mathbf{x}} \mathrm{d}^3 k.$$

We now introduce a matrix (kk) that plays a central role in many aspects of anisotropic elasticity. It is where all the information about the elastic constants is stored:

$$(kk)_{jm} = c_{jlms} k_l k_s, \tag{4.14}$$

Note that (kk) is a symmetric matrix. The 3×3 matrix of Fourier transforms is just the inverse of the matrix (kk):

$$\tilde{G}_{mp}(\mathbf{k}) = \left[(kk)^{-1}\right]_{mp},$$

where we put square brackets around $(kk)^{-1}$ to make clear that $\tilde{G}_{mp}(\mathbf{k})$ is the mp-element of the inverse matrix of (kk) not the inverse of the matrix element $(kk)_{mp}$. Taking the inverse transform we obtain

$$G_{mp}(\mathbf{x}) = \frac{1}{(2\pi)^3} \int \left[(kk)^{-1}\right]_{mp} e^{-i\mathbf{k}\cdot\mathbf{x}} \mathrm{d}^3 k. \tag{4.15}$$

We introduce some additional vectors to help us evaluate this triple integral. Let the unit vectors $\hat{\mathbf{e}}_1, \hat{\mathbf{e}}_2, \hat{\mathbf{e}}_3$ be aligned with the Cartesian axes used to define the elastic constant tensor. We may express $\mathbf{x}$ in spherical polar coordinates defined with respect to $\hat{\mathbf{e}}_1$, $\hat{\mathbf{e}}_2$, $\hat{\mathbf{e}}_3$ as follows:

$$\mathbf{x} = x\hat{\rho} = x[\sin\theta\cos\phi\,\hat{\mathbf{e}}_1 + \sin\theta\sin\phi\,\hat{\mathbf{e}}_2 + \cos\theta\,\hat{\mathbf{e}}_3],$$

where we have also introduced x as the length of $\mathbf{x}$ and $\hat{\rho}$ as the direction of $\mathbf{x}$. It is useful to introduce two more orthonormal vectors $\hat{\mathbf{a}}$ and $\hat{\mathbf{b}}$ which are perpendicular to $\mathbf{x}$, such that $\hat{\mathbf{a}}, \hat{\mathbf{b}}$ and $\hat{\rho}$ form a right-handed set, $\hat{\mathbf{a}} \times \hat{\mathbf{b}} = \hat{\rho}$:

$$\hat{\mathbf{a}} = \sin\phi\,\hat{\mathbf{e}}_1 - \cos\phi\,\hat{\mathbf{e}}_2$$
$$\hat{\mathbf{b}} = \cos\theta\cos\phi\,\hat{\mathbf{e}}_1 + \cos\theta\sin\phi\,\hat{\mathbf{e}}_2 - \sin\theta\,\hat{\mathbf{e}}_3$$
$$\hat{\rho} = \sin\theta\cos\phi\,\hat{\mathbf{e}}_1 + \sin\theta\sin\phi\,\hat{\mathbf{e}}_2 + \cos\theta\,\hat{\mathbf{e}}_3.$$

Let k be the length of the vector $\mathbf{k}$ and let $\hat{\xi}$ be the direction of $\mathbf{k}$. Then $\mathbf{k}$ may be expressed in the coordinate system $\hat{\mathbf{a}}, \hat{\mathbf{b}}, \hat{\rho}$ as follows:

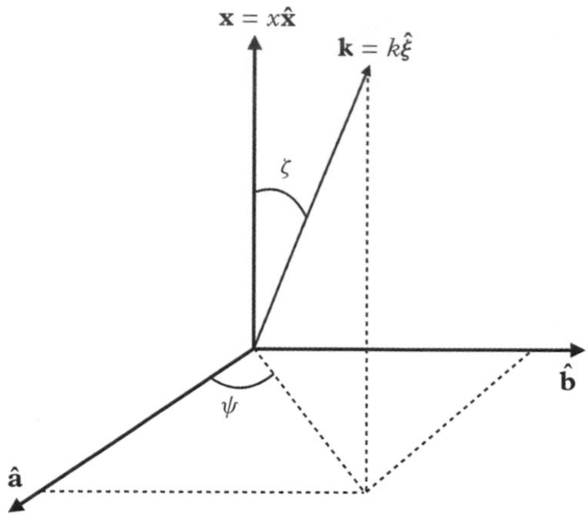

Figure 4.1 *To illustrate the vectors and angles used to define* **x** *and* **k**.

$$\mathbf{k} = k\hat{\boldsymbol{\xi}} = k\left[\sin\zeta\cos\psi\hat{\mathbf{a}} + \sin\zeta\sin\psi\hat{\mathbf{b}} + \cos\zeta\hat{\rho}\right],$$

where ζ is the angle between **x** and **k**, see Fig. 4.1.

At this point we recover the separation of variables in eqn. 4.9. Since $(kk) = k^2(\xi\xi)$ then the Fourier transform $\tilde{G}_{mp}(\mathbf{k})$ is given by

$$\tilde{G}_{mp}(\mathbf{k}) = [(kk)^{-1}]_{mp} = \frac{1}{k^2}\left[(\xi\xi)^{-1}\right]_{mp},$$

where we recognise $1/k^2$ as being proportional to the Fourier transform of $1/x$ and the directional dependence of g_{mp} is contained in the inverse transform of $\left[(\xi\xi)^{-1}\right]_{mp}$. We note that since (kk) is symmetric, so is its inverse, and therefore $G_{mp}(\mathbf{x})$ is also symmetric, as required by the Maxwell reciprocity theorem. Since $\mathrm{d}^3k = k^2\sin\zeta\,\mathrm{d}k\,\mathrm{d}\zeta\,\mathrm{d}\psi$ and $\mathbf{k}\cdot\mathbf{x} = kx\cos\zeta$, eqn. 4.15 becomes

$$G_{mp}(\mathbf{x}) = \frac{1}{(2\pi)^3}\int_0^{2\pi}\mathrm{d}\psi\int_0^{\pi}\mathrm{d}\zeta\,\sin\zeta\left[(\xi\xi)^{-1}\right]_{mp}\int_0^{\infty}\mathrm{d}k\frac{1}{\cancel{k^2}}\cancel{k^2}\cos(kx\cos\zeta).$$

The integration over k may be evaluated as follows:

$$\int_0^{\infty}\mathrm{d}k\cos(kx\cos\zeta) = \frac{1}{2}\int_{-\infty}^{\infty}\mathrm{d}k e^{ikx\cos\zeta} = \pi\delta(x\cos\zeta) = \frac{\pi}{x\sin\zeta}\delta(\zeta - \pi/2).$$

This simplifies the evaluation of $G_{mp}(\mathbf{x})$ considerably:

$$G_{mp}(\mathbf{x}) = \frac{1}{(2\pi)^3}\int_0^{2\pi} d\psi \int_0^{\pi} d\zeta \cancel{\sin\zeta}\left[(\xi\xi)^{-1}\right]_{mp}\frac{\pi}{x\cancel{\sin\zeta}}\delta(\zeta-\pi/2)$$

$$= \frac{1}{8\pi^2 x}\int_0^{2\pi}\left[(\xi\xi)^{-1}\right]_{mp} d\psi. \tag{4.16}$$

The anticipated separation into radial $(1/x)$ and orientational dependences is now explicit. The delta function $\delta(\zeta - \pi/2)$ requires that the integration with respect to ψ is around the unit circle perpendicular to the vector $\mathbf{x}$, with its centre at the origin. On this circle $\hat{\xi} = \hat{\mathbf{a}}\cos\psi + \hat{\mathbf{b}}\sin\psi$. Thus, the orientational dependence of the Green's function resides in this line integral. In most cases of elastic anisotropy the line integral may be evaluated only numerically. But it is only one integral to be evaluated for each direction $\hat{\rho}$, in contrast to the three integrals in the inverse Fourier transform of eqn. 4.15, and it is numerically stable. Finally, it is stressed that eqn. 4.16 is valid in *all* crystal symmetries. This remarkable result was first shown by Synge.[9] The derivation of eqn. 4.16 appears in his book *The hypercircle in mathematical physics*, Cambridge University Press: Cambridge (1957), p.411–3.

4.4 The Green's function in an isotropic elastic medium

In an isotropic medium the Green's function[10] can be determined directly by taking the inverse Fourier transform in eqn. 4.15.

Using eqn. 3.11 for the elastic constants in an isotropic elastic medium the matrix element $(kk)_{jm}$ is as follows:

$$(kk)_{jm} = \mu k^2\delta_{jm} + (\lambda+\mu)k_j k_m. \tag{4.17}$$

As we have seen in the previous section, the inverse of this matrix is the matrix of Fourier transforms $(kk)_{jm}\tilde{G}_{mp}(\mathbf{k}) = \delta_{jp}$. Rather than taking the brute force approach and evaluating the inverse of (kk) directly, we may multiply both sides of $(kk)_{jm}\tilde{G}_{mp}(\mathbf{k}) = \delta_{jp}$ by k_j and obtain an equation for $k_m\tilde{G}_{mp}(\mathbf{k})$:

$$k_m\tilde{G}_{mp}(\mathbf{k}) = \frac{k_p}{(\lambda+2\mu)k^2}.$$

[9] John Lighton Synge FRS 1897–1995, Irish mathematician and physicist.

[10] The solution in an isotropic medium for the displacement field of a point force was first derived by Lord Kelvin FRS in 1848, in a short paper entitled *Note on the integration of the equations of equilibrium of an elastic solid*, which can be found in Article 37 of Volume 1 of his *Mathematical and Physical Papers*, p.97, https://archive.org/details/mathematicaland01kelvgoog

When this is substituted into $(\mu k^2 \delta_{jm} + (\lambda+\mu)k_j k_m)\tilde{G}_{mp}(\mathbf{k}) = \delta_{jp}$ we find

$$\tilde{G}_{jp}(\mathbf{k}) = \frac{\delta_{jp}}{\mu k^2} - \frac{(\lambda+\mu)k_j k_p}{\mu(\lambda+2\mu)k^4}. \tag{4.18}$$

It may be shown by contour integration that

$$\int \frac{1}{k^2} e^{-i\mathbf{k}\cdot\mathbf{x}} \mathrm{d}^3 k = \frac{2\pi^2}{x}$$

$$\int \frac{k_j k_p}{k^4} e^{-i\mathbf{k}\cdot\mathbf{x}} \mathrm{d}^3 k = \pi^2 \left(\frac{\delta_{jp}}{x} - \frac{x_j x_p}{x^3} \right),$$

where $x = |\mathbf{x}|$. Using these integrals to evaluate the inverse Fourier transform we obtain

$$\begin{aligned} G_{jp}(\mathbf{x}) &= \frac{1}{8\pi\mu(\lambda+2\mu)x} \left((\lambda+3\mu)\delta_{jp} + (\lambda+\mu)\frac{x_j x_p}{x^2} \right) \\ &= \frac{1}{16\pi\mu(1-\nu)x} \left((3-4\nu)\delta_{jp} + \frac{x_j x_p}{x^2} \right). \end{aligned} \tag{4.19}$$

We shall use this result extensively.

4.5 The multipole expansion

Consider a point defect in a crystal such as a foreign atom occupying a site that would normally be occupied by a host atom, that is, a substitutional point defect. Let $\mathbf{f}^{(n)}$ be the excess force exerted on the nth host atom when the site is occupied by the foreign atom as compared to when the site is occupied by a host atom. At equilibrium the net force on each atom is zero. But as we saw in section 2.4 the excess forces exerted by the defect on the host atoms in this equilibrium state are the source of the stress field it creates in the host. We call the excess force that the defect exerts on a host atom a 'defect force'. In this equilibrium state the defect may be viewed as a collection of defect forces that sum to zero because their sum is the negative of the net force on the defect atom, which is zero at equilibrium. The elastic displacement field generated by this collection of defect forces may be computed using the superposition principle and the Green's function. This does involve approximations because the Green's function and the superposition principle assume linear elasticity is valid. But beyond some distance from the defect the predicted displacement, stress and strain fields are expected to be accurate. Indeed, when we wish to calculate interaction energies between defects over distances much larger than the spatial extent of their defect forces, elasticity theory is very accurate. It provides also unrivalled physical insight into such interactions.

Let the foreign atom be located at $\mathbf{d}$. Let the position of the nth neighbour relative to the foreign atom be $\mathbf{X}^{(n)}$, and let the defect force it experiences be $\mathbf{f}^{(n)}$. Therefore the displacement at $\mathbf{x}$ caused by the defect is

$$u_i(\mathbf{x}) = \sum_n G_{ij}(\mathbf{x} - \mathbf{d} - \mathbf{X}^{(n)}) f_j^{(n)}. \tag{4.20}$$

Within linear elasticity this is exact, but the approximations of linear elasticity are such that this expression becomes accurate only when $|\mathbf{x} - \mathbf{d}| \gg |\mathbf{X}^{(n)}|$. When this condition is satisfied we can expand the Green's function as a Taylor series:

$$G_{ij}(\mathbf{x} - \mathbf{d} - \mathbf{X}^{(n)}) = G_{ij}(\mathbf{x} - \mathbf{d}) - G_{ij,k}(\mathbf{x} - \mathbf{d}) X_k^{(n)} + \frac{1}{2} G_{ij,kl}(\mathbf{x} - \mathbf{d}) X_k^{(n)} X_l^{(n)} - \dots.$$

Inserting this expansion into eqn. 4.20 we obtain the following series:

$$\begin{aligned} u_i(\mathbf{x}) = {} & G_{ij}(\mathbf{x} - \mathbf{d}) \sum_n f_j^{(n)} \\ & - G_{ij,k}(\mathbf{x} - \mathbf{d}) \sum_n X_k^{(n)} f_j^{(n)} \\ & + \frac{1}{2} G_{ij,kl}(\mathbf{x} - \mathbf{d}) \sum_n X_k^{(n)} X_l^{(n)} f_j^{(n)} \\ & - \frac{1}{6} G_{ij,klm}(\mathbf{x} - \mathbf{d}) \sum_n X_k^{(n)} X_l^{(n)} X_m^{(n)} f_j^{(n)} \dots. \end{aligned} \tag{4.21}$$

This equation is called the multipole expansion of the displacement field of a point defect. At equilibrium the first term on the right hand side of eqn. 4.21 is zero because the defect forces sum to zero. The next term involves the first moment of the defect forces, which is called the elastic dipole tensor: $p_{kj} = \sum_n X_k^{(n)} f_j^{(n)}$. The next term involves the second moment of the defect forces, $q_{klj} = \sum_n X_k^{(n)} X_l^{(n)} f_j^{(n)}$ and is called the elastic quadrupole tensor. The next term involves the third moment of the defect forces $o_{klmj} = \sum_n X_k^{(n)} X_l^{(n)} X_m^{(n)} f_j^{(n)}$, and is called the elastic octupole tensor, and so on.

Exercise 4.3

Equilibrium requires not only that the net force on the point defect is zero but also that the net torque exerted on the defect atom is zero. Show that this condition is satisfied provided the elastic dipole tensor is symmetric: $p_{ij} = p_{ji}$.

The multipole expansion reveals how the elastic displacement field of a point defect decays with distance x. The dipole tensor gives rise to a displacement field that decays as $1/x^2$, the quadrupole as $1/x^3$, the octupole as $1/x^4$ and so on. The displacement fields from higher order poles become increasingly significant as the defect is approached. The orientational dependence of each of these contributions to the displacement field is contained in the derivatives of the Green's functions.

The multipole expansion reflects the symmetry of the atomic environment surrounding the point defect. This is very significant for those point defects that have particular symmetries, such as crowdions which have a distinct axial symmetry.

Finally, we note that eqn. 4.21 is an exemplar of 'multi-scale modelling'. It takes information from the atomic scale, namely the defect forces, and applies the theory of elasticity to determine the elastic field of the point defect at long range. It is a bridge between the atomic and continuum length scales, and perhaps this is why it was described by Leibfried and Breuer as the most important new concept in defect physics.[11] Kröner[12] was the first to introduce a systematic description of point defects in terms of force multipoles.[13]

The multipole expansion in electrostatics is often used to determine the electric field of a cluster of charges at distances larger than the cluster itself. The same idea applies to the multipole expansion in eqn. 4.21: it may be used to calculate the displacement field of a cluster of point defects at distances larger than the cluster itself. The sums in eqn. 4.21 are taken over all atoms within the cluster of point defects and the atoms with which they interact surrounding the cluster. In this way dipole, quadrupole, octupole,... tensors may be defined for point defect clusters of any size and used to calculate their long range elastic fields.

4.6 Relation between the Green's functions for an elastic continuum and a crystal lattice

Consider a crystal containing N lattice sites, with one atom at each lattice site. In harmonic lattice theory the potential energy of the distorted crystal is expanded to second order in the displacements of atoms from their perfect crystal positions. Using a slightly different notation from eqn. 3.28 the potential energy is as follows:

$$E = -\sum_n f_i(\mathbf{X}^{(n)})u_i(\mathbf{X}^{(n)}) + \frac{1}{2}\sum_n\sum_p S_{ij}(\mathbf{X}^{(n)} - \mathbf{X}^{(p)})u_i(\mathbf{X}^{(n)})u_j(\mathbf{X}^{(p)}). \tag{4.22}$$

We have replaced $u_i^{(n)}$ and $S_{ij}^{(np)}$ in eqn. 3.28 with $u_i(\mathbf{X}^{(n)})$ and $S_{ij}(\mathbf{X}^{(n)} - \mathbf{X}^{(p)})$, respectively. The force $f_i(\mathbf{X}^{(n)})$ is not zero this time because now there is a crystal defect that

[11] Leibfried, G and Breuer, N, *Point defect in metals I*, Springer-Verlag: Berlin (1978), p.146. ISBN 978-3662154489. Günther Leibfried 1915–77, German physicist.

[12] Ekkehart Kröner 1919–2000, German physicist.

[13] Kröner, E, *Kontinuumstheorie der Versetzungen und Eigenspannungen*, Springer-Verlag: Berlin (1958). ISBN 978-3540022619.

displaces the atom from its perfect crystal position at $\mathbf{X}^{(n)}$. It is equivalent to the body forces we have discussed in elasticity. This is different from eqn. 3.28, where the forces were all set to zero, because the perfect crystal configuration was being considered with no defects present.

By minimising the energy in eqn. 4.22 with respect to the displacements we obtain the following equation:

$$f_m(\mathbf{X}^{(n)}) = \sum_p S_{mj}(\mathbf{X}^{(n)} - \mathbf{X}^{(p)}) u_j(\mathbf{X}^{(p)}). \tag{4.23}$$

We want the inverse of this relationship, that is, we want an expression for the displacements in terms of the forces:

$$u_l(\mathbf{X}^{(n)}) = \sum_p G_{lh}(\mathbf{X}^{(n)} - \mathbf{X}^{(p)}) f_h(\mathbf{X}^{(p)}). \tag{4.24}$$

This equation is equivalent in the discrete lattice to eqn. 4.6 in the continuum, and it defines the crystal lattice Green's function. In this section we investigate the relationship between this crystal lattice Green's function and the Green's function in linear elasticity. To invert eqn. 4.23 we use discrete Fourier transforms defined by

$$\tilde{W}(\mathbf{k}) = \sum_n W(\mathbf{X}^{(n)}) e^{i\mathbf{k}\cdot\mathbf{X}^{(n)}} \tag{4.25}$$

$$W(\mathbf{X}^{(p)}) = \frac{1}{N} \sum_{\mathbf{k}} \tilde{W}(\mathbf{k}) e^{-i\mathbf{k}\cdot\mathbf{X}^{(p)}}. \tag{4.26}$$

Born von Karman periodic boundary conditions are applied to the crystal faces, and there are N wave vectors $\mathbf{k}$ in the Brillouin zone. Taking the discrete Fourier transform of eqn. 4.23 we obtain

$$\tilde{f}_m(\mathbf{k}) = \tilde{S}_{mj}(\mathbf{k}) \tilde{u}_j(\mathbf{k}), \tag{4.27}$$

where $\tilde{S}_{mj}(\mathbf{k})$ is a 3×3 matrix for each wave vector $\mathbf{k}$ in the Brillouin zone. Inverting this equation we obtain

$$\tilde{u}_l(\mathbf{k}) = \left[\tilde{S}^{-1}(\mathbf{k})\right]_{lh} \tilde{f}_h(\mathbf{k}). \tag{4.28}$$

Taking the discrete inverse Fourier transform we find

$$u_l(\mathbf{X}^{(n)}) = \frac{1}{N}\sum_{\mathbf{k}}\left[\tilde{S}^{-1}(\mathbf{k})\right]_{lh}\tilde{f}_h(\mathbf{k})e^{-i\mathbf{k}\cdot\mathbf{X}^{(n)}}$$

$$= \frac{1}{N}\sum_{\mathbf{k}}\left[\tilde{S}^{-1}(\mathbf{k})\right]_{lh}\left(\sum_p f_h(\mathbf{X}^{(p)})e^{i\mathbf{k}\cdot\mathbf{X}^{(p)}}\right)e^{-i\mathbf{k}\cdot\mathbf{X}^{(n)}}$$

$$= \sum_p\left(\frac{1}{N}\sum_{\mathbf{k}}\left[\tilde{S}^{-1}(\mathbf{k})\right]_{lh}e^{-i\mathbf{k}\cdot(\mathbf{X}^{(n)}-\mathbf{X}^{(p)})}\right)f_h(\mathbf{X}^{(p)}).$$

Comparing this with eqn. 4.24 we obtain the desired expression for the crystal lattice Green's function:[14]

$$G_{lh}(\mathbf{X}^{(n)}-\mathbf{X}^{(p)}) = \frac{1}{N}\sum_{\mathbf{k}}\left[\tilde{S}^{-1}(\mathbf{k})\right]_{lh}e^{-i\mathbf{k}\cdot(\mathbf{X}^{(n)}-\mathbf{X}^{(p)})}. \tag{4.29}$$

To aid comparison we quote here the Green's function from eqn. 4.15:

$$G_{mp}(\mathbf{x}) = \frac{1}{(2\pi)^3}\int\left[(kk)^{-1}\right]_{mp}e^{-i\mathbf{k}\cdot\mathbf{x}}\,\mathrm{d}^3k.$$

Since $S_{mj}(\mathbf{X}^{(n)}) = S_{mj}(-\mathbf{X}^{(n)})$, and $\sum_n S_{mj}(\mathbf{X}^{(n)}) = 0$, we may write[15]

$$\tilde{S}_{mj}(\mathbf{k}) = \sum_n S_{mj}(\mathbf{X}^{(n)})e^{i\mathbf{k}\cdot\mathbf{X}^{(n)}}$$

$$= \frac{1}{2}\sum_n S_{mj}(\mathbf{X}^{(n)})\left(e^{i\mathbf{k}\cdot\mathbf{X}^{(n)}} + e^{-i\mathbf{k}\cdot\mathbf{X}^{(n)}} - 2\right)$$

$$= \frac{1}{2}\sum_n S_{mj}(\mathbf{X}^{(n)})\left(2\cos\left(\mathbf{k}\cdot\mathbf{X}^{(n)}\right) - 2\right)$$

$$= -2\sum_n S_{mj}(\mathbf{X}^{(n)})\sin^2\left(\frac{\mathbf{k}\cdot\mathbf{X}^{(n)}}{2}\right).$$

[14] This Green's function has been used to calculate the displacement fields of defects in discrete lattices where atoms interact by harmonic potentials. See Tewary, VK, *Adv. Phys.* **22**, 757–810 (1973). https://doi.org/10.1080/00018737300101389

[15] See *Solid state physics* by Ashcroft, NW and Mermin, ND, pp. 437–40 (1976). ISBN 978-0030839931.

In the limit of long wavelengths the magnitude of the wave vector **k** becomes very small. In that limit we may write

$$\tilde{S}_{mj}(\mathbf{k}) \to -\frac{1}{2}\sum_{n} S_{mj}(\mathbf{X}^{(n)})\left(\mathbf{k}\cdot\mathbf{X}^{(n)}\right)^2$$
$$= \left[-\frac{1}{2}\sum_{n} S_{mj}(\mathbf{X}^{(n)})X_h^{(n)}X_l^{(n)}\right]k_h k_l.$$

Comparing with the expression for the elastic constant tensor c_{ikjl} in eqn. 3.31 we recognise the term in square brackets as being Ωc_{mhjl}, and therefore in the limit of long wavelengths $\tilde{S}_{mj}(\mathbf{k}) \to \Omega(kk)_{mj}$.

Each **k**-point in the Brillouin zone is associated with a volume in **k**-space of $(2\pi)^3/(N\Omega)$. Replacing the sum over **k**-points in eqn. 4.29 with a continuous integral over all **k**-space we find the crystal lattice Green's function, in the limit of long wavelengths, becomes

$$G_{mp}(\mathbf{X}^{(n)}) \to \frac{1}{N\Omega}\frac{N\Omega}{(2\pi)^3}\int \mathrm{d}^3k\left[(kk)^{-1}\right]_{mp} e^{-i\mathbf{k}\cdot\mathbf{X}^{(n)}}$$
$$= \frac{1}{(2\pi)^3}\int \mathrm{d}^3k\left[(kk)^{-1}\right]_{mp} e^{-i\mathbf{k}\cdot\mathbf{X}^{(n)}}, \tag{4.30}$$

which is identical to the Green's function in linear elasticity. We conclude that the elastic and crystal lattice Green's functions are equivalent only in the limit of long wavelengths compared to the distance from the force creating the displacement, so that $\mathbf{k}\cdot\mathbf{x} \ll 1$. At smaller wavelengths the lattice Green's function may be expected to deviate significantly from the Green's function in linear elasticity. The implication is that when we are studying the elastic displacement fields of defects with the Green's function in linear elasticity the fields nearer to the defect are less accurate than those further away. At larger distances from the defect the smaller wavelength contributions to the crystal lattice Green's function tend to cancel out, leaving only the longer wavelength components.

4.7 Eshelby's ellipsoidal inclusion

One of the most useful applications of the Green's function is to the solution of the elastic fields of inclusions. It is also a rare example of an analytic solution to a three-dimensional problem in elasticity. A region, which will become the inclusion, in an infinite, homogeneous, isotropic, elastic medium undergoes a transformation which changes its shape and or size. If the region were not constrained by the surrounding medium it would undergo a homogeneous 'transformation strain' e_{ij}^T, where the superscript T signifies it is the transformation strain. The transformation strain is sometimes called an eigenstrain or 'stress-free strain'. But given that the inclusion *is* constrained by the surrounding medium in reality, what is the final elastic state of the inclusion and the surrounding medium? Eshelby[16] formulated a set of operations for solving the problem, and a detailed solution for an ellipsoidal inclusion,[17] where the inclusion and the surrounding matrix have the same isotropic elastic constants. Applications include mechanical twinning and slip bands (see section 10.2.6), where the 'inclusion' is a region of the host material that undergoes a simple shear in response to an applied shear stress. It has also been applied to martensitic transformations where a region undergoes a spontaneous phase change involving a shape change with shear and dilation components in general. Mura has solved the inclusion problem in anisotropic elasticity.[18]

Eshelby's elegant thought-experiment to formulate and solve the ellipsoidal inclusion problem may be summarised as follows. Cut out from the medium the ellipsoidal region $\mathcal{R}$ that is to transform. Allow it to undergo the homogeneous transformation strain e^T_{ij} without any constraint applied to its surface $\mathcal{S}$. At this point there is no stress in the inclusion or in the matrix. The region $\mathcal{R}$ no longer fits back into the hole from which it was removed. Return the transformed region to its original shape by applying tractions $-\sigma^T_{ij}n_j$ to the surface $\mathcal{S}$, where $\sigma^T_{ij} = 2\mu e^T_{ij} + \lambda\delta_{ij}e^T_{kk}$. The ellipsoidal region $\mathcal{R}$ is then inserted back into the hole and the bonds across the interface are reformed. At this point there is no strain in the matrix or inclusion but there are tractions $-\sigma^T_{ij}n_j$ in the interface. These tractions are annihilated by applying an equal and opposite distribution $+\sigma^T_{ij}n_j$ which produce the constrained displacements u^C_i in the inclusion and the surrounding medium, from which strains e^C_{ij} and stresses σ^C_{ij} may be calculated. The final constrained stresses in the surrounding medium are σ^C_{ij}. But since the inclusion was already stressed by $-\sigma^T_{ij}$ the final constrained stresses in the inclusion are $\sigma^I_{ij} = \sigma^C_{ij} - \sigma^T_{ij}$.

The Green's function is used to calculate u^C_i:

$$u^C_i(\mathbf{x}) = \int_{\mathcal{S}} G_{ij}(\mathbf{x}-\mathbf{x}')\sigma^T_{jk}n_k \mathrm{d}S(\mathbf{x}'), \tag{4.31}$$

where the integral is over the surface $\mathcal{S}$ of the inclusion before it is transformed. The field point $\mathbf{x}$ may be inside or outside the inclusion. Eshelby showed that when this formulation is applied to an *ellipsoidal* inclusion the total final stress and strain are homogeneous inside the inclusion. Markenscoff [19] has proved[20] in isotropic and anisotropic elasticity that ellipsoids are the only shapes of inclusions in which the total stress and strain fields are homogeneous throughout them. The versatility of Eshelby's analysis stems in part from the wide range of shapes that can be described by ellipsoids, including needles, discs, prolate and oblate spheroids and spheres.

16 John Douglas Eshelby, FRS 1916–81, British physicist.

17 Eshelby, JD, *Proc. R. Soc. A* **241**, 376–96 (1957). https://doi.org/10.1098/rspa.1957.0133. This is one of the most highly cited papers in the mechanics of materials, possibly *the* most highly cited.

18 See Chapter 3 of Mura, T, *Micromechanics of defects in solids*, Kluwer Academic Publishers: Dordrecht (1991). ISBN 90-247-3256-5.

19 Xanthippi Markenscoff 1947–, Greek born, US theoretical materials scientist and engineer.

20 Markenscoff, X, *J. Elast.* **49**, 163–6 (1998). https://doi.org/10.1023/A:1007474108433

The surface integral in eqn. 4.31 is transformed into a volume integral over the interior $\mathcal{R}$ of the inclusion using the divergence theorem:

$$u_i^C(\mathbf{x}) = \int_{\mathcal{R}} G_{ij,k'}(\mathbf{x}-\mathbf{x}')\sigma_{jk}^T \, dV'$$

$$= -\int_{\mathcal{R}} G_{ij,k}(\mathbf{x}-\mathbf{x}')\sigma_{jk}^T \, dV'. \quad (4.32)$$

In these integrals the integration variable is $\mathbf{x}'$, as signified by the prime on the dV. In the first line the Green's function $G_{ij}(\mathbf{x}-\mathbf{x}')$ is differentiated with respect to $x_{k'}$. In the second line it is differentiated with respect to x_k, which introduces the minus sign.

Using eqn. 4.70 for the derivative of the Green's function the constrained displacement field becomes

$$u_i^C(\mathbf{x}) = \frac{\sigma_{jk}^T}{16\pi\mu(1-\nu)} \int_{\mathcal{R}} \frac{1}{|\mathbf{x}-\mathbf{x}'|^2} \left((3-4\nu)\rho_k\delta_{ij} - (\rho_j\delta_{ik} + \rho_i\delta_{jk}) + 3\rho_i\rho_j\rho_k\right) dV', \quad (4.33)$$

where ρ_i is the direction cosine $(x_i - x_i')/|\mathbf{x}-\mathbf{x}'|$, which specifies the direction of the vector from $\mathbf{x}'$ to $\mathbf{x}$. Exploiting the symmetry of the stress tensor $\sigma_{jk}^T = \sigma_{kj}^T$ this integral may be rewritten as follows:

$$u_i^C(\mathbf{x}) = \frac{\sigma_{jk}^T}{16\pi\mu(1-\nu)} \int_{\mathcal{R}} \frac{1}{|\mathbf{x}-\mathbf{x}'|^2} \left((1-2\nu)\left(\rho_k\delta_{ij} + \rho_j\delta_{ik}\right) - \rho_i\delta_{jk} + 3\rho_i\rho_j\rho_k\right) dV'$$

$$= \frac{\sigma_{jk}^T}{16\pi\mu(1-\nu)} \int_{\mathcal{R}} \frac{1}{|\mathbf{x}-\mathbf{x}'|^2} f_{ijk}(\boldsymbol{\rho}) dV', \quad (4.34)$$

where

$$f_{ijk}(\boldsymbol{\rho}) = (1-2\nu)\left(\rho_k\delta_{ij} + \rho_j\delta_{ik}\right) - \rho_i\delta_{jk} + 3\rho_i\rho_j\rho_k. \quad (4.35)$$

Using Hooke's law the integral in eqn. 4.34 may also be expressed in terms of the transformation strain tensor e_{jk}^T:

$$u_i^C(\mathbf{x}) = \frac{e_{jk}^T}{8\pi(1-\nu)} \int_{\mathcal{R}} \frac{1}{|\mathbf{x}-\mathbf{x}'|^2} \left((1-2\nu)\left(\rho_k\delta_{ji} + \rho_j\delta_{ik} - \rho_i\delta_{jk}\right) + 3\rho_i\rho_j\rho_k\right) \mathrm{d}V'$$

$$= \frac{e_{jk}^T}{8\pi(1-\nu)} \int_{\mathcal{R}} \frac{1}{|\mathbf{x}-\mathbf{x}'|^2} g_{ijk}(\boldsymbol{\rho})\,\mathrm{d}V', \tag{4.36}$$

where

$$g_{ijk}(\boldsymbol{\rho}) = (1-2\nu)\left(\rho_k\delta_{ij} + \rho_j\delta_{ik} - \rho_i\delta_{jk}\right) + 3\rho_i\rho_j\rho_k. \tag{4.37}$$

From afar the inclusion looks like a point defect. In that case, with the origin of the coordinate system at the centre of the inclusion, we have $|\mathbf{x}| \approx |\mathbf{x}-\mathbf{x}'|$. The displacement field of the inclusion is then

$$u_i^C(\mathbf{x}) = \frac{\sigma_{jk}^T V}{16\pi\mu(1-\nu)x^2} f_{ijk}(\boldsymbol{\rho}) \tag{4.38}$$

$$= \frac{e_{jk}^T V}{8\pi(1-\nu)x^2} g_{ijk}(\boldsymbol{\rho}), \tag{4.39}$$

where $\mathbf{x} = x\boldsymbol{\rho}$ and V is the volume of the inclusion. As we shall see in the next chapter the decay of the displacement field as the inverse square of the distance from the inclusion is consistent with the radial dependence of the displacement field of a point defect. The angular dependence of $u_i^C(\mathbf{x})$ is contained in $\sigma_{jk}^T f_{ijk}(\boldsymbol{\rho})$ and $e_{jk}^T g_{ijk}(\boldsymbol{\rho})$.

Consider a field point $\mathbf{x}$ *inside* the ellipsoid. It is convenient to redefine ρ_i as $(x_i' - x_i)/|\mathbf{x}-\mathbf{x}'|$. Since $g_{ijk}(\boldsymbol{\rho})$ is an odd function of $\boldsymbol{\rho}$ this redefinition will introduce a minus sign into eqn. 4.36. To evaluate the volume integral in eqn. 4.36 we divide $\mathcal{R}$ into an infinite number of infinitesimal cones, with their apices located at $\mathbf{x}$. Each cone axis is along a particular direction $\boldsymbol{\rho}$, and it terminates at the surface $\mathcal{S}$ of the ellipsoid. Let the length of the cone from $\mathbf{x}$ along the direction $\boldsymbol{\rho}$ to $\mathcal{S}$ be $L = L(\boldsymbol{\rho})$. If $\mathrm{d}\omega$ is the infinitesimal solid angle subtended at the cone apex the infinitesimal area of the base of a right circular cone is $L^2\mathrm{d}\omega$.

To evaluate the contribution of each infinitesimal cone to $\int_{\mathcal{R}} \mathrm{d}V'/|\mathbf{x}'-\mathbf{x}|^2$ consider a right circular cone with its apex at the origin of a Cartesian coordinate system, its axis along the x_1-axis, its length L, and the radius of its base R. We evaluate $\int \mathrm{d}V/|\mathbf{x}|^2$ for this cone and take the limit that R becomes infinitesimal. Consider an element of area $\mathrm{d}x_1\mathrm{d}x_2$ at $(x_1, x_2, 0)$ inside the cone. When this element of area is rotated about the x_1-axis it traces a ring of volume $2\pi x_2 \mathrm{d}x_1\mathrm{d}x_2$. The distance of all points on this ring from the origin is $|\mathbf{x}| = \sqrt{x_1^2 + x_2^2}$. Therefore, for this finite cone we have:

$$\int \frac{\mathrm{d}V}{|\mathbf{x}|^2} = \int_{x_1=0}^{L} \mathrm{d}x_1 \int_{x_2=0}^{Rx_1/L} \mathrm{d}x_2 \frac{2\pi x_2}{x_1^2 + x_2^2}$$

$$= \pi L \ln(1 + R^2/L^2). \tag{4.40}$$

Taking the limit $R \to 0$ this integral becomes $(\mathrm{d}S)/L$ where $\mathrm{d}S = \pi R^2$ is the area of the cone base. Since $\mathrm{d}S = L^2 \mathrm{d}\omega$ the contribution of an infinitesimal cone to $\int_{\mathcal{R}} \mathrm{d}V'/|\mathbf{x}' - \mathbf{x}|^2$ is $L(\boldsymbol{\rho})\mathrm{d}\omega$.

Remembering the minus sign arising from the redefinition of $\boldsymbol{\rho}$ the constrained displacement field inside the ellipsoid, eqn. 4.36, becomes the following:

$$u_i^C(\mathbf{x}) = -\frac{e_{jk}^T}{8\pi(1-\nu)} \int L(\boldsymbol{\rho}) g_{ijk}(\boldsymbol{\rho})\, \mathrm{d}\omega, \tag{4.41}$$

where the integral is over 4π steradians.[21]

To find $L(\boldsymbol{\rho})$ let $\mathbf{r} = \mathbf{x} + \xi\boldsymbol{\rho}$ be the vector equation of the straight line through $\mathbf{x}$ in the direction of $\boldsymbol{\rho}$, where ξ is an arbitrary real number. Let (X_1, X_2, X_3) be a point on the surface of the ellipsoid satisfying the equation:

$$\frac{X_1^2}{a_1^2} + \frac{X_2^2}{a_2^2} + \frac{X_3^2}{a_3^2} = 1, \tag{4.42}$$

where a_1, a_2, a_3 are the semi-axes of the ellipsoid. Then $L(\boldsymbol{\rho})$ is determined by the positive solution of the following quadratic equation:

$$\frac{(x_1 + L\rho_1)^2}{a_1^2} + \frac{(x_2 + L\rho_2)^2}{a_2^2} + \frac{(x_3 + L\rho_3)^2}{a_3^2} = 1. \tag{4.43}$$

The positive root is given by

$$L(\boldsymbol{\rho}) = \sqrt{\frac{f^2}{g^2} + \frac{e}{g}} - \frac{f}{g}, \tag{4.44}$$

[21] A steradian is a unit of solid angle.

where

$$e = 1 - \left(\frac{x_1^2}{a_1^2} + \frac{x_2^2}{a_2^2} + \frac{x_3^2}{a_3^2} \right)$$

$$f = \frac{x_1 \rho_1}{a_1^2} + \frac{x_2 \rho_2}{a_2^2} + \frac{x_3 \rho_3}{a_3^2}$$

$$g = \frac{\rho_1^2}{a_1^2} + \frac{\rho_2^2}{a_2^2} + \frac{\rho_3^2}{a_3^2}. \tag{4.45}$$

When eqn. 4.44 is inserted in eqn. 4.41, the square root, which is even in $\boldsymbol{\rho}$, multiplies $g_{ijk}(\boldsymbol{\rho})$, which is odd in $\boldsymbol{\rho}$. The contributions to the integrand in eqn. 4.41 from the square root for $\pm\boldsymbol{\rho}$ cancel.

Defining the vector $(\lambda_1, \lambda_2, \lambda_3) = (\rho_1/a_1^2, \rho_2/a_2^2, \rho_3/a_3^2)$ the constrained displacement field of eqn. 4.41 then becomes

$$u_i^C(\mathbf{x}) = \frac{e_{jk}^T x_m}{8\pi(1-\nu)} \int \frac{\lambda_m g_{ijk}}{g} \, \mathrm{d}\omega. \tag{4.46}$$

The constrained strains inside the ellipsoid are then as follows:

$$e_{il}^C = \frac{1}{2}\left(\frac{\partial u_i^C}{\partial x_l} + \frac{\partial u_l^C}{\partial x_i} \right) = \frac{e_{jk}^T}{16\pi(1-\nu)} \int \frac{\lambda_i g_{ijk} + \lambda_l g_{ijk}}{g} \, \mathrm{d}\omega. \tag{4.47}$$

We see the constrained strains, and hence the stresses, inside the ellipsoid are homogeneous throughout the ellipsoid. They depend only on its shape. They remain homogeneous also in anisotropic elasticity.

It is convenient to rewrite eqn. 4.47 as a tensor equation as follows:

$$e_{il}^C = S_{iljk} e_{jk}^T, \tag{4.48}$$

where S_{iljk} is often called the Eshelby tensor:

$$S_{iljk} = \frac{1}{16\pi(1-\nu)} \int \frac{\lambda_i g_{ijk} + \lambda_l g_{ijk}}{g} \, \mathrm{d}\omega. \tag{4.49}$$

It is not difficult to see that for a general ellipsoid terms of the form S_{1122} and S_{2211} are not equal. All terms S_{iljk} containing any component of $\boldsymbol{\rho}$ raised to an odd power are zero. Thus all terms coupling extensions to shears are zero, such as S_{2212}. Similarly all terms coupling different shear components are zero, such as S_{1213}. The non-zero components of S_{iljk} are those coupling extensions to extensions, such as S_{1111} and S_{1133}, and those coupling shear components to the same shear components, such as S_{1212}. In detail:

$$
\begin{aligned}
S_{1111} &= \frac{1}{8\pi(1-\nu)}\int \frac{\lambda_1 g_{111}}{g}\,\mathrm{d}\omega \\
&= \frac{(1-2\nu)}{8\pi(1-\nu)}\int \frac{\rho_1^2}{a_1^2 g}\mathrm{d}\omega + \frac{3a_1^2}{8\pi(1-\nu)}\int \frac{\rho_1^4}{a_1^4 g}\mathrm{d}\omega \\
&= \frac{(1-2\nu)}{8\pi(1-\nu)}I_1 + \frac{3a_1^2}{8\pi(1-\nu)}I_{11}
\end{aligned}
\tag{4.50}
$$

$$
\begin{aligned}
S_{1122} &= \frac{1}{8\pi(1-\nu)}\int \frac{\lambda_1 g_{122}}{g}\,\mathrm{d}\omega \\
&= \frac{a_2^2}{8\pi(1-\nu)}\int \frac{3\rho_1^2\rho_2^2}{a_1^2 a_2^2 g}\,\mathrm{d}\omega - \frac{(1-2\nu)}{8\pi(1-\nu)}\int \frac{\rho_1^2}{a_1^2 g}\mathrm{d}\omega \\
&= \frac{a_2^2}{8\pi(1-\nu)}I_{12} - \frac{(1-2\nu)}{8\pi(1-\nu)}I_1
\end{aligned}
\tag{4.51}
$$

$$
\begin{aligned}
S_{1212} &= \frac{1}{16\pi(1-\nu)}\int \frac{\lambda_1 g_{212} + \lambda_2 g_{112}}{g}\mathrm{d}\omega \\
&= \frac{(1-2\nu)}{16\pi(1-\nu)}\int \frac{\rho_1^2}{a_1^2 g} + \frac{\rho_2^2}{a_2^2 g}\mathrm{d}\omega + \frac{(a_1^2+a_2^2)}{16\pi(1-\nu)}\int \frac{3\rho_1^2\rho_2^2}{a_1^2 a_2^2 g}\mathrm{d}\omega \\
&= \frac{(1-2\nu)}{16\pi(1-\nu)}(I_1+I_2) + \frac{(a_1^2+a_2^2)}{16\pi(1-\nu)}I_{12}.
\end{aligned}
\tag{4.52}
$$

Other components of the Eshelby tensor are found by cyclic permutation of the indices in eqns. 4.50 to 4.52. The integrals I_1, I_{11} and I_{12} are defined as follows:

$$
\begin{aligned}
I_1 &= \int \frac{\rho_1^2}{a_1^2 g}\,\mathrm{d}\omega = 2\pi a_1 a_2 a_3 \int_0^\infty \frac{\mathrm{d}u}{(a_1^2+u)\Delta(u)} \\
I_{11} &= \int \frac{\rho_1^4}{a_1^4 g}\,\mathrm{d}\omega = 2\pi a_1 a_2 a_3 \int_0^\infty \frac{\mathrm{d}u}{(a_1^2+u)^2\Delta(u)} \\
I_{12} &= \int \frac{3\rho_1^2\rho_2^2}{a_1^2 a_2^2 g}\mathrm{d}\omega = 2\pi a_1 a_2 a_3 \int_0^\infty \frac{\mathrm{d}u}{(a_1^2+u)(a_2^2+u)\Delta(u)}.
\end{aligned}
\tag{4.53}
$$

The ω-integrations are over 4π steradians. The expressions on the right of each line in eqn. 4.53 were derived by Routh.[22] The function $\Delta(u) = \sqrt{(a_1^2+u)(a_2^2+u)(a_3^2+u)}$. Using eqn. 4.53 the following relations between the integrals, and their cyclic permutations, are easily deduced:

$$\begin{aligned} I_1 + I_2 + I_3 &= 4\pi \\ 3I_{11} + I_{12} + I_{13} &= 4\pi/a_1^2 \\ 3a_1^2 I_{11} + a_2^2 I_{12} + a_3^2 I_{13} &= 3I_1 \\ \frac{(I_1 - I_2)}{(a_2^2 - a_1^2)} &= I_{12}. \end{aligned} \tag{4.54}$$

The fourth line in eqn. 4.54 is obtained by splitting the factor $(a_1^2+u)^{-1}(a_2^2+u)^{-1}$ in I_{12} in eqn. 4.53 into partial fractions.

I_1, I_2 and I_3 may be expressed in terms of elliptic integrals.[23] Assuming $a_1 > a_2 > a_3$ we find

$$\begin{aligned} I_1 &= \frac{4\pi a_1 a_2 a_3}{(a_1^2 - a_2^2)\sqrt{(a_1^2 - a_3^2)}} \left[K(\theta, k^2) - E(\theta, k^2)\right] \\ I_2 &= \frac{4\pi a_1 a_2 a_3}{(a_1^2 - a_2^2)(a_2^2 - a_3^2)\sqrt{(a_1^2 - a_3^2)}} \\ &\quad \times \left[(a_1^2 - a_3^2) E(\theta, k^2) - (a_2^2 - a_3^2) K(\theta, k^2) - \frac{a_3}{a_1 a_2}(a_1^2 - a_2^2)\sqrt{(a_1^2 - a_3^2)}\right] \\ I_3 &= \frac{4\pi a_1 a_2 a_3}{(a_2^2 - a_3^2)\sqrt{(a_1^2 - a_3^2)}} \left[\frac{a_2\sqrt{(a_1^2 - a_3^2)}}{a_1 a_3} - E(\theta, k^2)\right], \end{aligned} \tag{4.55}$$

where the amplitude $\theta = \sin^{-1}\sqrt{(1 - a_3^2/a_1^2)}$ and the modulus $k^2 = (a_1^2 - a_2^2)/(a_1^2 - a_3^2)$. $K(\theta, z)$ is the incomplete elliptic integral of the first kind:

[22] Routh, EJ, *A treatise on analytical statics*, Vol. 2, Cambridge University Press: London (1892) p. 106–8. https://archive.org/details/treatiseonanalyt02routiala/page/n8. Edward John Routh FRS 1831-1907, British mathematician.

[23] Kellogg, OD, *Foundations of potential theory*, Dover Publications: New York (1954) p.192–6. ISBN 0486601447. Oliver Dimon Kellogg 1878–1932, US mathematician.

The integrals appear more explicitly in Gradshteyn, IS and Ryzhik, IM *Table of integrals, series and products*, 5th edn. corrected and enlarged by Alan Jeffrey, Academic Press Inc.: Orlando (1980). ISBN 0-12-294760-6. Expressions for I_1, I_2 and I_3 appear in 3.133, numbers 1, 7 and 13 respectively on p.220-222. Izrail Solomonovich Gradshteyn 1899–1958, Russian mathematician. Iosif Moiseevich Ryzhik ?–1941, Russian mathematician.

$$K(\theta, z) = \int_0^{\theta} \frac{d\phi}{\sqrt{1 - z\sin^2\phi}}. \tag{4.56}$$

$E(\theta, z)$ is the incomplete elliptic integral of the second kind:

$$E(\theta, z) = \int_0^{\theta} \sqrt{1 - z\sin^2\phi}\, d\phi. \tag{4.57}$$

Having evaluated any two of I_1, I_2 and I_3 the third follows immediately using the first line of eqn 4.54. The integrals I_{12}, I_{23} and I_{31} then follow from the last line of eqn. 4.54 and its cyclic permutations. I_{11}, I_{22} and I_{33} may then be deduced from the second or third lines of eqn. 4.54 and their cyclic permutations. Thus, once any two of I_1, I_2 and I_3 have been evaluated using eqn. 4.55 all the other integrals can be deduced without explicit evaluation.

The constrained stresses σ_{ij}^C inside the ellipsoid are deduced from the strains e_{kl}^C using Hooke's law and eqn. 4.48:

$$\sigma_{ij}^C = 2\mu S_{ijkl} e_{kl}^T + \delta_{ij} \frac{2\mu\nu}{1 - 2\nu} S_{mmkl} e_{kl}^T. \tag{4.58}$$

As explained at the beginning of this section the total stress inside the ellipsoid is $\sigma_{ij}^I = \sigma_{ij}^C - \sigma_{ij}^T$. In question 6 of the problem set of this chapter it is shown that the total elastic energy of the ellipsoidal inclusion and matrix is $-(V/2)\sigma_{ij}^I e_{ij}^T$. The elastic interaction energy between an ellipsoidal inclusion and an applied strain field is derived in question 10 of problem set 6, where it is also expressed as a volume integral taken over the inclusion only. Eshelby showed[24] how the elastic field outside an ellipsoidal inclusion may be calculated in detail. In that paper and the 1957 paper he also showed how his analysis may be adapted to determine how a uniform stress field is disturbed by an ellipsoidal inhomogeneity, where the local elastic constants differ from those of the surrounding medium, such as an ellipsoidal crack. In section 10.2.6 we model slip bands as ellipsoidal inclusions to develop a theory of work hardening.

4.8 The equation of motion and elastic waves

We shall use Hamilton's principle of classical mechanics[25] to derive the equation of motion of the continuum. Hamilton's principle states that the following integral is

[24] Eshelby, JD, *Proc. R. Soc. A* **252**, 561–9 (1959). https://doi.org/10.1098/rspa.1959.0173

[25] For example, see Goldstein, H, *Classical mechanics*, Addison-Wesley: Reading, MA (1980). ISBN 0-201-02969-3

stationary with respect to variations δu_i of the displacement field which vanish at times $t = t_1$ and $t = t_2$ and on the surface of the volume V:

$$\delta \int_{t_1}^{t_2} \mathrm{d}t \int_V \mathrm{d}V \mathcal{L}(u_i, \dot{u}_i, u_{i,j}) = 0. \tag{4.59}$$

$\mathcal{L}$ is the Lagrangian density, which is the difference between the kinetic energy and the potential energy densities:

$$\mathcal{L} = \frac{1}{2}\rho \dot{u}_i^2 - \left(-f_i u_i + \frac{1}{2} c_{ijkl} u_{i,j} u_{k,l}\right).$$

In this expression ρ is the mass density, $\dot{u}_i$ is the velocity of the medium at $\mathbf{x}$ and we have included a contribution from body forces f_i in the potential energy density. Substituting this Lagrangian density into eqn. 4.59 and performing the variation we obtain

$$\delta \int_{t_1}^{t_2} \mathrm{d}t \int_V \mathrm{d}V \mathcal{L}(u_i, \dot{u}_i, u_{i,j}) = \int_{t_1}^{t_2} \mathrm{d}t \int_V \mathrm{d}V\, \rho \dot{u}_i \delta \dot{u}_i + f_i \delta u_i - c_{ijkl} u_{i,j} \delta u_{k,l}$$

$$= \int_V \mathrm{d}V [\rho \dot{u}_i \delta u_i]_{t=t_1}^{t_2} - \int_{t_1}^{t_2} \mathrm{d}t \int_V \mathrm{d}V \rho \ddot{u}_i \delta u_i - f_i \delta u_i + \left(c_{ijkl} u_{i,j} \delta u_k\right)_{,l} - c_{ijkl} u_{i,jl} \delta u_k$$

$$= -\int_{t_1}^{t_2} \int_S c_{ijkl} u_{i,j} \delta u_k n_l \mathrm{d}S - \int_{t_1}^{t_2} \mathrm{d}t \int_V \mathrm{d}V (\rho \ddot{u}_i - f_i - \sigma_{ij,j}) \delta u_i \tag{4.60}$$

We have used $\delta u_i = 0$ at $t = t_1$ and $t = t_2$, and on the surface S of the body at all times. We have also used the divergence theorem to convert a volume integral into a surface integral. In order for the variation to be zero the integrand must be zero:

$$\rho \ddot{u}_i = f_i + \sigma_{ij,j}, \tag{4.61}$$

which is the equation of motion of the continuum.

4.8.1 Elastic waves

In the absence of forces in the equation of motion, eqn. 4.61, we try a wave solution of the form $u_i(\mathbf{x}, t) = A_i(\mathbf{k}, \omega) e^{i(\mathbf{k}\cdot\mathbf{x} - \omega t)}$, which leads to the following equation:

$$(\rho \omega^2 \delta_{ij} - (kk)_{ij}) A_j(\mathbf{k}, \omega) = 0. \tag{4.62}$$

This is a set of three equations in the wave amplitudes $A_i(\mathbf{k},\omega)$, and the condition for non-trivial solutions is that the following 3×3 determinant is zero:

$$\begin{vmatrix} (kk)_{11}-\rho\omega^2 & (kk)_{12} & (kk)_{13} \\ (kk)_{12} & (kk)_{22}-\rho\omega^2 & (kk)_{23} \\ (kk)_{13} & (kk)_{23} & (kk)_{33}-\rho\omega^2 \end{vmatrix} = 0. \tag{4.63}$$

In general this provides a set of three solutions for $\omega^2=\omega^2(\mathbf{k})$, which are known as the three branches of the dispersion relations. By writing $(kk)=k^2(\xi\xi)$ and by making ω/k the quantity to be determined the determinant can be rewritten as follows:

$$\begin{vmatrix} (\xi\xi)_{11}-\rho\omega^2/k^2 & (\xi\xi)_{12} & (\xi\xi)_{13} \\ (\xi\xi)_{12} & (\xi\xi)_{22}-\rho\omega^2/k^2 & (\xi\xi)_{23} \\ (\xi\xi)_{13} & (\xi\xi)_{23} & (\xi\xi)_{33}-\rho\omega^2/k^2 \end{vmatrix} = 0. \tag{4.64}$$

This makes it clear that ω is a linear function of k and that ω varies in general with the direction of $\mathbf{k}$. Thus, in general, the group velocity of the wave, $\mathbf{v}_g=\nabla_{\mathbf{k}}\omega(\mathbf{k})$, depends on the direction of the wave but not on its frequency. In particular, we note that in general the wave does not travel in the direction of its wave vector because in general $\mathbf{v}_g$ is not parallel to $\mathbf{k}$. Since the $(\xi\xi)$ matrix is symmetric its eigenvalues are always real and its eigenvectors are always perpendicular to each other. The eigenvectors are the amplitudes $\mathbf{A}(\mathbf{k},\omega)$ in eqn. 4.62 and their directions relative to the wave vector $\mathbf{k}$ determine whether the waves are longitudinal or transverse or a mixture of the two.

In the isotropic elastic approximation (kk) is given by eqn. 4.17. One solution of the dispersion equations, eqn. 4.63, is $c_l^2=\omega^2/k^2=(\lambda+2\mu)/\rho$, and is independent of the direction of $\mathbf{k}$, as expected in an isotropic medium. When this solution is substituted back into eqn. 4.62 it is found the wave amplitude is parallel to the wave vector and it is called the longitudinal wave or 'P-wave'. This wave produces alternating compression and dilation of the medium along the direction of propagation of the wave. The other two solutions are degenerate with $c_t^2=\omega^2/k^2=\mu/\rho$, and they are also independent of the direction of the wave vector. In these cases the wave amplitudes are perpendicular to each other and to the wave vector, and they are called transverse waves, or 'S-waves'. These waves produce shears of the medium and are not associated with local volume changes. For all three waves the group velocity is always parallel to the wave vector, and there is always one purely longitudinal wave and two purely transverse waves for all wave vectors. This combination of features of the propagation of elastic waves in isotropic materials is unique, and may be found in anisotropic materials only when the wave vector coincides with a direction of high rotational symmetry.

In section 4.6 we saw that the crystal lattice Green's function converged to the elastic Green's function in the limit of long wavelengths. For the same reason the dispersion relations for the crystal lattice converge to those of the elastic continuum in the limit of long wavelengths, that is, as $|\mathbf{k}| \to 0$.

4.9 The elastodynamic Green's function

Having analysed the properties of elastic waves we are in a position to tackle the time-dependent Green's function of elastodynamics. The time-dependent generalisation of eqn. 4.8 for the elastostatic Green's function is to consider a point force that is applied as an infinitely abrupt pulse at $t = 0$ that generates a displacement field that propagates away. In an isotropic medium we can expect a spherical longitudinal wave to travel outwards from the point pulse followed by two degenerate spherical shear waves. Thus in the equation of motion, eqn. 4.61, we set $f_i = \delta_{im}\delta(\mathbf{x})\delta(t)$ to describe a pulse of force of unit magnitude along the x_m-axis at the origin at time $t = 0$. The elastodynamic Green's function in an infinite, homogeneous elastic medium is then defined by the following partial differential equation:

$$c_{ijkl}G_{km,lj}(\mathbf{x}, t) + \delta_{im}\delta(\mathbf{x})\delta(t) = \rho\ddot{G}_{im}(\mathbf{x}, t), \tag{4.65}$$

where as usual the two dots over the Green's function signify the second derivative with respect to time.

To solve this equation we exploit the translational symmetry of the medium in space and time through four-dimensional Fourier transforms:

$$\tilde{f}(\mathbf{k}, \omega) = \int\int f(\mathbf{x}, t)\, e^{i(\mathbf{k}\cdot\mathbf{x}-\omega t)}\, \mathrm{d}^3x\, \mathrm{d}t \tag{4.66}$$

$$f(\mathbf{x}, t) = \frac{1}{(2\pi)^4}\int\int \tilde{f}(\mathbf{k}, \omega)\, e^{-i(\mathbf{k}\cdot\mathbf{x}-\omega t)}\, \mathrm{d}^3k\, \mathrm{d}\omega. \tag{4.67}$$

Applying the four-dimensional Fourier transform to eqn. 4.65 we obtain

$$\left((kk)_{ik} - \rho\omega^2\delta_{ik}\right)\tilde{G}_{km}(\mathbf{k}, \omega) = \delta_{im}, \tag{4.68}$$

from which it follows that

$$G_{km}(\mathbf{x}, t) = \int\int \left[((kk) - \rho\omega^2 I)^{-1}\right]_{km} e^{-i(\mathbf{k}\cdot\mathbf{x}-\omega t)}\, \mathrm{d}^3k\, \mathrm{d}\omega,$$

where I is the 3×3 identity matrix.

As with the elastostatic Green's function we resort to the isotropic elastic approximation to make further progress,[26] whereupon we have

$$((kk) - \rho\omega^2 I)_{ik} = (\mu k^2 - \rho\omega^2)\delta_{ik} + (\lambda + \mu)k_i k_k,$$

Exercise 4.4

Using the same procedure to invert (kk) following eqn. 4.17 show that

$$\tilde{G}_{im}(\mathbf{k},\omega) = \left[((kk) - \rho\omega^2 I)^{-1}\right]_{im} = \frac{1}{\rho}\left[\frac{\delta_{im} - k_i k_m/k^2}{c_t^2 k^2 - \omega^2} + \frac{k_i k_m/k^2}{c_l^2 k^2 - \omega^2}\right].$$

Taking the inverse Fourier transform we obtain the following Green's function for $t > 0$ for the point force pulse at the origin:

$$\begin{aligned} 4\pi\rho x G_{ij}(\mathbf{x},t) = {}& \frac{\delta(x - c_l t)}{c_l}\frac{x_i x_j}{x^2} + \frac{\delta(x - c_t t)}{c_t}\left(\delta_{ij} - \frac{x_i x_j}{x^2}\right) \\ & + [H(x - c_l t) - H(x - c_t t)]\frac{t}{x^2}\left(\delta_{ij} - \frac{3x_i x_j}{x^2}\right), \end{aligned} \qquad (4.69)$$

where $H(x)$ is the Heaviside step function: $H(x) = 1$ if $x > 0$, $H(x) = 0$ if $x < 0$. We see in this solution two spherical waves emanating from the point force pulse at the origin, with the longitudinal wave expanding faster than the shear wave. The material beyond the longitudinal wave front is undeformed.

4.10 Problem set 4

1. Derive the Green's function in linear isotropic elasticity using the line integral in eqn. 4.16.
2. A *dumb-bell* interstitial defect[27] has axial symmetry and is located at the origin of a Cartesian coordinate system in an elastically isotropic medium. With its axis along the unit vector $\hat{\mathbf{e}}$ it exerts a defect force $f\hat{\mathbf{e}}$ at $h\hat{\mathbf{e}}$ and another defect force $-f\hat{\mathbf{e}}$ at $-h\hat{\mathbf{e}}$. Write down the dipole tensor for the defect.

[26] This was first done by Stokes in *Trans. Camb. Phil. Soc.* **9** 1–62 (1849). https://archive.org/stream/transactionsofca09camb. Sir George Gabriel Stokes PRS 1819–1903, Irish mathematician and physicist.

[27] An interstitial defect occupies a site between those of the host crystal. The defect force for an interstitial defect is therefore the entire force it exerts on a neighbour in the relaxed configuration.

Show that in an isotropic medium

$$G_{ij,k}(\mathbf{x}) = \frac{-1}{16\pi\mu(1-\nu)x^2}\left((3-4\nu)\rho_k\delta_{ij} - (\rho_j\delta_{ik} + \rho_i\delta_{jk}) + 3\rho_i\rho_j\rho_k\right), \tag{4.70}$$

where $\rho_i = x_i/x$.

Hence show that for $x \gg h$ the displacement field at $\mathbf{x}$ of the dumb-bell defect is given by

$$u_i(\mathbf{x}) = \frac{2fh}{16\pi\mu(1-\nu)x^2}\left[(3\cos^2\alpha - 1)\rho_i + 2(1-2\nu)(\cos\alpha)e_i\right]$$

where $\cos\alpha = e_j\rho_j$.

Hence show that at a given distance x from the point defect the elastic displacement along the axis of the defect is $4(1-\nu)$ larger than the displacement along any direction perpendicular to the defect axis, and of opposite sign.

3. Consider a cubic crystal where the elastic anisotropy ratio is close to 1. Then $D = c_{11} - c_{12} - 2c_{44}$ is small compared to c_{12} and c_{44}. Equation 3.22 for the elastic constant tensor in a cubic crystal may always be expressed as

$$c_{ijkl} = c^0_{ijkl} + \delta_{ij}\delta_{jk}\delta_{kl}D,$$

where c^0_{ijkl} is the elastic constant tensor for an isotropic crystal where we choose $c^0_{12} = c_{12} = \lambda$, $c^0_{44} = c_{44} = \mu$. Let $(KK)_{jm} = c_{jlms}k_lk_s$ and $(kk)_{jm} = c^0_{jlms}k_lk_s$. Show that

$$(KK)^{-1} = (kk)^{-1} - (kk)^{-1}V(KK)^{-1},$$

where $V_{jm} = D(k_1^2\delta_{j1}\delta_{m1} + k_2^2\delta_{j2}\delta_{m2} + k_3^2\delta_{j3}\delta_{m3})$. This equation[28] is exact whatever the magnitude of D. Although it can be solved for $(KK)^{-1}$ the inverse Fourier transform requires the solution of a sextic equation to locate the poles, which can be done only numerically.

However, since D/μ is small it suggests a perturbation expansion:

$$(KK)^{-1} = (kk)^{-1} - (kk)^{-1}V(kk)^{-1} + (kk)^{-1}V(kk)^{-1}V(kk)^{-1} - \ldots$$

[28] This is a Dyson equation—see Economou, EN, *Green's functions in quantum physics*, Springer-Verlag: Berlin (1983). ISBN 978-3642066917.

Using eqn. 4.18 for $\left[(kk)^{-1}\right]_{jp}$ show that to first order in D/c_{44} the Green's function tensor in the cubic crystal in k-space becomes

$$
\begin{aligned}
\tilde{G}_{ij}(\mathbf{k}) &= \left[(KK)^{-1}\right]_{ij} \\
&= \frac{1}{\mu}\frac{\delta_{ij}}{k^2} - \frac{1}{\mu}\left(\frac{\lambda+\mu}{\lambda+2\mu}\right)\frac{k_i k_j}{k^4} \\
&+ \frac{D}{\mu}\frac{1}{\mu}\frac{(\delta_{i1}k_1^2\delta_{j1} + \delta_{i2}k_2^2\delta_{j2} + \delta_{i3}k_3^2\delta_{j3})}{k^4} \\
&- \frac{D}{\mu}\frac{1}{\mu}\left(\frac{\lambda+\mu}{\lambda+2\mu}\right)\frac{(k_1^3(\delta_{i1}k_j + k_i\delta_{j1}) + k_2^3(\delta_{i2}k_j + k_i\delta_{j2}) + k_3^3(\delta_{i3}k_j + k_i\delta_{j3}))}{k^6} \\
&+ \frac{D}{\mu}\frac{1}{\mu}\left(\frac{\lambda+\mu}{\lambda+2\mu}\right)^2 k_i k_j \frac{(k_1^4 + k_2^4 + k_3^4)}{k^8}
\end{aligned}
\tag{4.71}
$$

Hence show that the first order corrections to the Green's functions in isotropic elasticity are

$$
\Delta\tilde{G}_{11}(\mathbf{k}) = \frac{D}{\mu}\frac{1}{\mu}\left(\frac{k_1^2}{k^4} - \chi\frac{2k_1^4}{k^6} + \chi^2\frac{k_1^2(k_1^4 + k_2^4 + k_3^4)}{k^8}\right)
$$

$$
\Delta\tilde{G}_{12}(\mathbf{k}) = \frac{D}{\mu}\frac{1}{\mu}\left(-\chi\frac{k_1^3 k_2 + k_1 k_2^3}{k^6} + \chi^2\frac{k_1 k_2(k_1^4 + k_2^4 + k_3^4)}{k^8}\right)
$$

where $\chi = (\lambda+\mu)/(\lambda+2\mu)$. To invert these Fourier transforms the following integral is useful (it may be derived by contour integration):

$$
\frac{1}{(2\pi)^3}\int \frac{1}{k^{2n}} e^{-i\mathbf{k}\cdot\mathbf{x}} \mathrm{d}^3 k = -\frac{1}{4\pi}\frac{(-1)^n x^{2n-3}}{(2n-2)!},
$$

where $n = 1, 2, 3, \ldots$. Other integrals needed to evaluate the inverse transforms may by obtained by differentiating this integral with respect to components of $\mathbf{x}$. Thus, we obtain the following inverse transforms

$$
\frac{1}{(2\pi)^3}\int \frac{k_1^2}{k^4} e^{-i\mathbf{k}\cdot\mathbf{x}} \mathrm{d}^3 k = \frac{1}{8\pi x}\left(1 - \frac{x_1^2}{x^2}\right)
$$

$$\frac{1}{(2\pi)^3}\int\frac{k_1^3k_j}{k^6}e^{-i\mathbf{k}\cdot\mathbf{x}}\mathrm{d}^3k=\frac{3}{32\pi x}\left(\delta_{1j}-\frac{x_1x_j}{x^2}\right)\left(1-\frac{x_1^2}{x^2}\right)$$

$$\frac{1}{(2\pi)^3}\int\frac{k_1^5k_j}{k^8}e^{-i\mathbf{k}\cdot\mathbf{x}}\mathrm{d}^3k=\frac{5}{64\pi x}\left(\delta_{1j}-\frac{x_1x_j}{x^2}\right)\left(1-\frac{x_1^2}{x^2}\right)^2$$

$$\frac{1}{(2\pi)^3}\int\frac{k_1^2k_2^4}{k^8}e^{-i\mathbf{k}\cdot\mathbf{x}}\mathrm{d}^3k=\frac{1}{64\pi x}\left(1-\frac{x_2^2}{x^2}\right)\left(\frac{x_3^2}{x^2}+5\frac{x_1^2x_2^2}{x^4}\right)$$

$$\frac{1}{(2\pi)^3}\int\frac{k_1k_2k_3^4}{k^8}e^{-i\mathbf{k}\cdot\mathbf{x}}\mathrm{d}^3k=-\frac{1}{64\pi x}\frac{x_1x_2}{x^2}\left(1-\frac{x_3^2}{x^2}\right)\left(1-5\frac{x_3^2}{x^2}\right).$$

Hence show that

$$\Delta G_{11}(\mathbf{x})=\frac{1}{\mu}\frac{D}{\mu}\frac{1}{8\pi x}\left[\left(1-\frac{x_1^2}{x^2}\right)-\frac{3\chi}{2}\left(1-\frac{x_1^2}{x^2}\right)^2+\frac{\chi^2}{8}\left(5\left(1-\frac{x_1^2}{x^2}\right)^3\right.\right.$$
$$\left.\left.+\left(1-\frac{x_2^2}{x^2}\right)\left(\frac{x_3^2}{x^2}+5\frac{x_1^2x_2^2}{x^4}\right)+\left(1-\frac{x_3^2}{x^2}\right)\left(\frac{x_2^2}{x^2}+5\frac{x_1^2x_3^2}{x^4}\right)\right)\right]$$

$$\Delta G_{12}(\mathbf{x})=\frac{1}{\mu}\frac{D}{\mu}\frac{1}{8\pi x}\frac{x_1x_2}{x^2}\left[\frac{3\chi}{4}\left(1+\frac{x_3^2}{x^2}\right)-\frac{\chi^2}{8}\left(5\left(\left(1-\frac{x_1^2}{x^2}\right)^2+\left(1-\frac{x_2^2}{x^2}\right)^2\right)\right.\right.$$
$$\left.\left.+\left(1-\frac{x_3^2}{x^2}\right)\left(1-5\frac{x_3^2}{x^2}\right)\right)\right].$$

The other elements of the Green's function are found by permuting the subscripts.

4. In a hexagonal crystal show that the (kk) matrix is as follows:

$$(kk)=\begin{bmatrix} c_{11}k_1^2+c_{66}k_2^2+c_{44}k_3^2 & \frac{1}{2}(c_{11}+c_{12})k_1k_2 & (c_{44}+c_{13})k_1k_3 \\ \frac{1}{2}(c_{11}+c_{12})k_1k_2 & c_{66}k_1^2+c_{11}k_2^2+c_{44}k_3^2 & (c_{44}+c_{13})k_2k_3 \\ (c_{44}+c_{13})k_1k_3 & (c_{44}+c_{13})k_2k_3 & c_{44}(k_1^2+k_2^2)+c_{33}k_3^2 \end{bmatrix},$$

where $c_{66}=\frac{1}{2}(c_{11}-c_{12})$.

Since the elastic constant tensor is invariant with respect to rotations about the hexagonal axis (which is the x_3-axis), we may choose $\mathbf{k}=(0,k_2,k_3)$. With this simplification the determinant in eqn. 4.63 can be factorised. Determine the three solutions for $\omega^2=\omega^2(0,k_2,k_3)$.

Determine the frequencies and polarisations of the three waves with **k** in the basal plane and normal to the basal plane.

5. Consider the Green's function, $\gamma_{km}(\mathbf{x}, t)$ in isotropic elasticity, for an oscillatory unit body force located at the origin described by $f_i(t) = \delta_{im}\delta(\mathbf{x})\cos(\omega t)$. The equation of motion for $\gamma_{km}(\mathbf{x}, t)$ is as follows:

$$c_{ijkl}\gamma_{km,lj}(\mathbf{x}, t) + \delta_{im}\delta(\mathbf{x})\cos(\omega t) = \rho\ddot{\gamma}_{im}(\mathbf{x}, t).$$

To satisfy the time dependence in this equation we must have[29] $\gamma_{im}(\mathbf{x}, t) = \chi_{im}(\mathbf{x})\cos\omega t$, where $\chi(\mathbf{x})$ satisfies following equation:

$$c_{ijkl}\chi_{km,lj}(\mathbf{x}) + \delta_{im}\delta(\mathbf{x}) = -\rho\omega^2\chi_{im}(\mathbf{x})$$

Show that the Fourier transform $\tilde{\chi}(\mathbf{k})$ satisfies

$$\left[(kk)_{ik} - \rho\omega^2\delta_{ik}\right]\tilde{\chi}_{km} = \delta_{im}$$

and therefore, in an isotropic medium,

$$\tilde{\chi}_{im} = \frac{1}{\rho}\left[\frac{\delta_{im} - k_i k_m/k^2}{c_t^2 k^2 - \omega^2} + \frac{k_i k_m/k^2}{c_l^2 k^2 - \omega^2}\right].$$

Hence show that

$$\chi_{im}(\mathbf{x}) = \frac{\delta_{im}}{4\pi\rho c_t^2}\frac{\cos(\omega x/c_t)}{x} + \frac{1}{4\pi\rho\omega^2 x}\frac{\partial}{\partial x_i}\frac{\partial}{\partial x_m}(\cos(\omega x/c_t) - \cos(\omega x/c_l)).$$

6. It was shown in section 4.7 that the stress in an inclusion, which would undergo a homogeneous strain e^T_{ij} if it were not constrained by the surrounding matrix, is given by $\sigma^I_{ij} = \sigma^C_{ij} - \sigma^T_{ij}$, where σ^T_{ij} and σ^C_{ij} are the stresses related respectively to the strain e^T_{ij} and the constrained strain e^C_{ij} by Hooke's law. The stress and strain fields in the matrix surrounding the inclusion are the constrained fields σ^C_{ij} and e^C_{ij}. Let the strain related to σ^I_{ij} through Hooke's law be $e^I_{ij} = e^C_{ij} - e^T_{ij}$. Then the total elastic energy is

$$E_{el} = \frac{1}{2}\int_{\mathcal{R}} \sigma^I_{ij}e^I_{ij}\mathrm{d}V + \frac{1}{2}\int_{matrix} \sigma^C_{ij}e^C_{ij}\mathrm{d}V,$$

where $\mathcal{R}$ is the region that has transformed and *matrix* is the surrounding matrix. Using the continuity of surface tractions and of constrained displacements at all

[29] If damping were included in the equation of motion there would be a phase difference between the vibrations of the medium and those of the oscillator.

points in the interface surrounding the inclusion, and the divergence theorem, prove that

$$\frac{1}{2}\int\limits_{matrix} \sigma_{ij}^{C} e_{ij}^{C} \mathrm{d}V = -\frac{1}{2}\int\limits_{\mathcal{R}} \sigma_{ij}^{I} e_{ij}^{C} \mathrm{d}V.$$

Hence show that the total elastic energy may be expressed as the following volume integral over the inclusion only, *regardless of its shape*:

$$E_{el} = -\frac{1}{2}\int\limits_{\mathcal{R}} \sigma_{ij}^{I} e_{ij}^{T} \mathrm{d}V.$$

Since the constrained stress and strain fields are homogeneous inside an *ellipsoidal* inclusion the elastic energy is $-(V/2)\sigma_{ij}^{I}e_{ij}^{T}$, where V is the volume of the ellipsoid.

In question 10 of Problem set 6 it is shown that the interaction energy between an ellipsoidal inclusion and an applied strain field e_{ij}^{A} is as follows:

$$E_{int} = -\int\limits_{\mathcal{R}} \sigma_{ij}^{T} e_{ij}^{A} \mathrm{d}V,$$

where again the volume integral is taken over the inclusion only.

7. Show that the Green's function in isotropic elasticity (eqn. 4.19) may be expressed as

$$G_{ij}(\mathbf{x}-\mathbf{x}') = \frac{1}{4\pi\mu}\frac{\delta_{ij}}{|\mathbf{x}-\mathbf{x}'|} - \frac{1}{16\pi\mu(1-\nu)}\frac{\partial}{\partial x_i}\frac{\partial}{\partial x_j}|\mathbf{x}-\mathbf{x}'|.$$

An inclusion would undergo a homogeneous transformation strain e_{jk}^{T} if it were not constrained by the surrounding matrix. Let σ_{jk}^{T} be the stress obtained by applying Hooke's law to the transformation strain e_{ij}^{T}. By applying the divergence theorem to eqn. 4.31 for the constrained displacement field $u_i^{C}(\mathbf{x})$, and using the above form of the Green's function, show that the displacement field when the inclusion is constrained by the matrix is given by

$$u_i^{C}(\mathbf{x}) = -\sigma_{jk}^{T}\left\{\frac{1}{4\pi\mu}\delta_{ij}\phi_{,k} - \frac{1}{16\pi\mu(1-\nu)}\psi_{,ijk}\right\},$$

where $\phi(\mathbf{x})$ and $\psi(\mathbf{x})$ are the potentials:

$$\phi(\mathbf{x}) = \int_{\mathcal{R}} \frac{d^3x'}{|\mathbf{x}-\mathbf{x}'|}$$

$$\psi(\mathbf{x}) = \int_{\mathcal{R}} |\mathbf{x}-\mathbf{x}'| \, d^3x'.$$

The integrations in these potentials are carried over the region $\mathcal{R}$ occupied by the inclusion before it transforms. Show that

$$\nabla^2\psi(\mathbf{x}) = 2\phi(\mathbf{x})$$

$$\nabla^2\phi(\mathbf{x}) = \begin{cases} 4\pi & \text{if } \mathbf{x} \text{ is in } \mathcal{R} \\ 0 & \text{otherwise} \end{cases}$$

Consider an inclusion in which the stress-free transformation strain is a dilation $e^T_{ij} = \frac{e^T}{3}\delta_{ij}$. Show that

$$u^C_i(\mathbf{x}) = -\frac{(1+\nu)}{12\pi(1-\nu)} e^T \phi_{,i}(\mathbf{x}).$$

Hence prove that the dilation in the matrix is zero, while the dilation in the inclusion is

$$e^C = \frac{e^T}{3}\left(\frac{1+\nu}{1-\nu}\right).$$

Show that the stress in the inclusion is

$$\sigma^I_{jk} = \sigma^C_{jk} - \sigma^T_{jk} = -\frac{4\mu e^C}{3}\delta_{jk} = -\frac{4\mu e^T}{9}\left(\frac{1+\nu}{1-\nu}\right)\delta_{jk}.$$

Remarkably, these results are independent of the shape of the inclusion, but they do assume the elastic constants inside the inclusion are the same as those outside, and that they are isotropic. For $\nu = \frac{1}{3}$ we find the matrix reduces the dilation of the inclusion by a factor of $\frac{1}{3}$.

5

Point defects

The theory perhaps suffers from the disadvantage that its limitations are more immediately obvious than are those of other approximate methods which have to be used in dealing with the solid state.... A simple treatment at cottage-industry level may sometimes provide a certain degree of insight into some phenomenon or other when an exact calculation, based on a theoretical model which apes reality precisely, will give accurate answers but may be too unsurveyable, or too numerical, to give much insight.

JD Eshelby in *Point defect behaviour and diffusional processes*, ed. RE Smallman and E Harris, The Metals Society: London (1977), pp.3–10. Copyright © Institute of Materials, Minerals and Mining, reprinted by permission of Taylor & Francis Ltd, http://www.tandfonline.com, on behalf of Institute of Materials, Minerals and Mining.

5.1 Introduction

Point defects are ubiquitous in crystalline materials. They are classified as either intrinsic or extrinsic. Intrinsic point defects are vacancies and self-interstitials. A vacancy is a vacant atomic site that would normally be occupied by a host atom. A self-interstitial is created when one of the atoms of the crystal is occupying a site in the space, called the 'interstice', between usual crystal sites. Extrinsic point defects are foreign atoms that have been incorporated into the crystal structure either on substitutional or interstitial sites.

Unlike linear and planar defects there may be populations of point defects in thermodynamic equilibrium in a crystal. For example, the equilibrium concentration of vacancies in a crystal at a given temperature and pressure is $\exp(-G^f/k_B T)$ where G^f is the Gibbs free energy of formation of the point defect. The thermodynamic equilibrium concentration of foreign atoms in a crystal depends on their chemical potentials, which in turn depend on the partial pressures of these impurities in the vapour phase in equilibrium with the crystal.

Thermodynamic equilibrium is seldom achieved in crystals, but its existence is important because it indicates the state of the system towards which it will naturally evolve under prescribed environmental conditions provided it is not being driven away from thermodynamic equilibrium by external influences such as mechanical deformation or irradiation. The diffusion of point defects leads to mass transport and possibly changes of phase in the solid state. For example, diffusion in pure crystals usually occurs through

Physics of elasticity and crystal defects. Adrian P. Sutton, Oxford University Press (2020). © Adrian P. Sutton.
DOI: 10.1093/oso/9780198860785.001.0001

the migration of vacancies, as they are the only sites in the crystal that can undergo a change of occupation through a host atom jumping into one of them. This is analogous to electronic conduction in a p-type semiconductor which occurs through the movement of electronic holes at the top of the valence band.

Point defects may be created in concentrations far in excess of those prescribed by thermodynamic equilibrium through processes such as plastic deformation and irradiation by high-energy particles. When this happens the point defects may aggregate and form small interstitial and vacancy clusters that can be very mobile, especially when they are small. Intrinsic point defects such as vacancies may be attracted to foreign atoms and form new combined defects called complexes. The variety of point defects can be huge.

In this chapter we will focus on how point defects interact with each other through their elastic fields. We begin with the model of a point defect as a misfitting sphere. We will go on to consider models of point defects that capture the symmetry of the atomic environment of the point defect.

5.2 The misfitting sphere model of a point defect

The simplest model of a point defect is a misfitting sphere[1] due to Bilby.[2] Cut out a spherical hole of radius r_h in an elastic continuum. Insert into the hole a sphere of radius $r_s \neq r_h$ and weld the interface. The sphere may be larger or smaller than the hole. We consider first the case where the inserted sphere is rigid and the continuum is infinite in extent. All the deformation occurs in the continuum surrounding the sphere.

Let $\mathbf{u}^{\infty}(\mathbf{r})$ be the displacement field in the continuum surrounding the sphere at radius r. The superscipt 'infinity' is to remind us that the continuum is infinite in extent. The displacement field will be purely radial so that $\mathbf{u}^{\infty}(\mathbf{r}) = \mathbf{u}^{\infty}(|\mathbf{r}|) = \mathbf{u}^{\infty}(r)$. In the isotropic approximation the displacement field must satisfy Navier's equation, eqn. 4.4, which in the absence of body forces becomes

$$\mu\nabla^2\mathbf{u}^{\infty} + (\lambda+\mu)\nabla(\nabla\cdot\mathbf{u}^{\infty}) = 0. \tag{5.1}$$

We have to be careful how we interpret this equation in spherical polar coordinates. If we naively substitute the usual expressions for the Laplacian, gradient and divergence in spherical polar coordinates (r,θ,ϕ) we will go wrong. When we rotate a Cartesian coordinate system (x_1,x_2,x_3) into another Cartesian coordinate system (x_1',x_2',x_3') the transformation is the same at all points in space. But when we use spherical polar coordinates (or other curvilinear coordinates) the transformation varies with position because the coordinates r, θ and ϕ vary with position. Consequently, $(du_r, du_\theta, du_\phi)$ does

[1] Bilby, BA, *Proc. Phys. Soc. A* **63**, 191 (1950). https://doi.org/10.1088/0370-1298/63/3/302

[2] Bruce Alexander Bilby FRS 1922–2013, British materials physicist.

not form a vector. The solution is to use tensor calculus, following Sokolnikoff.[3] Then it is found that the strain tensor in spherical polar coordinates is as follows:

$$
\begin{aligned}
e_{rr} &= \frac{\partial u_r}{\partial r}\\
e_{\theta\theta} &= \frac{1}{r}\frac{\partial u_\theta}{\partial \theta} + \frac{u_r}{r}\\
e_{\phi\phi} &= \frac{1}{r\sin\theta}\frac{\partial u_\phi}{\partial \phi} + \frac{u_r}{r} + u_\theta \frac{\cot\theta}{r}\\
e_{r\phi} &= \frac{1}{2}\left(\frac{1}{r\sin\theta}\frac{\partial u_r}{\partial \phi} - \frac{u_\phi}{r} + \frac{\partial u_\phi}{\partial r}\right)\\
e_{r\theta} &= \frac{1}{2}\left(\frac{1}{r}\frac{\partial u_r}{\partial \theta} - \frac{u_\theta}{r} + \frac{\partial u_\theta}{\partial r}\right)\\
e_{\phi\theta} &= \frac{1}{2}\left(\frac{1}{r}\frac{\partial u_\phi}{\partial \theta} - \frac{u_\phi \cot\theta}{r} + \frac{1}{r\sin\theta}\frac{\partial u_\theta}{\partial \phi}\right).
\end{aligned}
$$

Hooke's law in isotropic elasticity in spherical polar coordinates is as follows:

$$
\begin{aligned}
\sigma_{rr} &= 2\mu e_{rr} + \lambda(e_{rr} + e_{\theta\theta} + e_{\phi\phi})\\
\sigma_{\theta\theta} &= 2\mu e_{\theta\theta} + \lambda(e_{rr} + e_{\theta\theta} + e_{\phi\phi})\\
\sigma_{\phi\phi} &= 2\mu e_{\phi\phi} + \lambda(e_{rr} + e_{\theta\theta} + e_{\phi\phi})\\
\sigma_{r\theta} &= 2\mu e_{r\theta}\\
\sigma_{r\phi} &= 2\mu e_{r\phi}\\
\sigma_{\theta\phi} &= 2\mu e_{\theta\phi}.
\end{aligned}
$$

The equations of equilibrium in isotropic elasticity in spherical polar coordinates are as follows:

$$
\frac{\partial \sigma_{rr}}{\partial r} + \frac{1}{r\sin\theta}\frac{\partial \sigma_{r\phi}}{\partial \phi} + \frac{1}{r}\frac{\partial \sigma_{r\theta}}{\partial \theta} + \frac{2\sigma_{rr} - \sigma_{\theta\theta} - \sigma_{\phi\phi} + \sigma_{r\theta}\cot\theta}{r} + f_r = 0
$$

[3] Sokolnikoff, IS, *Mathematical theory of elasticity*, 2nd edn., section 48, McGraw-Hill: New York (1956). ISBN 978-0070596290. Ivan Stephan Sokolnikoff 1901–76 Russian-born, US mathematician.

$$\frac{\partial \sigma_{\phi r}}{\partial r} + \frac{1}{r\sin\theta}\frac{\partial \sigma_{\phi\phi}}{\partial \phi} + \frac{1}{r}\frac{\partial \sigma_{\phi\theta}}{\partial \theta} + \frac{3\sigma_{\phi r} + 2\sigma_{\phi\theta}\cot\theta}{r} + f_\phi = 0$$

$$\frac{\partial \sigma_{\theta r}}{\partial r} + \frac{1}{r\sin\theta}\frac{\partial \sigma_{\theta\phi}}{\partial \phi} + \frac{1}{r}\frac{\partial \sigma_{\theta\theta}}{\partial \theta} + \frac{3\sigma_{\theta r} + (\sigma_{\theta\theta} - \sigma_{\phi\phi})\cot\theta}{r} + f_\theta = 0,$$

where (f_r, f_θ, f_ϕ) is the body force at (r, θ, ϕ).

Exercise 5.1

Consider a radially symmetric elastic displacement field. In spherical polar coordinates the displacement field $\mathbf{u}$ is then purely radial, $\mathbf{u} = [u_r, 0, 0]$ and u_r is a function of r only. Using the equations above for the strain tensor, Hooke's law and the equations of equilibrium in spherical polar coordinates show that the equations of equilibrium become just one equation:

$$\frac{\partial \sigma_{rr}}{\partial r} + \frac{2\sigma_{rr} - \sigma_{\theta\theta} - \sigma_{\phi\phi}}{r} + f_r = 0.$$

Hence derive a differential equation for u_r in isotropic elasticity, in the absence of body force, and show that the general solution is

$$u_r(r) = Ar + D/r^2, \tag{5.2}$$

where A and D are arbitrary constants.

Returning to the misfitting sphere in an infinite medium we note that since $\mathbf{u}^\infty$ must be finite at $r = \infty$ we choose $u^\infty(r) = D/r^2$. The constant D is determined by the boundary condition at the surface of the hole: $u^\infty(r_h) = r_s - r_h$. Therefore $D = (r_s - r_h)r_h^2$.

One of the surprising features of this model is that the dilation is zero everywhere. Inside the sphere there is no elastic strain because the sphere is assumed to be rigid. Outside the sphere the strain in the radial direction is $e_{rr} = \mathrm{d}u_r^\infty/\mathrm{d}r = -2D/r^3$, while $e_{\theta\theta} = e_{\phi\phi} = u_r/r = D/r^3$, so that the trace of the strain tensor is zero. But there is an overall volume change ΔV^∞ and it is all located in the misfitting sphere: $\Delta V^\infty = (4\pi/3)(r_s^3 - r_h^3) \approx 4\pi r_h^2(r_s - r_h) = 4\pi D$.

In reality all bodies are finite. Consider a finite body with a free surface $\mathcal{S}$. The solution for the misfitting sphere in an infinite medium has to be corrected by applying surface tractions $-\sigma_{ij}^\infty n_j$ over the surface $\mathcal{S}$, where $\sigma_{ij}^\infty = 2\mu\partial u_i^\infty/\partial x_j$, and in Cartesian coordinates $u_i^\infty = Dx_i/r^3$. These tractions cancel those that would exist on the surface if the infinite solution were not corrected. To calculate the change ΔV^I in the volume of the body caused by this distribution of surface tractions consider the integral:

$$\int_S \sigma_{ij}^{\infty} x_i n_j \mathrm{d}S = \int_V (\sigma_{ij}^{\infty} x_i)_{,j} \mathrm{d}V = \int_V \sigma_{ij,j}^{\infty} x_i + \sigma_{ij}^{\infty} \delta_{ij} \mathrm{d}V = \int_V -f_i x_i \mathrm{d}V + V\langle\sigma_{ii}\rangle,$$

where the divergence theorem and the equilibrium condition have been used. Since there is no body force in V the average trace of the stress tensor is $\langle\sigma_{ii}\rangle$ inside the body when the distribution of surface tractions $-\sigma_{ij}^{\infty} n_j$ is applied is given by

$$\begin{aligned} V\langle\sigma_{ii}\rangle &= -\int_S \sigma_{ij}^{\infty} x_i n_j \mathrm{d}S \\ &= -\int_S 2\mu D\left(\frac{\delta_{ij}}{r^3} - \frac{3x_i x_j}{r^5}\right) x_i n_j \mathrm{d}S \\ &= 4\mu D \int_S \frac{x_i n_i}{r^3} \mathrm{d}S, \end{aligned}$$

where $(x_i n_i / r^3)\mathrm{d}S$ is an element of solid angle, and the final surface integral is therefore 4π. The change in volume, ΔV^I, caused by the elimination of the surface tractions is given by $3B\Delta V^I = \langle\sigma_{ii}\rangle V = 16\pi\mu D = 4\mu\Delta V^{\infty}$. The total volume change caused by the misfitting sphere in a finite body is given by

$$\Delta V = \Delta V^{\infty} + \Delta V^I = 4\pi D\left(\frac{3B+4\mu}{3B}\right) = 3\Delta V^{\infty}\left(\frac{1-\nu}{1+\nu}\right), \tag{5.3}$$

where we have used $B = 2\mu(1+\nu)/(3(1-2\nu))$, and ν is Poisson's ratio. For $\nu = 1/3$, which is typical of many metals, the increase in the overall volume of the body as a result of the presence of free surfaces is approximately 50%. It is remarkable that this result does not depend on the size of the body, or its shape, or where the defect is located within it. This is an example of an insight from a simple model into a universal property of point defects that would be virtually impossible to obtain from a purely atomistic model.

Exercise 5.2

Consider a misfitting *deformable* sphere at the centre of a much larger sphere of radius R on the surface of which there are no tractions. The radius of the hole into which the misfitting sphere is inserted is r_h. The radius of the unconstrained (i.e. stress-free) deformable sphere is r_s. The bulk modulus and the shear modulus of the large sphere of radius R are B and μ respectively, and the bulk modulus of the misfitting sphere is B'.

To first order in the difference $r_s - r_h$ show that the volume change of the sphere of radius R is

continued

Exercise 5.2 Continued

$$\Delta V = \left(\frac{1 + \frac{4\mu}{3B}}{1 + \frac{4\mu}{3B'}} \right) \delta v,$$

where $\delta v = 4\pi r_h^2 (r_s - r_h)$ is the misfit volume.

Show that the dilation in the large sphere is

$$e_{ii} = \frac{\frac{4\mu}{3B}}{1 + \frac{4\mu}{3B'}} \frac{\delta v}{V},$$

where $V = 4\pi R^3/3$ is the volume of the sphere of radius R.

5.3 Interaction energies

The misfitting sphere model has been used widely owing to its attractive simplicity. But its central assumption that the defect has spherical symmetry limits its applicability principally to substitutional defects at sites of cubic symmetry. It is less applicable to interstitial defects which may display highly non-spherical symmetries, such as split interstitials and crowdion defects. We saw in section 4.5 how the multipole expansion may be applied to derive the displacement field of a substitutional point defect. It may also be applied to the displacement field of an interstitial defect, with the sole difference that the defect forces are the entire forces, rather than the excess forces, exerted by the foreign atom on the surrounding neighbours in the equilibrium state. The multipole expansion captures the symmetry of the site occupied by the defect. In this section we consider the interaction energy between two point defects using the multipole expansion. This is a hybrid approach using interatomic forces to describe the sources of the elastic fields and elasticity theory to describe the interactions over distances much larger than the atomic scale. The analysis of this section follows Siems.[4]

We begin by thinking about the problem entirely from an atomistic viewpoint. In eqn. 3.28 we wrote down the harmonic expansion of the potential energy of the crystal. Here we write it slightly differently:

$$E = -\sum_n f_i^{(n)} u_i^{(n)} + \frac{1}{2} \sum_n \sum_p \frac{\partial^2 E}{\partial u_i^{(n)} \partial u_j^{(p)}} u_i^{(n)} u_j^{(p)}, \tag{5.4}$$

[4] Siems, R, *Phys. Stat. Sol.* **30**, 645–58 (1968). https://doi.org/10.1002/pssb.19680300226

where we interpret the forces $f_i^{(n)}$ as being forces due to an existing defect which create the displacement field $u_i^{(n)}$. Let the vector of defect forces be **f**, the vector of atomic displacements be **u** and the matrix of second derivatives be **S**. Then we can rewrite eqn. 5.4 for the potential energy of the defect as follows:

$$E = -\mathbf{f}\cdot\mathbf{u} + \frac{1}{2}\mathbf{uSu} = -\frac{1}{2}\mathbf{f}\cdot\mathbf{u} = -\frac{1}{2}\mathbf{uSu},$$

where E is minimised when $\mathbf{f} = \mathbf{Su}$. Notice that this energy is negative. It is the energy of *relaxing* the system of defect forces and harmonic atomic interactions. The defect forces do work when the atoms on which they act are displaced thereby reducing the potential energy by $-\mathbf{f}\cdot\mathbf{u}$. These displacements induce further displacements that extend throughout the crystal and which raise the potential energy by $\frac{1}{2}\mathbf{uSu}$. A balance between these two competing terms is established when $\mathbf{f} = \mathbf{Su}$, and the reduction in the potential energy is then only half of $-\mathbf{f}\cdot\mathbf{u}$.

If there is a second defect, with defect force vector $\tilde{\mathbf{f}}$ and vector of atomic displacements $\tilde{\mathbf{u}}$, with $\tilde{\mathbf{f}} = \mathbf{S}\tilde{\mathbf{u}}$ and $\tilde{E} = -\frac{1}{2}\tilde{\mathbf{f}}\cdot\tilde{\mathbf{u}}$, we may use the linear superposition principle to write the total potential energy as follows:

$$\begin{aligned} E^T &= -(\mathbf{f}+\tilde{\mathbf{f}})\cdot(\mathbf{u}+\tilde{\mathbf{u}}) + \frac{1}{2}(\mathbf{u}+\tilde{\mathbf{u}})\mathbf{S}(\mathbf{u}+\tilde{\mathbf{u}}) \\ &= E + \tilde{E} + E_{int}, \end{aligned}$$

where the interaction energy E_{int} is given by

$$\begin{aligned} E_{int} &= -(\mathbf{f}\cdot\tilde{\mathbf{u}} + \tilde{\mathbf{f}}\cdot\mathbf{u}) + \frac{1}{2}(\mathbf{uS}\tilde{\mathbf{u}} + \tilde{\mathbf{u}}\mathbf{Su}) \\ &= -\mathbf{f}\cdot\tilde{\mathbf{u}} = -\tilde{\mathbf{f}}\cdot\mathbf{u}. \end{aligned} \tag{5.5}$$

The last equality follows from Maxwell's reciprocity theorem. This result for the interaction energy may be understood in the following way. Suppose the first defect already exists and we introduce the second defect. The atoms on which the defect forces **f** of the first defect act are displaced further by the displacement field $\tilde{\mathbf{u}}$ of the second defect. The additional work done is $-\mathbf{f}\cdot\tilde{\mathbf{u}}$. If we repeat the argument but with the second defect existing before the first the work done is $-\tilde{\mathbf{f}}\cdot\mathbf{u}$.

To describe the interaction energy over large separations between the defects we may use defect forces from an atomistic calculation and elasticity theory to evaluate the displacement field. That is why we call this a hybrid approach.

With the first point defect at the origin of the Cartesian coordinate system the interaction energy with a second point defect at **x** is

$$E_{int} = -\sum_n \tilde{f}_i(\mathbf{R}^{(n)})u_i(\mathbf{x}+\mathbf{R}^{(n)}), \tag{5.6}$$

where $\tilde{f}_i(\mathbf{R}^{(n)})$ is the force exerted by the defect at $\mathbf{x}$ on its neighbour at the relative position $\mathbf{R}^{(n)}$ and $u_i(\mathbf{x}+\mathbf{R}^{(n)})$ is the displacement of this neighbour due to the defect at the origin. Provided $|\mathbf{x}| \gg |\mathbf{R}^{(n)}|$ we may expand $u_i(\mathbf{x}+\mathbf{R}^{(n)})$ about $u_i(\mathbf{x})$:

$$u_i(\mathbf{x}+\mathbf{R}^{(n)}) = u_i(\mathbf{x}) + R_j^{(n)} u_{i,j}(\mathbf{x}) + \frac{1}{2} R_j^{(n)} R_k^{(n)} u_{i,jk}(\mathbf{x}) + \frac{1}{6} R_j^{(n)} R_k^{(n)} R_l^{(n)} u_{i,jkl}(\mathbf{x}) + \ldots \quad (5.7)$$

Substituting this Taylor expansion into the interaction energy, eqn. 5.6, we obtain

$$E_{int} = -\left(\sum_n \tilde{f}_i(\mathbf{R}^{(n)})\right) u_i(\mathbf{x}) - \tilde{p}_{ij} u_{i,j}(\mathbf{x}) - \frac{1}{2} \tilde{q}_{ijk} u_{i,jk}(\mathbf{x}) - \frac{1}{6} \tilde{o}_{ijkl} u_{i,jkl}(\mathbf{x}) - \ldots, \quad (5.8)$$

where the sum of the defect forces is zero provided the defect at $\mathbf{x}$ is relaxed and $\tilde{p}_{ij}$, $\tilde{q}_{ijk}$ and $\tilde{o}_{ijkl}$ are the dipole, quadrupole and octupole moments respectively of the forces exerted by the defect at $\mathbf{x}$.

We can now use the multipole expansion, eqn. 4.21, to expand the displacement field due to the defect at the origin:

$$\begin{aligned} u_i(\mathbf{x}) &= \sum_m G_{ia}(\mathbf{x}-\mathbf{R}^{(m)}) f_a(\mathbf{R}^{(m)}) \\ &= \sum_m \left(G_{ia}(\mathbf{x}) - R_b^{(m)} G_{ia,b}(\mathbf{x}) + \frac{1}{2} R_b^{(m)} R_c^{(m)} G_{ia,bc}(\mathbf{x}) - \frac{1}{6} R_b^{(m)} R_c^{(m)} R_d^{(m)} G_{ia,bcd}(\mathbf{x}) + \ldots \right) f_a(\mathbf{R}^{(m)}) \\ &= G_{ia}(\mathbf{x}) \sum_m f_a(\mathbf{R}^{(m)}) - G_{ia,b}(\mathbf{x}) p_{ab} + \frac{1}{2} G_{ia,bc}(\mathbf{x}) q_{abc} - \frac{1}{6} G_{ia,bcd}(\mathbf{x}) o_{abcd} + \ldots \end{aligned} \quad (5.9)$$

The sum of the forces exerted by the defect at the origin on its neighbours is zero provided the defect is relaxed. The dipole, quadrupole and octupole moments of the forces exerted by the defect at the origin are p_{ab}, q_{abc} and o_{abcd} respectively. Inserting the displacement field $u_i(\mathbf{x})$ into the interaction energy of eqn. 5.8, and collecting terms involving the same order of differentiation of the Green's function we obtain finally:

$$\begin{aligned} E_{int} &= \tilde{p}_{ij} G_{ia,bj}(\mathbf{x}) p_{ab} \\ &\quad + \frac{1}{2} \left(\tilde{q}_{ijk} G_{ia,bjk}(\mathbf{x}) p_{ab} - \tilde{p}_{ij} G_{ia,bcj}(\mathbf{x}) q_{abc} \right) \\ &\quad + \frac{1}{12} \left(2\tilde{o}_{ijkl} G_{ia,bjkl}(\mathbf{x}) p_{ab} + 2\tilde{p}_{ij} G_{ia,bcdj}(\mathbf{x}) o_{abcd} - 3\tilde{q}_{ijk} G_{ia,bcjk}(\mathbf{x}) q_{abc} \right) + \ldots \end{aligned} \quad (5.10)$$

The first line is the dipole–dipole interaction and it decays as $1/|\mathbf{x}|^3$. The second line is the dipole–quadrupole interaction and it decays as $1/|\mathbf{x}|^4$. The third line describes interactions that depend on the fourth derivative of the Green's function, which decays as $1/|\mathbf{x}|^5$, and involves quadrupole–quadrupole interactions and dipole–octupole interactions.

5.4 The λ-tensor

In the previous section we saw that a point defect at $\mathbf{x}$ has an interaction energy with another elastic field according to eqn. 5.8. Thus, if the strain field with which it interacts is homogeneous then the interaction energy involves only the dipole tensor, a strain gradient at $\mathbf{x}$ will interact with the quadrupole tensor, the second derivative of the strain at $\mathbf{x}$ will interact with the octupole tensor of the defect and so on. If there is a dilute solution of these point defects their dipole tensors can drive a change of shape of the crystal. If c is the atomic concentration of the point defect then the λ-tensor describes the rate of change of the spontaneous homogeneous strain of the crystal with c in the absence of any forces applied to the surface of the crystal:

$$\lambda_{kl} = \frac{\mathrm{d}e^h_{kl}}{\mathrm{d}c} \tag{5.11}$$

To expose the relationship between the λ-tensor and the dipole tensor consider the elastic energy of the crystal containing a solution of the defects. It is assumed that the defects are sufficiently far apart that their defect forces do not overlap, and therefore the dipole tensors are the same as those for an isolated defect. In a volume V let there be N defects. If the atomic volume of the host atoms is Ω then $c = N\Omega/V$ in the dilute limit. If the crystal undergoes a homogeneous strain e^h_{ij} the elastic energy of the volume V becomes

$$E = -N\rho_{ij}e^h_{ij} + \frac{V}{2}c_{ijkl}e^h_{ij}e^h_{kl}.$$

Minimising this elastic energy with respect to the homogeneous strain we obtain

$$N\rho_{ij} = Vc_{ijkl}e^h_{kl}.$$

Substituting $N = Vc/\Omega$ and differentiating the resulting expression with respect to the concentration c we obtain

$$\rho_{ij} = \Omega c_{ijkl}\lambda_{kl}. \tag{5.12}$$

This is a very useful relationship because the λ-tensor is experimentally measurable using X-ray diffraction for example. By measuring the dependence of the homogeneous strain on concentration c of the solute atoms, in the limit of small concentrations, the λ-tensor may be deduced using eqn. 5.11.

5.5 Problem set 5

1. Consider two point defects occupying sites of cubic point group symmetry in a slightly anisotropic cubic crystal, such as aluminium for which the anisotropy ratio is 1.2. Their elastic dipole tensors are $\tilde{p}_{ij} = \tilde{d}\delta_{ij}$ and $p_{ij} = \delta_{ij}d$. The elastic interaction energy at large separations is given by

$$E_{int} = \tilde{p}_{ij} G_{ia,jb}(\mathbf{x}) p_{ab} = \tilde{d} d G_{ij,ji}(\mathbf{x}).$$

Show that in an elastically isotropic medium $G_{ij,ji} = 0$.

Using eqn. 4.71 for the first order correction to the Fourier transform of the Green's function in a cubic crystal show that

$$G_{ij,ji}(\mathbf{x}) = -\left(\frac{c_{11} - c_{12} - 2c_{44}}{(c_{12} + 2c_{44})^2}\right) \frac{1}{(2\pi)^3} \int \frac{k_1^4 + k_2^4 + k_3^4}{k^4} e^{-i\mathbf{k}\cdot\mathbf{x}} \mathrm{d}^3 k$$

$$= -\left(\frac{3}{8\pi}\right)\left(\frac{c_{11} - c_{12} - 2c_{44}}{(c_{12} + 2c_{44})^2}\right) \frac{1}{x^3}\left[5\frac{x_1^4 + x_2^4 + x_3^4}{x^4} - 3\right]. \qquad (5.13)$$

2. Consider a substitutional point defect located at the origin of a Cartesian coordinate system in an infinite simple cubic crystal structure with lattice constant a. The forces exerted by the defect on each of the six nearest neighbours have magnitude f, and they are directed along the bonds. Show that the displacement field $u_i(\mathbf{x})$ at $\mathbf{x}$ is approximately $-2afG_{ij,j}(\mathbf{x})$, and state the nature of the approximation.

Show that in the isotropic elastic approximation:

$$u_i(\mathbf{x}) = \frac{(1-2\nu)af}{4\pi\mu(1-\nu)} \frac{x_i}{x^3}.$$

Hence calculate the volume change ΔV^∞ associated with the point defect in an infinite medium. By comparing your answer with ΔV^∞ in the model of a point defect as a misfitting sphere in an infinite medium obtain an expression for $D = \Delta V^\infty/(4\pi)$ in the misfitting sphere model in terms of the bond length a and defect force f.

3. Consider two vacancies in tungsten, which has a body-centred cubic crystal structure, and its elastic properties are well approximated as isotropic.[5] Aligning the coordinate axes with the sides of the cubic unit cell show that the dipole tensor is $p_{ij} = p\delta_{ij}$.

[5] This problem is inspired by section 23.2 of Teodosiu's book.

In the first problem of this set it was shown that dipole–dipole interaction energy is zero because $G_{ij,ji} = 0$. The symmetry of atomic sites in the bcc crystal dictates that the quadrupole tensor is zero.

Show that there are only two types of non-zero independent elements of the octupole tensor: o_{1111} and o_{1122}. Note that $o_{1122} = o_{1212} = o_{1221}$. Hence show that the octupole tensor may be expressed as

$$o_{ijkl} = o_{1111}\delta_{ij}\delta_{jk}\delta_{kl} + o_{1122}\left(\delta_{ij}\delta_{kl} + \delta_{ik}\delta_{jl} + \delta_{il}\delta_{jk} - 3\delta_{ij}\delta_{jk}\delta_{kl}\right).$$

Show that the first non-zero term in the interaction energy between the defects is

$$E_{int} = \frac{1}{6}G_{ij,jiii}\left[p(\tilde{o}_{1111} - 3\tilde{o}_{1122}) + \tilde{p}(o_{1111} - 3o_{1122})\right] + \frac{1}{2}G_{ij,jikk}(p\tilde{o}_{1122} + \tilde{p}o_{1122}),$$

where $G_{ij,jiii}$ means $G_{1j,j111} + G_{2j,j222} + G_{3j,j333}$.

Hence show that

$$E_{int} = \frac{7(1-2\nu)}{8\pi\mu(1-\nu)x^5}\left[p(\tilde{o}_{1111} - 3\tilde{o}_{1122})\right]\left(\frac{5(x_1^4 + x_2^4 + x_3^4)}{x^4} - 3\right).$$

Thus, in an infinite isotropic cubic crystal like tungsten two vacancies interact through the dipole–octupole interaction. This interaction energy displays the same angular dependence,[6] as was found in the first problem of this set for the dipole–dipole interaction energy between two point defects occupying sites of cubic symmetry in a weakly anisotropic cubic crystal. However, the dipole–dipole interaction energy in a cubic crystal varies with separation x as $1/x^3$, as compared with $1/x^5$ for the variation of the dipole–octupole interaction energy in an isotropic crystal. Therefore, whenever there is a departure of the anisotropy ratio from unity in a cubic crystal the dipole–dipole interaction will dominate at long range.

[6] The angular dependence is the cubic harmonic $K_{4,1} = (\sqrt{21}/4)[5((x_1/x)^4 + (x_2/x)^4 + (x_3/x)^4) - 3]$, which is normalised as follows: $\int (K_{4,1})^2 \mathrm{d}\Omega = 4\pi$ where $\mathrm{d}\Omega = \sin\theta \mathrm{d}\theta \mathrm{d}\phi$ is an element of solid angle. Cubic harmonics are combinations of spherical harmonics with cubic symmetry. See Fehlner, WR and Vosko, SH, *Can. J. Phys.* **54**, 2159–69 (1976). https://doi.org/10.1139/p76-256

6

Dislocations

6.1 Introduction

When crystals are deformed there is a limit to which the deformation remains elastic or reversible. Beyond this limit the crystal fractures if it is brittle. If it is ductile further deformation takes place, but the additional deformation is irreversible or 'plastic'. When the load is removed from a ductile crystal that has been deformed plastically the shape of the crystal has changed permanently. It is a characteristic property of many metals and alloys that they can undergo plastic deformation before they fracture, which enables them to be extruded, pressed, rolled and forged into everyday objects from car bodies to drink cans and furniture to girders for bridges and skyscrapers.

In the early twentieth century there was a great deal of interest to discover what happened inside a crystal when it ceased to deform elastically and began to deform plastically. It was a mystery why some crystals such as copper and gold were extremely ductile whereas others such as diamond were brittle, at least at room temperature. Equally baffling was the observation that some body-centred cubic (bcc) metals that were ductile at room temperature became brittle at low temperatures displaying almost no ductility before they fractured, while some face-centred cubic (fcc) metals remained ductile at temperatures as low as a few kelvin. The rate at which a crystal was deformed was also found to have a strong influence on the ductile/brittle behaviour of many crystalline materials. The purity of the crystal was also found to affect the stress required to make it deform plastically. Beginning to address these questions brought about the birth of the science of materials in the 1930s. Today our understanding of all these phenomena and processes is based on dislocations and how they interact with other defects in the material. It is the movement of dislocations that leads to plastic deformation of crystalline matter, and in general this happens more readily in metals than non-metals. During the 1950s transmission electron microscopy made it possible to observe dislocations moving inside very thin crystals undergoing plastic deformation. The complexity of many processes involved in plasticity is so great that our understanding is still far from complete, and we shall return to some of these issues in Chapter 10.

Physics of elasticity and crystal defects. Adrian P. Sutton, Oxford University Press (2020). © Adrian P. Sutton.
DOI: 10.1093/oso/9780198860785.001.0001

6.2 Dislocations as the agents of plastic deformation

To a very good approximation crystals deform plastically at constant volume. Permanent changes of shape are brought about by shearing processes in which planes of atoms slide over each other. The sliding creates steps at the surface of the crystal, which creates surface roughness that can be felt with a finger if a metal paperclip is straightened. In 1926 Frenkel[1] produced a startling back-of-an-envelope calculation demonstrating that the stress required to slide a plane of atoms en masse over another was orders of magnitude larger than the stresses observed to initiate plastic deformation of many metals.

Consider two adjacent planes of atoms in a crystal spaced d apart. Let the crystal periodicity in the plane in the direction of sliding be b. Frenkel sought to calculate the maximum stress required to slide one entire plane rigidly over the other. Let x be the relative displacement of the two atomic planes in the direction of sliding. Let $\gamma(x)$ be the energy per unit area of the plane associated with this relative displacement. What do we know about $\gamma(x)$? We know it must be periodic, with a period of b, and that $\gamma(x)$ is minimised when $x = nb$ where n in an integer. Therefore we can write

$$\gamma(x) = A_0 + \sum_{n=1}^{\infty} A_n \cos\left(\frac{2n\pi x}{b}\right) + B_n \sin\left(\frac{2n\pi x}{b}\right), \tag{6.1}$$

which is just a Fourier expansion of $\gamma(x)$. If we make the simplest approximation and assume $\gamma(x)$ is an even function, taking just the first term of the cosine series we obtain $\gamma(x) = A\sin^2(\pi x/b)$, where A is a positive constant with the dimensions of energy per unit area. Here we have chosen $x = 0$ to be at a minimum of $\gamma(x)$. The slope of $\gamma(x)$ is the negative of the stress required to sustain the relative displacement x. Frenkel argued that as $x \to 0$ this slope is determined by the elastic shear modulus μ because $\mathrm{d}\gamma/\mathrm{d}x = (A\pi/b)\sin(2\pi x/b) \to A2\pi^2 x/b^2$ and this has to be equated to $\mu x/d$. Therefore, $A = (\mu b^2)/(2\pi^2 d)$ and the stress required to slide one plane of atoms rigidly and en masse over another is of order $\mu b/(2\pi d)$.

The stress required to transition from elastic to plastic deformation is called the yield stress. Frenkel's estimate of the yield stress is of order $\mu/10$. Experimentally observed values of the yield stress are typically between three and five orders of magnitude less than this. We have to conclude that entire planes of atoms do not slide en masse over each other during plastic deformation.

The resolution of this paradox came with the independent but almost simultaneous insights of Orowan[2] in Budapest, Polanyi in Manchester (UK)[3] and Taylor[4] in

[1] Jacov Frenkel 1894–1952, Soviet condensed matter physicist.

[2] Egon Orowan FRS 1902–89, Hungarian/British/US physicist and metallurgist, in Z. Physik **89**, 634. https://doi.org/10.1007/BF01341480

[3] Michael Polanyi FRS 1891–1976, Hungarian/British polymath, in Z. Physik **89**, 660. https://doi.org/10.1007/BF01341481

[4] Sir Geoffrey Ingram Taylor FRS OM 1891–1976, British physicist and mathematician, in Proc. R. Soc. A **145**, 388. https://doi.org/10.1098/rspa.1934.0106

Cambridge (UK) in 1934 who suggested that planes 'slip' past each other through the movement of linear defects called dislocations separating slipped and unslipped regions. Taylor's paper is especially remarkable as it includes a detailed discussion of the crystallographic nature of slip, namely the observation that it occurs on particular planes in particular directions, and of the hardening of crystals that occurs with increasing plastic deformation, which is called work hardening, and elastic interactions between dislocations leading to stable and unstable configurations of dislocations. Taylor drew on the earlier mathematical treatment of dislocations by Volterra[5] who was interested in the elastic equilibrium of bodies rendered multiply connected through the existence of dislocations.[6]

The sliding of one crystal plane over another, which is called slip, begins with the nucleation of a small dislocation loop. The plane where slip occurs is called the slip plane. Inside the loop the planes have slipped past each other by a vector **b**, which is called the Burgers vector. Outside the loop the planes have not slipped. The dislocation is the line separating the slipped and unslipped regions of the slip plane—see Fig. 6.1. We will show that under the influence of an applied shear stress on the slip plane in the direction of **b** the loop expands converting more of the unslipped region into the slipped

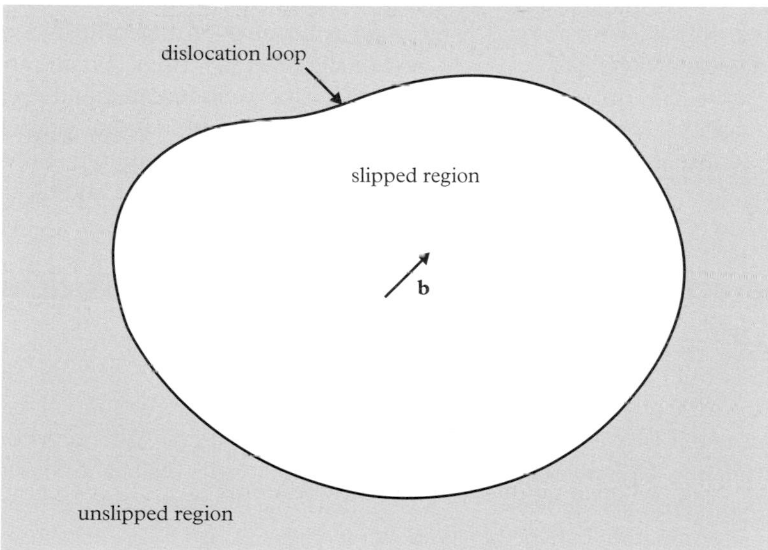

Figure 6.1 *A dislocation loop lying in a slip plane separating slipped and unslipped regions. The dislocation is the line separating the slipped and unslipped regions of the slip plane. Inside the loop the material beneath the slip plane has been translated with respect to material above it by the Burgers vector* **b**.

[5] Vito Volterra 1860–1940, Italian mathematician and physicist.

[6] Volterra, V, *Annales scientifiques de l'École Normale Supérieure* **24**, 401–517 (1907). http://www.numdam.org/item?id=ASENS_1907_3_24__401_0

region, enabling the applied shear stress to do more work thereby reducing the potential energy of the whole system including an external loading mechanism if there is one. If the loop is in a single crystal then when it reaches the surfaces of the crystal the entire crystal on one side of the slip plane has slipped by **b** with respect to the other side, and there is a step on the surface with unit normal $\hat{\mathbf{n}}$ of height $(\mathbf{b} \cdot \hat{\mathbf{n}})$. If the Burgers vector **b** is a crystal lattice vector in the slip plane then as the dislocation moves it recreates the same crystal structure in its wake. In principle **b** may be any lattice vector in the slip plane, but we shall see that the elastic energy of the dislocation varies as $|\mathbf{b}|^2$ so that usually only the smallest lattice vectors are found. The smallest lattice vectors tend to occur in planes with the largest spacing.

The key physical insight of Orowan, Polanyi and Taylor in 1934 is that slip by dislocation motion localises the inevitable bond breaking and making, when one plane slides over another, to the very much smaller region where the dislocation line is located. In contrast, when an entire plane slides en masse over another the bond breaking and making occurs everywhere in the slip plane simultaneously. Therefore, the stress required to move a dislocation is orders of magnitude less than that required to slide a plane of a macroscopic crystal over another en masse. This was a giant step forward. But fundamental questions remained, such as why some crystals seem to undergo very limited slip, if any, before they fracture, why some are brittle at low temperatures and ductile at higher temperatures, why small concentration of impurities can have a seemingly disproportionate effect on the ease of slip, the choice of slip plane and Burgers vector, the stress to move a dislocation, how dislocations are created and even whether they do in fact exist. Answering these questions over the intervening years has led to some of the most interesting experimental, theoretical and computational condensed matter physics, and to have profound consequences for engineering and technology.

6.3 Characterisation of dislocations: the Burgers circuit

Consider a dislocation loop lying in a slip plane. Within the loop there is a constant relative displacement of the crystals above and below the slip plane equal to the Burgers vector. Because the relative displacement is constant throughout the slipped region inside the loop it does not vary with the direction of the dislocation line. Where the direction of the dislocation line is perpendicular to the Burgers vector the dislocation is said to have 'edge' character. Perhaps the reason for this name is that the dislocation line is then along the edge of a terminating half-plane, as shown in Fig. 6.2. An edge dislocation is a long straight dislocation where the dislocation line is everywhere perpendicular to the Burgers vector. The edge dislocation appeared in Taylor's paper of 1934. When the dislocation line is parallel to the Burgers vector the dislocation is said to have 'screw' character. A screw dislocation is a long straight dislocation where the dislocation line is everywhere parallel to the Burgers vector. The screw dislocation was introduced by JM Burgers[7] in

[7] Johannes Martinus Burgers 1895–1981, Dutch physicist, whose brother, Wilhelm Gerard Burgers 1897–1988, also a Dutch physicist, also worked on dislocations.

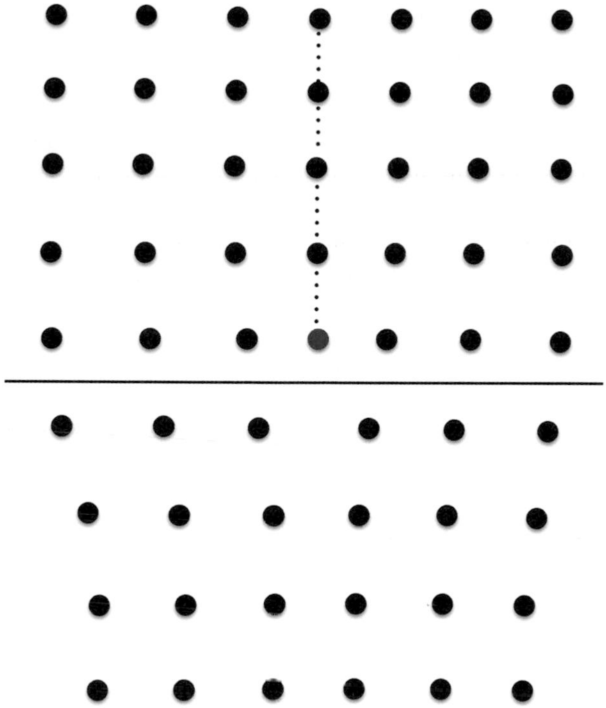

Figure 6.2 *A schematic illustration of an edge dislocation viewed along the dislocation line. Each dot represent a column of atoms normal to the page. The horizontal line is the trace of the slip plane. Above the slip plane there is an extra half plane shown by dotted lines. The edge dislocation is located at the termination of this extra half plane, which is shown by the red column of atoms.*

1939.[8] The reason for this name is that the crystal lattice planes normal to the dislocation line become a helicoidal surface like the thread of a screw, as shown in Fig. 6.3. If the angle between the line direction and the Burgers vector is ϕ then the dislocation may be regarded as a superposition of an edge dislocation with Burgers vector $\mathbf{b}\sin\phi$ and a screw dislocation with Burgers vector $\mathbf{b}\cos\phi$. Such a dislocation is said to have 'mixed' character, and its atomic structure changes from screw to edge type as ϕ varies from 0 to $\pi/2$.

Suppose we see some defect in a crystal. How do we know whether it is a dislocation or some other defect? If it is a dislocation how do we determine its Burgers vector? The answer to both questions is the Burgers circuit construction. This is a geometrical construction that was introduced by Frank[9] in a paper[10] which defined rigorously many

[8] Burgers, JM, *Koninklijke Nederlandsche Akademie van Wetenschappen* **42**, 293 (1939), http://www.dwc.knaw.nl/toegangen/digital-library-knaw/?pagetype=publDetail&pId=PU00014649

[9] Sir (Frederick) Charles Frank FRS 1911–98, British physicist.

[10] Frank, FC, *Phil. Mag.* **42**, 809 (1951), http://dx.doi.org/10.1080/14786445108561310

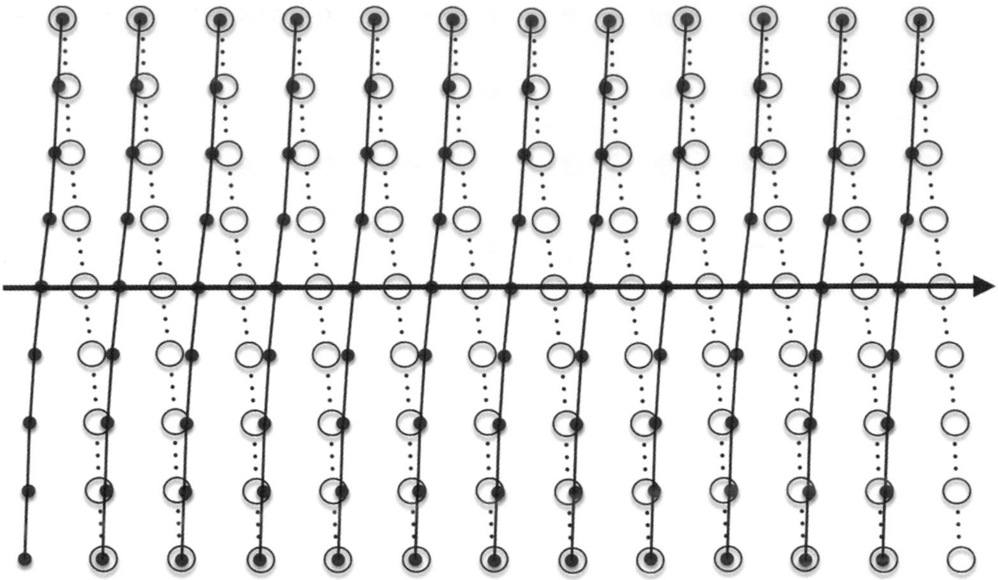

Figure 6.3 *A schematic illustration of a screw dislocation. The dislocation line is shown by the horizontal arrow. Solid circles and solid lines are below the page. Open circles and dotted lines are above the page. Planes of atoms are converted into a continuous spiral by the dislocation. This drawing was adapted from Hull, D, and Bacon, DJ,* Introduction to dislocations, *3rd edn., Pergamon Press: Oxford (1984). ISBN 0-08-028720-4.*

of the terms in use today when discussing dislocations and plasticity of crystals. The construction is illustrated in Fig. 6.4 for an edge dislocation. We draw a closed circuit in the elastically distorted crystal surrounding the dislocation which begins and ends at the same site. The circuit comprises steps between lattice sites of the elastically distorted crystal. It is important that the circuit passes through material that is recognisable as perfect crystal that has only been elastically strained, and that it does not go near the core of the dislocation where the local atomic environment cannot be mapped onto the perfect crystal. It does not matter how far the circuit goes from the dislocation as long as it does not enclose any other dislocations and as long as it keeps away from the dislocation core. The circuit is then mapped onto a perfect lattice, as shown in Fig. 6.4. If the defect is a dislocation the circuit mapped onto the perfect lattice will not close. By convention, the closure failure from the finish to the start of the circuit mapped onto the perfect crystal is the Burgers vector when the circuit is taken in a clockwise sense when looking along the dislocation line, that is, into the page in Fig. 6.4. But if the circuit is started somewhere else or follows a different closed path in the dislocated crystal the Burgers vector will always be the same provided the circuit does not enclose any other dislocations and that it does not pass through the dislocation core. The Burgers vector is an invariant property

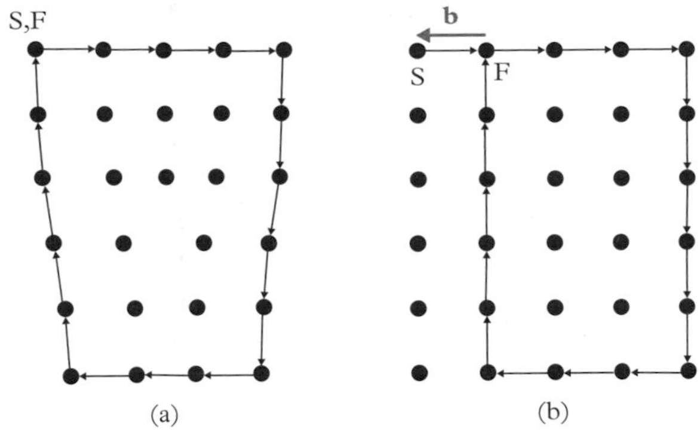

Figure 6.4 *Illustration of the Burgers circuit construction for an edge dislocation. The line sense of the edge dislocation in (a) is defined to be positive looking into the page. A right handed circuit is drawn around the dislocation line, starting at S and finishing at the same site F. The circuit is mapped, step by step, onto the perfect crystal in (b). It is found that S and F are no longer coincident, and therefore the circuit has a closure failure. The vector joining F to S, shown by the arrow in red, is defined by the FS/RH convention as the Burgers vector* **b** *of the dislocation.*

of the dislocation. The existence of the closure failure when the circuit is mapped into the perfect crystal is the defining property of a dislocation.[11] There is no closure failure if the circuit encloses a point defect only.

In a continuum there is no lattice to define a Burgers circuit in the manner of Fig. 6.4. If **u** is the elastic displacement field of the dislocation then the Burgers vector is defined in a continuum by the line integral:

$$b_i = \oint_C \frac{\partial u_i}{\partial x_k} \mathrm{d}x_k, \tag{6.2}$$

where C is any clockwise circuit taken around the dislocation line viewed along its positive direction. A word of caution: this convention is followed by many authors but some define the Burgers vector as the negative of this integral.

Both the Burgers circuit construction for a dislocation in a crystal and the line integral of eqn. 6.2 for a dislocation in a continuum depend on the direction of the dislocation line. If the line direction reverses then the Burgers vector changes sign. A dislocation is therefore defined by two vectors: the direction of its line $\hat{\mathbf{t}}$ and its Burgers vector $\mathbf{b}$. Together these two vectors define the sign of a straight dislocation. If just one of $\hat{\mathbf{t}}$ or $\mathbf{b}$ changes sign then the sign of the dislocation changes. If both $\hat{\mathbf{t}}$ and $\mathbf{b}$ change sign then the dislocation retains the same sign. Dislocations with the same sign repel while dislocations

[11] The Burgers circuit is an example of anholonomy which arises elsewhere in physics including the Aharonov–Bohm effect, Foucault's pendulum and the Berry phase.

of opposite sign attract each other. We will see in this chapter that dislocations are often created in the form of loops. At all points of a dislocation loop the Burgers vector is the same but the line direction changes. Dislocation segments in the loop with opposite line directions have opposite signs and if they come together they annihilate. We will see an example of this in the operation of a Frank–Read source.

A bit more terminology: when the Burgers vector is a lattice vector the dislocation is said to be a 'perfect' dislocation. When the Burgers vector is a fraction of a lattice vector the dislocation is called a 'partial' dislocation, or sometimes an 'imperfect', dislocation.

6.4 Glide, climb and cross-slip

When dislocations move in their slip plane no diffusion of atoms is required. This kind of motion is called *glide* or conservative because the number of atoms at the dislocation as it moves along is conserved. The normal to the slip plane of an edge dislocation is $\hat{\mathbf{n}} = \hat{\mathbf{b}} \times \hat{\mathbf{t}}$, where $\hat{\mathbf{b}}$ is a unit vector parallel to the Burgers vector and $\hat{\mathbf{t}}$ is a unit vector along the line direction. As long as an edge dislocation moves within its slip plane the number of atoms in the extra half-plane associated with the dislocation does not change, and this is why such motion is conservative.

If an edge dislocation moves out of its slip plane then the extra half plane either grows or shrinks requiring atoms to be added or removed from it. This is called *climb* or non-conservative motion. Because diffusion is involved the speed of climb is generally much less than the speed of glide. Climb may enable an edge dislocation to overcome an obstacle blocking glide on its slip plane, but since diffusion is involved this is a thermally activated process. The extra half plane of an edge dislocation grows when atoms are added to it, which is equivalent to saying that vacancies are emitted from it. In this way edge dislocations are sources and sinks for vacancies and self-interstitials, and thus they can regulate the populations of these point defects when the populations deviate from those in thermal equilibrium, for example due to irradiation.

In contrast to an edge dislocation the slip plane of a screw dislocation is not uniquely defined because $\mathbf{b}$ and $\hat{\mathbf{t}}$ are parallel or anti-parallel to each other. Consequently a screw dislocation gliding on one plane may switch to gliding on another plane, which is a process called cross-slip. For example, a screw dislocation in a bcc crystal with Burgers vector $\mathbf{b} = 1/2[111]$ gliding on a $(1\bar{1}0)$ plane may cross-slip onto the planes $(10\bar{1})$ or $(0\bar{1}1)$. As a result screw dislocations move only conservatively. Cross-slip provides a mechanism for screw dislocations to overcome barriers to glide on a slip plane by gliding on an inclined plane over or under the obstacle. Although cross-slip is a process that happens during glide it is thermally activated. In fcc crystals this is because perfect dislocations may be dissociated into partial dislocations separated by stacking faults (see the next chapter), and before cross-slip can take place the partial dislocations have to be recombined which is a process requiring energy. But even when perfect dislocations cross-slip the atomic structure of the dislocation core has to adjust from that on one slip plane to that on another, and this is also a process requiring energy.

Exercise 6.1

During irradiation with high energy neutrons a highly non-equilibrium population of vacancies is created in a metal. In a bcc crystal some of the vacancies cluster together on a $\{111\}$-type plane forming hexagonal dislocation loops with sides along $\langle 1\bar{1}0\rangle$ directions. The Burgers vector of the dislocation loop is $1/2\langle 111\rangle$ type. Show that such a loop may move in a conservative manner in the direction of its Burgers vector, tracing out a hexagonal prism as it moves. Such a loop is called 'prismatic' for this reason. This example illustrates how individual defects, vacancies in this case, may come together to form a new defect, a prismatic dislocation loop in this case, and the mechanism of their motion changes from diffusion of individual vacancies to glide of the loop as a whole involving no diffusion at all.

6.5 The interaction energy between a dislocation and another source of stress

This section and the following section are based on the analysis presented in section 5 of a review article by Eshelby.[12] Consider a dislocation D in a body with external surface S_o. There may be constant tractions acting on the surface of the body, and other defects D′ inside the body creating internal stresses. Define a closed surface S inside the body enclosing D only so that it separates D from other sources of internal stress and the external load. The position of S is otherwise arbitrary. The internal surface S divides the body into two regions: region I contains D and region II is the rest of the body, as illustrated in Fig. 6.5.

Let the elastic fields created by D be u_i^D, e_{ij}^D and σ_{ij}^D. Let u_i^A, e_{ij}^A and σ_{ij}^A be the elastic fields created by the constant loads applied to the external surface and by sources of internal stress other than D. Using linear superposition the total elastic fields in the body are $\sigma_{ij} = \sigma_{ij}^D + \sigma_{ij}^A$, $e_{ij} = e_{ij}^D + e_{ij}^A$ and $u_i = u_i^D + u_i^A$. The total elastic energy density is $\frac{1}{2}\sigma_{ij}e_{ij} = \frac{1}{2}(\sigma_{ij}^D + \sigma_{ij}^A)(e_{ij}^D + e_{ij}^A)$. The elastic interaction energy density involves only the cross terms between D and A, so that the total interaction energy is the following integral:

$$E_{int} = \frac{1}{2}\int_V \left(\sigma_{ij}^D e_{ij}^A + \sigma_{ij}^A e_{ij}^D\right) \mathrm{d}V. \qquad (6.3)$$

In region I u_i^D cannot be defined everywhere owing to the elastic singularity D, but u_i^A is well defined throughout region I. Similarly u_i^A cannot be defined everywhere in region II, but u_i^D is well defined throughout region II. It follows from Hooke's law that $\sigma_{ij}^A e_{ij}^D = \sigma_{ij}^D e_{ij}^A$. Therefore eqn. 6.3 can be rewritten as

[12] Eshelby, JD, *Solid State Phys.*, **3**, 79–144 (1956). https://doi.org/10.1016/S0081-1947(08)60132-0

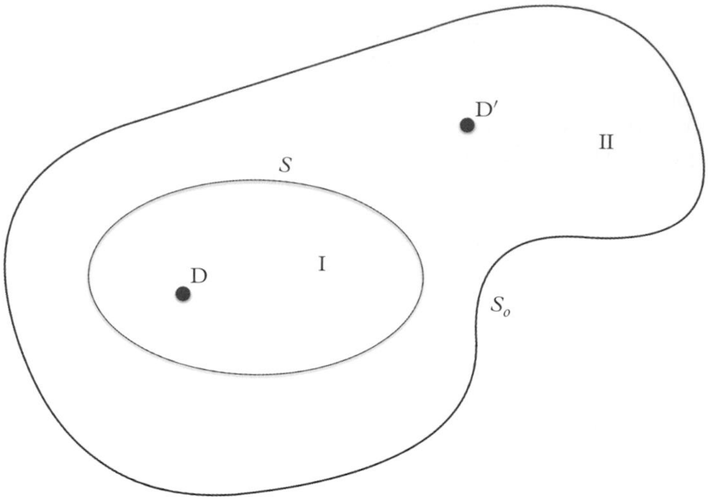

Figure 6.5 *A schematic illustration of a body with an external surface S_o containing a dislocation D and other defects represented by D′. The internal closed surface S separates D from the other defects in the body, and divides the body into region I containing D only and region II which is the remainder of the body.*

$$E_{int} = \int_I \sigma_{ij}^D u_{i,j}^A \mathrm{d}V + \int_{II} \sigma_{ij}^A u_{i,j}^D \mathrm{d}V. \tag{6.4}$$

Furthermore, $\sigma_{ij,j}^D = 0$ and $\sigma_{ij,j}^A = 0$ in the absence of body forces and therefore $\sigma_{ij}^D u_{i,j}^A = (\sigma_{ij}^D u_i^A)_{,j}$ and $\sigma_{ij}^A u_{i,j}^D = (\sigma_{ij}^A u_i^D)_{,j}$. The divergence theorem then transforms eqn. 6.4 into the following surface integrals:

$$E_{int} = \int_S \sigma_{ij}^D u_i^A n_j \mathrm{d}S + \int_{S_0} \sigma_{ij}^A u_i^D n_j \mathrm{d}S - \int_S \sigma_{ij}^A u_i^D n_j \mathrm{d}S. \tag{6.5}$$

If there is no external loading mechanism then the external surface is free of tractions and $\sigma_{ij}^A n_j = 0$. The integral over S_0 then vanishes. But if there is a load applied to the surface S_0 of the body then the integral over S_0 is the work done by the external loading mechanism and the change of the potential energy of the external loading mechanism must be included in the total interaction energy.

Suppose D moves by an *additional* infinitesimal amount. Then the displacement field u_i^D will change by δu_i^D and the external load will do *additional* work given by $\int_{S_0} \sigma_{ij}^A \delta u_i^D n_j \mathrm{d}S$. The potential energy of the loading mechanism will then change by $-\int_{S_0} \sigma_{ij}^A \delta u_i^D \mathrm{d}S$ because it has expended energy to do this work. For a constant external load this is exactly the same as we would calculate for the change in the potential energy of the external loading mechanism if the potential energy of the external loading mechanism were given by the following relationship:

$$E_{ext} = -\int_{S_0} \sigma_{ij}^{A} u_i^{D} n_j \mathrm{d}S. \tag{6.6}$$

Thus the total interaction energy, including the potential energy of the loading mechanism, is as follows:

$$E_{int}^{T} = E_{int} + E_{ext} = \int_{S} \left(\sigma_{ij}^{D} u_i^{A} - \sigma_{ij}^{A} u_i^{D} \right) n_j \mathrm{d}S. \tag{6.7}$$

The reduction of the potential energy of any external loading mechanism is equal to the work done by the external load on the body. But this is merely a redistribution of energy within the system as whole, where the system comprises the body and an external loading mechanism; it does not contribute to the change of the energy of the system as a whole. We see that the integral over the external surface of the body in eqn. 6.5 vanishes in the total interaction energy irrespective of whether there is an external load applied to the body or not.

As it stands eqn. 6.7 has limited utility because σ_{ij}^{D} and u_i^{D} are the stress and displacement field of the defect D that satisfy the boundary conditions at the surface S_0 of the body. But suppose we replace σ_{ij}^{D} and u_i^{D} with $\sigma_{ij}^{D} + \sigma_{ij}^{W}$ and $u_i^{D} + u_i^{W}$ where σ_{ij}^{W} and u_i^{W} are any stress and displacement fields that have no singularities within S. Then a straightforward application of the divergence theorem shows that the additional terms do not change the integral in eqn. 6.7. It follows that σ_{ij}^{D} and u_i^{D} may be replaced by any stress and displacement fields that have the same singularities within S. In particular they may be replaced by $\sigma_{ij}^{D\infty}$ and $u_i^{D\infty}$ where $\sigma_{ij}^{D\infty}$ and $u_i^{D\infty}$ are the stress and displacement fields of D in an *infinite* medium. With this change eqn. 6.7 is very useful indeed.

If D is a dislocation loop enclosing a surface Σ then Σ may be taken as the cut on either side of which the displacement by the Burgers vector **b** exists. Then the surface S may be taken as the positive and negative sides of Σ together with a tube surrounding the dislocation line, as shown in Fig. 6.6. The positive side of the loop is defined by the right-hand screw rule: progressing around the loop along the positive direction of the dislocation line advances a right-hand screw in the direction of the positive loop normal. As the radius of the tube shrinks to zero it contributes nothing to the integral in eqn. 6.7 because the dislocation is not associated with a resultant line of force. The term

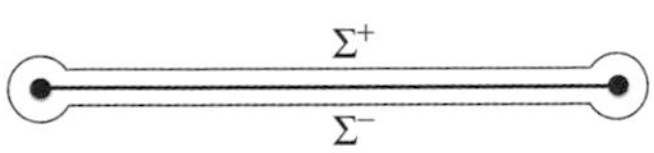

Figure 6.6 *Illustration of the surface of the integral in eqn. 6.7 for a dislocation loop. The loop is viewed edge on and the plane of the loop is the thicker horizontal line. The positive sense of the dislocation line is coming out of the page on the left and into the page on the right. The positive loop normal points upwards. The surface S comprises the surfaces Σ^+ and Σ^- above and below the plane of the loop and a tube surrounding the dislocation. The positive loop normal points from the negative surface Σ^- to the positive surface Σ^+.*

$\sigma_{ij}^D u_i^A$ is the same at opposing points on Σ^+ and Σ^-, and therefore this term contributes nothing to the integral. But there is a discontinuity in the displacement u_i^D on either side of the cut with u_i^D on Σ^+ minus u_i^D on Σ^- equal to $-b_i$. Therefore the total interaction energy becomes

$$E_{int}^T = +b_i \int_{\Sigma^+} \sigma_{ij}^A n_j \, \mathrm{d}S. \tag{6.8}$$

This equation has a clear physical interpretation. If we imagine the dislocation loop is created in the presence of the stress field σ_{ij}^A then $-E_{int}^T$ represents the work done by the stress field σ_{ij}^A when the surface Σ^- is translated by the Burgers vector with respect to the surface Σ^+. For example, in the case of an interstitial loop formed in a tensile stress field with the tensile axis along the loop normal $E_{int}^T < 0$ because $\mathbf{b} \cdot \hat{\mathbf{n}} < 0$.

Exercise 6.2

Using Maxwell's reciprocity theorem (eqn. 4.7) show that the interaction energy between two dislocation loops A and B is as follows:

$$E_{int} = b_i^A \int_{\Sigma_A} \sigma_{ij}^B n_j \, \mathrm{d}S = b_i^B \int_{\Sigma_B} \sigma_{ij}^A n_j \, \mathrm{d}S. \tag{6.9}$$

6.6 The Peach–Koehler force on a dislocation

Now that we have the interaction energy in eqn. 6.8 we may derive a general expression for the force on a dislocation due to an applied stress field. This force was first derived[13] by Peach and Koehler[14] and it is known as the Peach–Koehler force.

Consider a dislocation where the local direction of the dislocation line is $\hat{\mathbf{t}}$. Suppose an infinitesimal segment $\mathrm{d}l$ of the dislocation line is displaced by a vector $\mathbf{ds}$, as shown in Fig. 6.7. Bearing in mind the convention for defining the positive loop normal, with the shaded area in Fig. 6.7 in the plane of the page, the positive loop normal points into the page. Then the vector area swept by the displaced segment is positive if it is defined as $\mathbf{dA} = \mathbf{ds} \times \mathrm{d}l\hat{\mathbf{t}}$. The displacement of the dislocation segment changes the surface Σ in the surface integral of eqn. 6.8. As in the previous section let σ_{ij}^A be the stress field acting locally on the dislocation segment, where this stress is the resultant of the stresses due to an external loading mechanism and sources of internal stresses. The work done by the

[13] Peach, MO and Koehler, JS, *Phys. Rev.* **80**, 436 (1950), https://doi.org/10.1103/PhysRev.80.436

[14] This work formed part of the doctoral thesis of MO Peach under the supervision of James Stark Koehler 1914–2006 undertaken at Carnegie Institute of Technology.

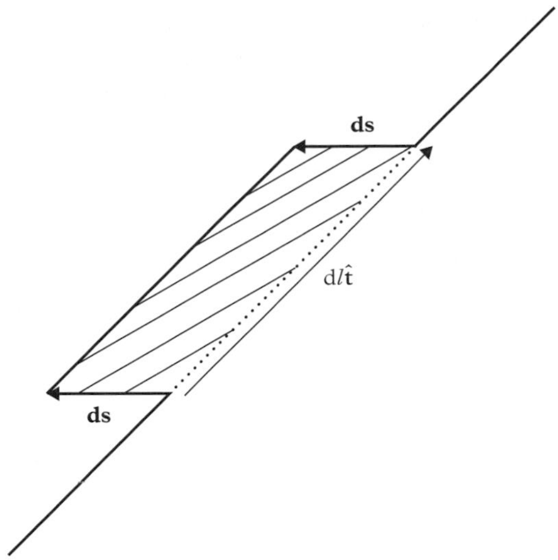

Figure 6.7 *A segment of dislocation of length dl moves by the vector* **ds** *sweeping the area shaded.*

applied stress is given by the change of $-E^T_{int}$ in eqn. 6.8:

$$dw = -dE^T_{int} = -b_i \sigma^A_{ij} \, dl \varepsilon_{jpq} \, ds_p \, t_q. \tag{6.10}$$

The force acting on the dislocation segment is $\partial w / \partial s_q = -\partial E^T_{int} / \partial s_q = F_q \, dl$, where F_q is the force per unit length:

$$F_q = b_i \sigma^A_{ij} \varepsilon_{jpq} t_p = \varepsilon_{qjp} \left(\sigma^A_{ji} b_i \right) t_p. \tag{6.11}$$

In vector notation this is $\mathbf{F} = (\boldsymbol{\sigma} \mathbf{b}) \times \hat{\mathbf{t}}$. This is the Peach–Koehler force. The force is always perpendicular to the dislocation line. For edge dislocations, where the slip plane is uniquely defined, it includes components that promote glide and climb of dislocations. If $\hat{\mathbf{n}}$ is the normal to the slip plane the climb component is $\mathbf{F}_c = (\mathbf{F} \cdot \hat{\mathbf{n}})\hat{\mathbf{n}}$ and the glide component is $\mathbf{F}_g = \mathbf{F} - \mathbf{F}_c = \hat{\mathbf{n}} \times (\mathbf{F} \times \hat{\mathbf{n}})$.

Exercise 6.3

Show that the glide component of the Peach–Koehler force per unit length on any dislocation is τb, where τ is the shear stress resolved on the slip plane in the direction of the Burgers vector and b is the magnitude of the Burgers vector.

Show that the climb component of the Peach–Koehler force on an edge dislocation is σb, where σ is the resolved normal stress in the direction of the Burgers vector.

6.7 Volterra's formula

We come now to the derivation of a general expression for the displacement field of a dislocation loop in anisotropic elasticity. This expression is useful because, as shown[15] by Nabarro,[16] the displacement field for an infinitesimal loop may be integrated to obtain solutions for loops of arbitrary shapes and sizes, which may then be differentiated to give strain fields and stress fields using Hooke's law. Assign a line sense to the dislocation loop. As before, the positive loop normal $\hat{\mathbf{n}}$ is defined by the right-hand screw rule applied to the positive line direction of the dislocation loop. Looking along the positive direction of the dislocation line draw a right-handed circuit around the dislocation to define the Burgers vector $\mathbf{b}$, as in eqn. 6.2. The surface inside the loop may be taken as the cut on either side of which the displacement field changes discontinuously by the Burgers vector. More precisely, the displacement field on the *negative* side of this cut minus the displacement field on the *positive* side of this cut is equal to the Burgers vector. The third line of eqn. 4.12 gives the displacement field in an infinite medium when displacements are prescribed on a surface $\mathcal{S}$. We may use this relation to write down the displacement field of a dislocation loop $\mathcal{L}$ with a displacement discontinuity equal to the Burgers vector between the negative and positive sides of a cut surface $\mathcal{S}$ bounded by $\mathcal{L}$. The result is known as Volterra's formula:

$$u_j(\mathbf{x}) = -c_{mikp}\, b_k \int_{\mathcal{S}^+} G_{ij,m'}(\mathbf{x}-\mathbf{x}')\, n_{p'}\, \mathrm{d}S' = c_{mikp}\, b_k \int_{\mathcal{S}^+} G_{ij,m}(\mathbf{x}-\mathbf{x}')\, n_{p'}\, \mathrm{d}S', \qquad (6.12)$$

where the integration is carried out on the positive side $\mathcal{S}^+$ of the cut only.

A dislocation loop created in this way is called a Volterra dislocation. It is a mathematical simplification of a real dislocation in the sense that it is a mathematical line where the displacement discontinuity equal to the Burgers vector appears infinitely abruptly. In other words a Volterra dislocation has no width, and the displacement by the Burgers vector appears as a step function at the dislocation line. Real dislocations in crystals have finite widths, which are called dislocation cores, where the relative displacement by the Burgers vector accumulates over several interatomic bond lengths. Nevertheless the concept of a Volterra dislocation describes the elastic field away from the core quite accurately. In the next chapter we will meet more realistic models of dislocations.

The stress and strain fields of the loop are more significant physically than the displacement field because they determine the energy of the loop and its interaction with other defects. In contrast to the stress and strain fields of the loop, the displacement field is inevitably dependent on the choice of the cut where the displacement by the Burgers vector is introduced. Differentiation of the displacement field obtained with Volterra's formula yields the strain field, which must be independent of the location of the cut. Later in this chapter we will derive Mura's formula for the *strain* field of the dislocation which

[15] Nabarro, FRN, *Phil. Mag.* **42**, 1224 (1951), http://dx.doi.org/10.1080/14786444108561379

[16] Frank Reginald Nunes Nabarro FRS 1916–2006, South African physicist born and educated in England.

depends only on the configuration of the dislocation *line*, and not on the cut bounded by it.

We have implied in this section that the choice of the cut plane is arbitrary. In an elastic continuum that is true. The dislocation is defined by the line where the cut terminates inside the medium. The location of the cut does affect the displacement field because it has to show the discontinuity by the Burgers vector on crossing the cut. But it does not affect the strain field and hence the stress field of the dislocation. For a crystal dislocation the location of the cut is defined by the slip plane of the dislocation when it glides. For an edge crystal dislocation the slip plane is uniquely defined by the cross product between the Burgers vector and the dislocation line direction. For a crystal screw dislocation the slip plane may change, as in cross-slip, but there is still only a finite number of choices of cut plane. Only in an elastic continuum is the cut plane truly arbitrary, and there it does not even need to be flat.

6.8 The infinitesimal loop

Consider a loop of infinitesimal area δA and unit normal n_p located at $\mathbf{x}'$. Volterra's formula, eqn. 6.12, provides the displacement field at $\mathbf{x}$:

$$\delta u_j(\mathbf{x}) = \delta A \, c_{mikp} \, b_k \, n_p \, G_{ij,m}(\mathbf{x}-\mathbf{x}'). \tag{6.13}$$

This expression is exact within linear anisotropic elasticity. If the loop is finite, with a characteristic size L, then it remains a good approximation provided $|\mathbf{x}-\mathbf{x}'|$ is more than about $2L$. To see that this is true consider a finite, planar, centrosymmetric loop, with its centre located at $\mathbf{R}$, where we take the cut to be the plane of area A bounded by the loop. Writing $\mathbf{x}' = \mathbf{R}+\boldsymbol{\rho}$ we have the Taylor expansion:

$$G_{ij,m}(\mathbf{x}-\mathbf{R}-\boldsymbol{\rho}) = G_{ij,m}(\mathbf{x}-\mathbf{R}) - \rho_p \, G_{ij,mp}(\mathbf{x}-\mathbf{R}) + \frac{1}{2}\rho_p \rho_q \, G_{ij,mpq}(\mathbf{x}-\mathbf{R}) - \ldots$$

Inserting this into Volterra's formula, eqn. 6.12, we obtain an expansion for the displacement field of the loop in terms of its areal moments:

$$u_j(\mathbf{x}) = c_{mikp} \, b_k \, n_p \left[G_{ij,m}(\mathbf{x}-\mathbf{R})A - G_{ij,mp}(\mathbf{x}-\mathbf{R}) \int_S \rho_p \, \mathrm{d}S + \frac{1}{2} G_{ij,mpq}(\mathbf{x}-\mathbf{R}) \int_S \rho_p \rho_q \, \mathrm{d}S - \ldots \right]. \tag{6.14}$$

For a centrosymmetric loop, such as a circle, ellipse, rectangle, hexagon, etc., only the even moments are non-zero. The first correction to eqn. 6.13 comes from second moments of the form $G_{ij,mpp}(\mathbf{x}-\mathbf{R}) \int_S \rho_p^2 \mathrm{d}S$, which is of order $L^4/|\mathbf{x}-\mathbf{R}|^4$. Thus, provided the distance from a finite centrosymmetric loop is more than about twice its size eqn. 6.13 is remarkably accurate.

In isotropic elasticity eqn. 6.13 becomes

$$\delta u_j(\mathbf{x}) = -\frac{\delta A\, b_k n_p}{8\pi(1-\nu)}\left[(1-2\nu)\frac{\delta_{kj}X_p + \delta_{pj}X_k - \delta_{kp}X_j}{X^3} + 3\frac{X_k X_p X_j}{X^5}\right], \tag{6.15}$$

where $\mathbf{X} = \mathbf{x} - \mathbf{x}'$. By integrating this expression the displacement fields of finite loops may be derived in isotropic elasticity. The displacement fields of infinite straight dislocations may be found by considering loops closed at infinity.

6.9 The dipole tensor of an infinitesimal loop

The leading term in the interaction energy between two point defects in the multipole expansion of eqn. 5.10 involves their dipole tensors. Since an infinitesimal dislocation loop is a point defect it should also be possible to define a dipole tensor for it. One way to derive the dipole tensor for an infinitesimal loop is to consider the interaction energy between two infinitesimal loops A and B using eqn. 6.9. If loop A is at the origin and loop B is at $\mathbf{x}$ then the interaction energy between them is as follows:

$$\delta E_{int} = b_f^B n_g^B (\delta A^B)(\delta\sigma_{fg}^A(\mathbf{x})), \tag{6.16}$$

where δA^B is the area of loop B and $\delta\sigma_{fg}^A(\mathbf{x})$ is the stress field of the infinitesimal loop A located at the origin evaluated at loop B. Taking only the first term of the moment expansion of the displacement field of a loop in eqn. 6.14, and differentiating it to obtain the strain field we obtain

$$\delta\sigma_{fg}^A(\mathbf{x}) = c_{fgjl} c_{mikp} b_k^A n_p^A (\delta A^A) G_{ij,ml}(\mathbf{x}). \tag{6.17}$$

Inserting this stress field into eqn. 6.16 we obtain an expression for the elastic interaction energy in terms of the dipole tensors $\delta\rho$ of the infinitesimal loops:

$$\delta E_{int} = \delta\rho_{jl}^B G_{ij,ml}(\mathbf{x}) \delta\rho_{mi}^A, \tag{6.18}$$

where

$$\begin{aligned}\delta\rho_{mi}^A &= c_{mikp} b_k^A n_p^A (\delta A^A)\\ \delta\rho_{jl}^B &= c_{jlfg} b_f^B n_g^B (\delta A^B).\end{aligned} \tag{6.19}$$

As with the interaction energy between two point defects eqn. 6.18 separates properties of the loops themselves contained in the dipole tensors from the radial and angular dependences of their interaction energy contained in the second derivatives of the Green's function. The dipole tensor of a loop in eqn. 6.19 has a simple physical interpretation. As first discussed[17] by Kroupa,[18] a dislocation loop may be thought of as a region that has undergone a transformation strain $e^T_{kp} = \frac{1}{2}(b_k n_p + b_p n_k)/\Delta$ where Δ is the thickness of the transformed region in the direction of the loop normal. (We will take the limit $\Delta \to 0$.) The transformation strain gives rise to a stress $\sigma^T_{im} = c_{imkp} e^T_{kp} = c_{imkp} b_k n_p/\Delta$. The surfaces of the loop are separated by the vector $\hat{\mathbf{n}}\Delta$. These surfaces are subjected to equal and opposite forces per unit area equal to $\sigma^T_{im} n_m$. The displacement at $\mathbf{x}$ is then given by

$$u_j(\mathbf{x}) = \int_{S'} G_{ji}(\mathbf{x} - \mathbf{x}')\sigma^T_{im} n_m \mathrm{d}S', \tag{6.20}$$

where the integration is over the surface of the transformed region constituting the loop. Inside the loop σ^T_{im} is constant. Applying the divergence theorem we obtain

$$u_j(\mathbf{x}) = \sigma^T_{im} \int_{V'} G_{ji,m}(\mathbf{x} - \mathbf{x}')\mathrm{d}V', \tag{6.21}$$

where the integral is over the volume of the transformed region. In the limit of an infinitesimal loop, and in the limit of $\Delta \to 0$, the volume integral becomes $G_{ji,m}(\mathbf{x})(\delta A)\Delta$ where δA is the area of the loop with normal $\hat{\mathbf{n}}$. The displacement field of the infinitesimal loop is therefore as follows:

$$u_j(\mathbf{x}) = c_{imkp}\frac{b_k n_p}{\Delta}\Delta(\delta A)G_{ji,m}(\mathbf{x}) = c_{imkp} b_k n_p(\delta A)G_{ji,m}(\mathbf{x}) = P_{im} G_{ji,m}(\mathbf{x}). \tag{6.22}$$

As discussed by Landau and Lifshitz[19] and the paper[20] by Burridge and Knopoff,[21] the displacement by the Burgers vector at a dislocation may be thought of as the response of the medium to a dipolar distribution of fictitious forces across the cut. These forces are fictitious in the sense that they are the forces required by linear elasticity to generate the displacement by the Burgers vector. They are equivalent to Kanzaki forces in harmonic lattice theory, which are the forces required to create a defect within a harmonic model of atomic interactions. The real forces are what we have called 'defect forces', and they are the true forces acting between atoms, as determined by quantum mechanics. But in a linear elastic theory it is the fictitious forces we need to generate a dislocation and they appear in the dipole tensor for the dislocation loop.

[17] Kroupa, F in *Theory of crystal defects*, Proceedings of the Summer School held in Hrazany in September 1964, Academia Publishing House of the Czechoslovak Academy of Sciences (1966), pp. 275–316.
[18] František Kroupa 1925–2009, Czech physicist.
[19] Landau, LD and Lifshitz, EM, *Theory of elasticity*, 3rd edn., Pergamon Press: Oxford (1986), section 27, p.111. ISBN 978-0750626330.
[20] Burridge, R and Knopoff, L, *Bull. Seismol. Soc. Am.* **54**, 1875–88 (1964), http://bssa.geoscienceworld.org/content/ssabull/54/6A/1875.full.pdf
[21] Leon Knopoff 1925–2011, US geophysicist and musicologist.

Volterra's formula, eqn. 6.12, follows directly from eqn. 6.21. That is because $\sigma^T_{im} = c_{imkp} b_k n_p / \Delta$ and $dV' = dS'\Delta$ so that eqn. 6.12 follows. This way of deriving Volterra's formula is arguably more satisfying than the rather formal presentation in section 6.7. A dislocation loop may thus be viewed in two equivalent ways. As discussed in section 6.7 the dislocation delineates a region in a plane which has undergone a displacement by the Burgers vector **b**. The second is that it delineates a region which has undergone a transformation strain $e^T_{kp} = \frac{1}{2}(b_k n_p + b_p n_k)/\Delta$, where Δ is the thickness of the transformed region in the direction of the loop normal and is very small compared to the diameter of the loop except in nanoscale loops. The transformed region inside the loop sets up closely spaced dipolar sheets of surface tractions inside the loop which generate the elastic field of the loop. This second way of viewing dislocations is particularly useful in the modelling of cracks as distributions of dislocations, as we shall see in Chapter 9.

Does this picture of terminating dipolar sheets of forces at a dislocation apply in an atomistic model? As a dislocation glides atoms on either side of the slip plane experience forces that introduce the relative displacement by the Burgers vector and enlarge the slipped region. Those forces generate vibrations of the crystal lattice, and as discussed in section 10.4.4 this radiation is a principal source of drag on the dislocation, and it gives rise to acoustic emission which can be detected experimentally: the forces are real and they have observable consequences. Once the dislocation has moved on the forces return to zero if the perfect crystal structure is recreated in the wake of the dislocation.[22] Since there is no crystal structure in a continuum the forces required to shear the medium to establish the relative displacement by the Burgers vector persist even after the dislocation has moved on. However, they produce no stress or strain field in the limit $\Delta \to 0$, and only the relative displacement by the Burgers vector on either side of the slip plane. Another difference is that the minimum value of Δ in a crystal is the spacing of atomic planes parallel to the slip plane, whereas in a continuum the limit $\Delta \to 0$ is usually taken.

Exercise 6.4

For a screw dislocation along the x_3-axis, with Burgers vector $b_k = b\delta_{k3}$, with the sheets occupying the half spaces $x_1 > 0, x_2 = \Delta/2$ and $x_1 > 0, x_2 = -\Delta/2$ show that in isotropic elasticity:

$$\sigma^T_{im} = \frac{\mu b}{\Delta}(\delta_{i3}\delta_{m2} + \delta_{i2}\delta_{m3}).$$

An edge dislocation along the x_3-axis may be created with $b_k = b\delta_{k1}$ and the same location of the sheets of force as in the screw dislocation. Show that for an edge dislocation in isotropic elasticity:

$$\sigma^T_{im} = \frac{\mu b}{\Delta}(\delta_{i1}\delta_{m2} + \delta_{i2}\delta_{m1}).$$

[22] This is always true for a perfect dislocation. More generally the forces return to zero for any dislocation separating regions of the slip plane that correspond to local minima in the γ-surface.

6.10 The infinitesimal loop in isotropic elasticity

The stress field of an infinitesimal loop is readily obtained in isotropic elasticity by differentiating the displacement field in eqn. 6.15 to get the distortion tensor and then applying Hooke's law:

$$\begin{aligned}\sigma_{ij}(\mathbf{x}) = &-\frac{\mu(\delta A)b}{4\pi(1-\nu)x^3}\\ &\times\{[3(1-2\nu)(\hat{\mathbf{b}}\cdot\hat{\mathbf{x}})(\hat{\mathbf{n}}\cdot\hat{\mathbf{x}})+(4\nu-1)(\hat{\mathbf{b}}\cdot\hat{\mathbf{n}})]\delta_{ij}\\ &+(1-2\nu)(\hat{b}_i n_j+n_i\hat{b}_j)+3\nu\left[(\hat{\mathbf{b}}\cdot\hat{\mathbf{x}})(n_i\hat{x}_j+\hat{x}_i n_j)+(\hat{\mathbf{n}}\cdot\hat{\mathbf{x}})(\hat{b}_i\hat{x}_j+\hat{x}_i\hat{b}_j)\right]\\ &+3(1-2\nu)(\hat{\mathbf{b}}\cdot\hat{\mathbf{n}})\hat{x}_i\hat{x}_j-15(\hat{\mathbf{b}}\cdot\hat{\mathbf{x}})(\hat{\mathbf{n}}\cdot\hat{\mathbf{x}})\hat{x}_i\hat{x}_j\}.\end{aligned} \tag{6.23}$$

where b is the magnitude of the Burgers vector, $\hat{\mathbf{b}}=\mathbf{b}/b$ is the unit vector parallel to $\mathbf{b}$ and $\hat{\mathbf{x}}=\mathbf{x}/x$.

In isotropic elasticity the second derivative of the Green's function is as follows:

$$\begin{aligned}G_{ik,jl}(x) = &\frac{1}{16\pi\mu(1-\nu)x^3}\\ &\times\{(3-4\nu)\delta_{ik}\left(3\hat{x}_l\hat{x}_j-\delta_{lj}\right)\\ &+15\hat{x}_i\hat{x}_j\hat{x}_k\hat{x}_l-3(\delta_{ij}\hat{x}_k\hat{x}_l+\delta_{il}\hat{x}_j\hat{x}_k+\delta_{jl}\hat{x}_i\hat{x}_k\\ &+\delta_{kj}\hat{x}_i\hat{x}_l+\delta_{kl}\hat{x}_i\hat{x}_j)+(\delta_{il}\delta_{kj}+\delta_{kl}\delta_{ij})\}.\end{aligned} \tag{6.24}$$

When this expression is inserted in eqn. 6.18 we obtain, after some tedious algebra, the following expression for the elastic interaction energy between two infinitesimal loops, A and B, in isotropic elasticity:[23]

$$\begin{aligned}E_{int}^{(AB)} = &\frac{\mu b^A b^B\left(\delta A^A\right)\left(\delta A^B\right)}{4\pi(1-\nu)x^3}\left[15(\hat{\mathbf{b}}^A\cdot\hat{\mathbf{x}})(\hat{\mathbf{b}}^B\cdot\hat{\mathbf{x}})(\hat{\mathbf{n}}^A\cdot\hat{\mathbf{x}})(\hat{\mathbf{n}}^B\cdot\hat{\mathbf{x}})\right.\\ &-3\nu\{(\hat{\mathbf{b}}^A\cdot\hat{\mathbf{b}}^B)(\hat{\mathbf{n}}^A\cdot\hat{\mathbf{x}})(\hat{\mathbf{n}}^B\cdot\hat{\mathbf{x}})+(\hat{\mathbf{b}}^A\cdot\hat{\mathbf{n}}^B)(\hat{\mathbf{n}}^A\cdot\hat{\mathbf{x}})(\hat{\mathbf{b}}^B\cdot\hat{\mathbf{x}})\\ &+(\hat{\mathbf{n}}^A\cdot\hat{\mathbf{b}}^B)(\hat{\mathbf{n}}^B\cdot\hat{\mathbf{x}})(\hat{\mathbf{b}}^A\cdot\hat{\mathbf{x}})+(\hat{\mathbf{n}}^A\cdot\hat{\mathbf{n}}^B)(\hat{\mathbf{b}}^A\cdot\hat{\mathbf{x}})(\hat{\mathbf{b}}^B\cdot\hat{\mathbf{x}})\}\end{aligned}$$

[23] Dudarev, SL and Sutton, AP, *Acta Mater.* **125**, 425–30 (2017). https://doi.org/10.1016/j.actamat.2016.11.060. Sergei Lvovich Dudarev 1960–, British materials physicist born in Belarus.

$$
\begin{aligned}
&-3(1-2\nu)(\hat{\mathbf{b}}^A\cdot\hat{\mathbf{n}}^A)(\hat{\mathbf{b}}^B\cdot\hat{\mathbf{x}})(\hat{\mathbf{n}}^B\cdot\hat{\mathbf{x}})\\
&-3(1-2\nu)(\hat{\mathbf{b}}^B\cdot\hat{\mathbf{n}}^B)(\hat{\mathbf{b}}^A\cdot\hat{\mathbf{x}})(\hat{\mathbf{n}}^A\cdot\hat{\mathbf{x}})\\
&-(1-2\nu)\{(\hat{\mathbf{b}}^A\cdot\hat{\mathbf{b}}^B)(\hat{\mathbf{n}}^A\cdot\hat{\mathbf{n}}^B)+(\hat{\mathbf{b}}^A\cdot\hat{\mathbf{n}}^B)(\hat{\mathbf{n}}^A\cdot\hat{\mathbf{b}}^B)\}\\
&-(4\nu-1)(\hat{\mathbf{b}}^A\cdot\hat{\mathbf{n}}^A)(\hat{\mathbf{b}}^B\cdot\hat{\mathbf{n}}^B)\big].
\end{aligned}
\tag{6.25}
$$

In this expression the loops are of arbitrary character, where the character is prismatic ($\hat{\mathbf{b}}\cdot\hat{\mathbf{n}}=1$ for a vacancy loop and $\hat{\mathbf{b}}\cdot\hat{\mathbf{n}}=-1$ for an interstitial loop), shear ($\hat{\mathbf{b}}\cdot\hat{\mathbf{n}}=0$) or a mixture ($0<|\hat{\mathbf{b}}\cdot\hat{\mathbf{n}}|<1$). The interaction energy separates into an inverse cube dependence on the separation between the loops and a term that depends on no less than ten angles. The ten angles are all the angles, taken in pairs, between the five unit vectors $\hat{\mathbf{x}},\hat{\mathbf{b}}^A,\hat{\mathbf{b}}^B,\hat{\mathbf{n}}^A,\hat{\mathbf{n}}^B$. Together with the magnitudes of the Burgers vectors, the loop areas and the distance between the loop centres, the interaction energy is a function of fifteen variables, and it is remarkable that this function has a closed form. Equation 6.25 may be applied to finite-sized planar loops provided the separation between them is more than about twice their size.

6.11 Mura's formula

We have already noted that the displacement field of a dislocation depends on the choice of the cut across which the relative displacement by the Burgers vector is introduced. This is explicit in Volterra's formula, eqn. 6.12. However, the strain and stress fields cannot depend on the choice of cut. Nevertheless, we can differentiate Volterra's formula to get the distortion field:

$$u_{j,g}(\mathbf{x})=c_{mikp}b_k\int_{S+}G_{ij,mg}(\mathbf{x}-\mathbf{x}')n_{p'}\,\mathrm{d}S', \tag{6.26}$$

but this also involves an integral over the cut surface. The distortion field should be independent of the choice of the cut and dependent only on the configuration of the dislocation line. Mura's formula makes this explicit by transforming the surface integral in eqn. 6.26 into a line integral along the dislocation line using a version of Stokes' theorem.

A familiar form of Stokes' theorem is

$$\int_S \varepsilon_{ijk}V_{k,j}(\mathbf{x}-\mathbf{x}')\,n_i\,\mathrm{d}S(\mathbf{x})=\oint_L V_p(\mathbf{x}-\mathbf{x}')\,\mathrm{d}x_p, \tag{6.27}$$

where L is the closed line where the surface S terminates. In this equation $\mathbf{x}'$ is constant and $\mathbf{x}$ ranges over the surface S and around the line L. In the following we revert to our usual notation $\mathrm{d}S(\mathbf{x})=\mathrm{d}S$. To obtain the version of Stokes' theorem needed here let $V_k(\mathbf{x}-\mathbf{x}') = f(\mathbf{x}-\mathbf{x}')\delta_{kq}$ so that the vector function $\mathbf{V}(\mathbf{x}-\mathbf{x}')$ has only one non-zero component and that is $V_q(\mathbf{x}-\mathbf{x}') = f(\mathbf{x}-\mathbf{x}')$. Then Stokes' theorem becomes

$$\varepsilon_{qij}\int_S f_{,j}(\mathbf{x}-\mathbf{x}')\,n_i\,\mathrm{d}S = \oint_L f(\mathbf{x}-\mathbf{x}')\,\mathrm{d}x_q.$$

Multiplying both sides of this equation by ε_{qpg} we obtain the following:

$$\begin{aligned}
\varepsilon_{qpg}\,\varepsilon_{qij}\int_S f_{,j}(\mathbf{x}-\mathbf{x}')\,n_i\,\mathrm{d}S &= \varepsilon_{qpg}\oint_L f(\mathbf{x}-\mathbf{x}')\,\mathrm{d}x_q \\
(\delta_{ip}\delta_{jg}-\delta_{jp}\delta_{ig})\int_S f_{,j}(\mathbf{x}-\mathbf{x}')\,n_i\,\mathrm{d}S &= \varepsilon_{qpg}\oint_L f(\mathbf{x}-\mathbf{x}')\,\mathrm{d}x_q \\
\int_S f_{,g}(\mathbf{x}-\mathbf{x}')\,n_p\,\mathrm{d}S &= \int_S f_{,p}(\mathbf{x}-\mathbf{x}')\,n_g\,\mathrm{d}S + \varepsilon_{qpg}\oint_L f(\mathbf{x}-\mathbf{x}')\,\mathrm{d}x_q.
\end{aligned} \tag{6.28}$$

Now we switch the integration variable from $\mathbf{x}$ to $\mathbf{x}'$, treating $\mathbf{x}$ as constant. Writing $\mathrm{d}S'$ for $\mathrm{d}S(\mathbf{x}')$ we obtain

$$\int_S f_{,g'}(\mathbf{x}-\mathbf{x}')\,n'_p\,\mathrm{d}S' = \int_S f_{,p'}(\mathbf{x}-\mathbf{x}')\,n'_g\,\mathrm{d}S' + \varepsilon_{qpg}\oint_L f(\mathbf{x}-\mathbf{x}')\,\mathrm{d}x'_q. \tag{6.29}$$

If we set $f(\mathbf{x}-\mathbf{x}') = G_{ij,m'}(\mathbf{x}-\mathbf{x}')$ then we obtain

$$\int_S G_{ij,m'g'}(\mathbf{x}-\mathbf{x}')\,n'_p\,\mathrm{d}S' = \int_S G_{ij,m'p'}(\mathbf{x}-\mathbf{x}')\,n'_g\,\mathrm{d}S' + \varepsilon_{qpg}\oint_L G_{ij,m'}(\mathbf{x}-\mathbf{x}')\,\mathrm{d}x'_q. \tag{6.30}$$

Multiplying both sides by c_{mikp} we obtain

$$\begin{aligned}
c_{mikp}\int_S G_{ij,m'g'}(\mathbf{x}-\mathbf{x}')\,n'_p\,\mathrm{d}S' &= c_{mikp}\int_S G_{ij,m'p'}(\mathbf{x}-\mathbf{x}')\,n'_g\,\mathrm{d}S' \\
&\quad + c_{mikp}\varepsilon_{qpg}\oint_L G_{ij,m'}(\mathbf{x}-\mathbf{x}')\,\mathrm{d}x'_q.
\end{aligned} \tag{6.31}$$

Recalling that $G_{ij,m'p'} = G_{ij,mp}$, it follows from the defining equation (eqn. 4.8) for the Green's function that $c_{mikp}G_{ij,m'p'} = 0$ at all points except $\mathbf{x}' = \mathbf{x}$. Since the surface S can always be chosen to avoid $\mathbf{x}$ the surface integral on the right hand side is zero. Inserting this result in eqn. 6.26 we obtain Mura's formula for the distortion tensor:

$$u_{j,g}(\mathbf{x}) = \varepsilon_{qpg}\,c_{mikp}\,b_k\oint_{L'} G_{ij,m'}(\mathbf{x}-\mathbf{x}')\,\mathrm{d}x_{q'}. \tag{6.32}$$

The stress field of the dislocation follows from Hooke's law: $\sigma_{ab}(\mathbf{x}) = c_{abjg}u_{j,g}(\mathbf{x})$. In isotropic elasticity this becomes

$$\sigma_{ab}(\mathbf{x}) = c_{abjg}\frac{\varepsilon_{qpg}b_k}{8\pi(1-\nu)}\oint_{L'}\left[(1-2\nu)\frac{\delta_{kj}X_p + \delta_{pj}X_k - \delta_{kp}X_j}{X^3} + 3\frac{X_kX_pX_j}{X^5}\right]\mathrm{d}x_{q'},$$

where $\mathbf{X} = \mathbf{x} - \mathbf{x}'$. Defining the line integral I_{kjpq} as

$$I_{kjpq} = \oint_{L'}\left[(1-2\nu)\frac{\delta_{kj}X_p + \delta_{pj}X_k - \delta_{kp}X_j}{X^3} + 3\frac{X_kX_pX_j}{X^5}\right]\mathrm{d}x_{q'}$$

the stress field may be conveniently expressed as follows:

$$\sigma_{ab}(\mathbf{x}) = \frac{\mu b_k}{8\pi(1-\nu)}\left[\frac{2\nu}{1-2\nu}\delta_{ab}\varepsilon_{qpj}I_{kjpq} + \varepsilon_{qpb}I_{kapq} + \varepsilon_{qpa}I_{kbpq}\right]. \tag{6.33}$$

6.12 The stress field of an edge dislocation in isotropic elasticity

To illustrate the application of Mura's formula the stress field of an infinitely long edge dislocation along the x_3-axis with $\mathbf{b} = [b, 0, 0]$ will be derived. In eqn. 6.33 we have $b_k = b\delta_{k1}$ and $q = 3$. We obtain

$$\sigma_{11} = \frac{\mu b}{4\pi(1-\nu)(1-2\nu)}(\nu I_{1213} - (1-\nu)I_{1123})$$

$$\sigma_{12} = \sigma_{21} = \frac{\mu b}{8\pi(1-\nu)}(I_{1113} - I_{1223})$$

$$\sigma_{13} = \sigma_{31} = -\frac{\mu b}{8\pi(1-\nu)}I_{1323}$$

$$\sigma_{22} = \frac{\mu b}{4\pi(1-\nu)(1-2\nu)}((1-\nu)I_{1213} - \nu I_{1123})$$

$$\sigma_{23} = \sigma_{32} = \frac{\mu b}{8\pi(1-\nu)}I_{1313}$$

$$\sigma_{33} = \frac{\mu b}{4\pi(1-\nu)(1-2\nu)}\nu(I_{1213} - I_{1123}) = \nu(\sigma_{11} + \sigma_{22}).$$

It is straightforward to evaluate the integrals:

$$I_{1213} = -\frac{2(1-2\nu)x_2}{(x_1^2+x_2^2)} + \frac{4x_1^2x_2}{(x_1^2+x_2^2)^2}$$

$$I_{1123} = \frac{2(1-2\nu)x_2}{(x_1^2+x_2^2)} + \frac{4x_1^2x_2}{(x_1^2+x_2^2)^2}$$

$$I_{1113} = \frac{2(1-2\nu)x_1}{(x_1^2+x_2^2)} + \frac{4x_1^3}{(x_1^2+x_2^2)^2}$$

$$I_{1223} = \frac{2(1-2\nu)x_1}{(x_1^2+x_2^2)} + \frac{4x_1x_2^2}{(x_1^2+x_2^2)^2}$$

$$I_{1313} = I_{1323} = 0.$$

We obtain the following stress components for an edge dislocation in isotropic elasticity:

$$\sigma_{11} = -\frac{\mu b}{2\pi(1-\nu)}x_2\frac{3x_1^2+x_2^2}{(x_1^2+x_2^2)^2}$$

$$\sigma_{12} = \frac{\mu b}{2\pi(1-\nu)}x_1\frac{x_1^2-x_2^2}{(x_1^2+x_2^2)^2}$$

$$\sigma_{13} = 0$$

$$\sigma_{22} = \frac{\mu b}{2\pi(1-\nu)}x_2\frac{x_1^2-x_2^2}{(x_1^2+x_2^2)^2}$$

$$\sigma_{23} = 0$$

$$\sigma_{33} = -\frac{\mu b\nu}{\pi(1-\nu)}\frac{x_2}{(x_1^2+x_2^2)}. \tag{6.34}$$

We see in eqn. 6.34 that the stress field is inversely proportional to the distance from the dislocation line. This is a general feature of straight dislocations. In particular the stress diverges as the dislocation line is approached. This follows from the assumption that the dislocation is a mathematical line with no width. In more realistic models the dislocation has a finite width and the stress does not become infinite.

Exercise 6.5

Sketch the stress component σ_{12} of eqn. 6.34 in the $x_1 - x_2$ plane. Sketch the hydrostatic stress $\frac{1}{3}\mathrm{Tr}\sigma$ for an edge dislocation in the $x_1 - x_2$ plane. Edge dislocations attract misfitting atoms. Where would a larger misfitting atom tend to locate itself to lower its elastic energy?

6.13 The elastic energy of a dislocation

The elastic energy of a dislocation is just the volume integral of the elastic energy density it creates:

$$E_{el} = \frac{1}{2}\int_V \sigma_{ij} u_{i,j} \mathrm{d}V.$$

It would be quite laborious to evaluate the elastic energy in this way as there are usually many non-zero stress and strain components. However, the task can be simplified significantly by applying the divergence theorem.

Consider an infinite straight dislocation parallel to the x_3-axis in an anisotropic elastic medium. Let the cut surface coincide with the half-plane $x_2 = 0$ where $x_1 \geq 0$, as shown in Fig. 6.8. Applying the divergence theorem to the elastic strain energy per unit length along x_3, and recalling that $\sigma_{ij,j} = 0$, we obtain

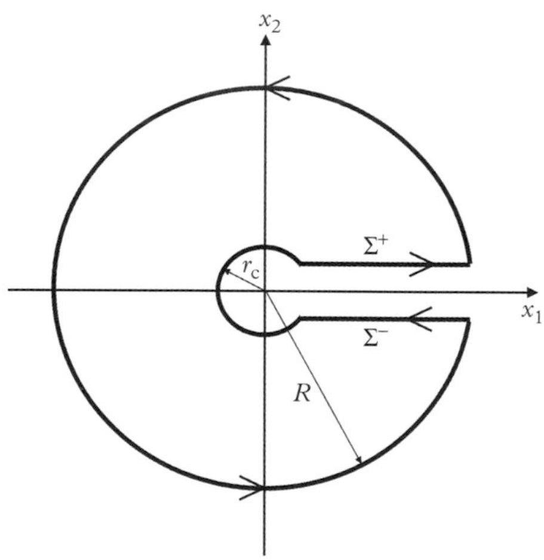

Figure 6.8 *To illustrate the evaluation of the surface integral in eqn. 6.35 for a straight dislocation lying along the x_3-axis. The positive line sense of the dislocation is coming out of the page. A right-handed circuit C is therefore anticlockwise, as shown by the arrows.*

$$E_{el} = \frac{1}{2}\int_C \sigma_{ij} u_i n_j \mathrm{d}S, \tag{6.35}$$

where the surface C comprises four segments of unit length along x_3. The first is the almost complete small cylinder of radius r_c around the dislocation line. In the limit the separation of the surfaces on either of the cut becomes infinitesimal the cylinder of radius r_c becomes complete. The integral of $\sigma_{ij}n_j$ around the cylinder is then zero because there is no resultant force associated with the dislocation line. The same argument applies to the segment around the outer cylinder of radius R, and it too is zero. That leaves the two contributions from the segments Σ^+ and Σ^- on either side of the cut. As the separation of the segments shrinks to zero they yield

$$E_{el} = \frac{1}{2}\int_{r_c}^{R} \sigma_{i2}(x_1, 0)\, b_i \mathrm{d}x_1. \tag{6.36}$$

For example, for an edge dislocation $b_i = b\delta_{i1}$ and the only stress component contributing to this integral is $\sigma_{12}(x_1, 0)$. Thus we obtain the following energy per unit length of an edge dislocation in isotropic elasticity:

$$E_{el}^{edge} = \frac{1}{2}\int_{r_c}^{R} \frac{\mu b^2}{2\pi(1-\nu)}\frac{\mathrm{d}x_1}{x_1} = \frac{\mu b^2}{4\pi(1-\nu)}\ln(R/r_c). \tag{6.37}$$

For a screw dislocation $b_i = b\delta_{i3}$ and the only stress component contributing to the integral in eqn. 6.36 is $\sigma_{32}(x_1, 0)$. In the isotropic elastic approximation the energy per unit length of a screw dislocation is as follows:

$$E_{el}^{screw} = \frac{\mu b^2}{4\pi}\ln(R/r_c). \tag{6.38}$$

The logarithmic divergence of the energy of a straight dislocation is common to other line singularities in physics, for example the electrostatic energy per unit length of an infinitely long thin line of charge or the magnetic energy per unit length of an infinitely long thin wire carrying a constant current. Of course dislocations exist in crystals of finite size, so their energy never becomes infinite. More significantly, dislocations often lower their elastic energies by organising themselves into configurations where their elastic fields have much shorter spatial extent than they would have if each dislocation were isolated. A classic example of such screening is the elastic field of a grain boundary comprising an array of dislocations with a spacing d. It is found (see the Problem set at the end of this chapter) that the elastic field decays exponentially from the grain boundary with a decay distance of d.

The surface C for the integral in eqn. 6.35 has to avoid the singularity at $x_1 = x_2 = 0$ where the stress tensor diverges. r_c is often called the core radius, but as we have argued the core has no radius in this theory. The meaning of r_c is more accurately described as the radius at which the elastic solution provides an acceptable description of the actual stresses near the dislocation. It is of order 0.5 nm in most crystalline materials. However, it is not necessarily the circular cross section that has been assumed here and many other places. For example, in fcc metals it is often found that the core is spread on the slip plane, so that it is much longer in that direction than it is normal to the slip plane by as much as a factor of ten. The total energy per unit length of the dislocation is the sum of the elastic energy and the core energy per unit length.

The surface tension of a liquid arises because the surface of the liquid has an energy per unit area. If the surface is stretched more surface area is created by diffusion of atoms from beneath the surface, which raises the energy of the system. The increase in energy may be thought of as the work done against the surface tension. Similarly a line tension may be associated with the energy per unit length of a dislocation: in isotropic elasticity if there are no other forces acting on them dislocations will always seek to minimise their lengths. But this is not necessarily true in elastic anisotropy because the energies per unit length of some line directions may be significantly less than those of other directions. In that case the total elastic energy may decrease even though there is an overall increase in the length of dislocation line.

The logarithmic singularity in the elastic energy per unit length of a dislocation is weak. Changing the outer radius R by an order of magnitude changes the elastic energy per unit length by only a factor of $\ln 10$. Sometimes we need a rough estimate of the energy of a dislocation per unit length for back-of-an-envelope calculations, and then we use $\mu b^2/2$. This is also a useful approximation for the line tension of a dislocation. It is of order 0.1 to 1 $\mathrm{eV\mathring{A}}^{-1}$. This is also the range of energies per unit length of the core.

The energy of a dislocation loop is also given by eqn. 6.35 where the surface C is that shown in Fig. 6.6. The energy of the loop is then

$$E_{el}^{loop} = \frac{1}{2}\int_{\Sigma+} \sigma_{ij} b_i n_j \mathrm{d}S, \tag{6.39}$$

where the integral is taken over the positive side of the cut surface inside the loop. This expression may be understood as the work done against the self-stress of the loop when it is created by increasing its Burgers vector from zero to the final b_i. By the conservation of energy this work becomes the elastic energy of the medium.

6.14 The Frank–Read source

For many years after it became accepted that dislocations are the agents of plastic deformation in metals it was not clear how they were created. To create a dislocation loop by homogeneous nucleation requires stresses approaching the theoretical shear strength, which are much higher than those observed when a metal yields at normal strain rates.

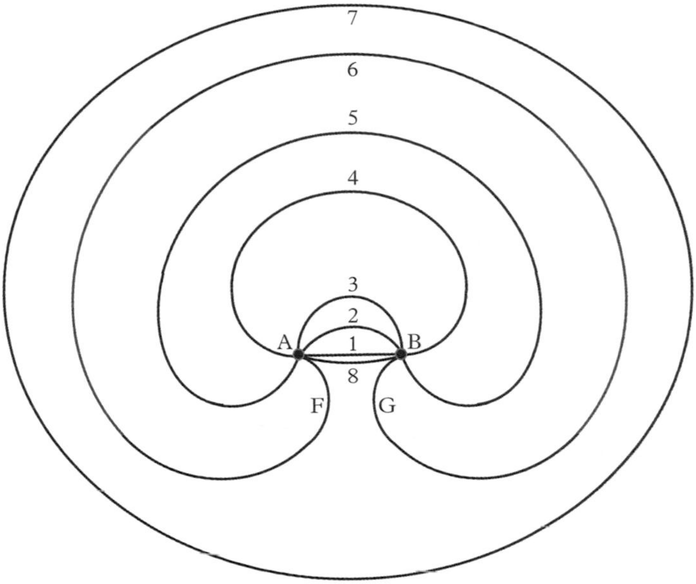

Figure 6.9 *Schematic illustration of the operation of a Frank–Read source. A segment of dislocation line is pinned at A and B (configuration 1). Under the action of an applied shear stress on the slip plane in the direction of the Burgers vector the dislocation segment bows (configuration 2) against the line tension tending to keep it straight. Eventually a critical configuration is reached, configuration 3, where the radius of curvature of the dislocation segment is a minimum. The segment continues to expand (configurations 4, 5). Segments F and G have opposite line senses and attract each other. When they meet, a loop is liberated and expands under the action of the applied stress, configuration 7. A segment is left behind, configuration 8, and returns to AB and the process starts again.*

A mechanism of generating dislocations was needed to explain these relatively small yield stresses. One of the most commonly observed mechanisms was proposed before it was observed by Frank and Read[24] and is known as a Frank–Read source.[25]

Figure 6.9 illustrates the operation of a Frank–Read source. We consider a segment of a dislocation pinned at A and B. There are many ways the dislocation could be pinned at two points. In the paper by Frank and Read they considered a rectangular prismatic loop ABCD with Burgers vector normal to the plane of the loop; see Fig. 6.10. A pure shear stress on the plane containing the Burgers vector and the line segment AB will make segments AB and CD move in opposite directions, but the segments AD and BC will experience no Peach–Koehler force and remain static. The result is that the segments AB and CD are pinned at their ends.

[24] Frank, FC and Read, WT, *Phys. Rev.* **79**, 722 (1950). https://doi.org/10.1103/PhysRev.79.722. W Thornton Read was a US physicist.

[25] There is an interesting account by Sir Charles Frank of how he and Thornton Read developed the idea of their dislocation source independently and precisely simultaneously in *Proc. R. Soc. A* **371**, 136 (1980). http://dx.doi.org/10.1098/rspa.1980.0069

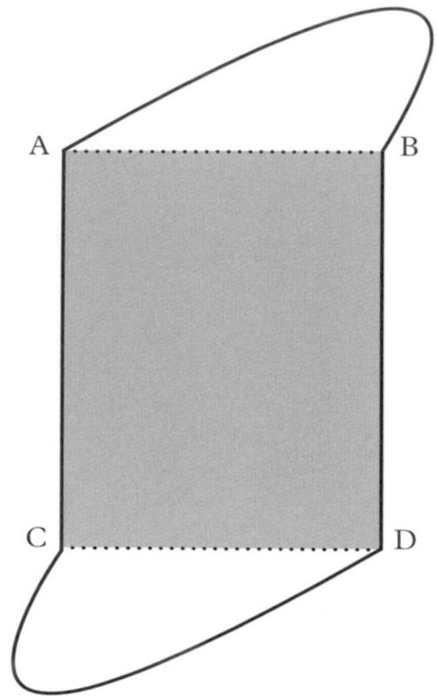

Figure 6.10 *A rectangular prismatic loop ABCD with Burgers vector normal to the page initially occupies the shaded area. A shear stress is applied in the direction of the Burgers vector on planes containing the Burgers vector and the lines AB and CD. Segments AB and CD bow under the influence of this applied shear stress. However, segments AC and BD experience no resolved shear stress and remain static. Segments AB and CD may then operate as Frank–Read sources of loops, as illustrated in Fig. 6.9. Jaehyun Cho has uploaded a movie of such a pair of Frank–Read sources operating at https://www.youtube.com/watch?v=jwK-TF7o2Oo.*

Let τ be the resolved shear stress on the slip plane in the direction of the Burgers vector of the segment AB. The Peach–Koehler force acting on the segment is then τb where, as usual, b is the magnitude of the Burgers vector. This force acts in a direction normal to the line and the dislocation bows out between the pinning points (see configuration 2 in Fig. 6.9). As a result the dislocation line length increases and this is opposed by the line tension, T. As the radius of curvature of the bowed segment decreases the stress required to make it bow further increases. Eventually a critical configuration is reached (configuration 3) where the radius of curvature is a minimum and the bowed segment expands with no further increase in the applied stress required. The segments F and G (see Fig. 6.9) have opposite line directions and they attract each other and annihilate. This reaction liberates a loop (configuration 7) which then expands freely under the influence of the applied stress, leaving behind a segment (configuration 8) which returns to the configuration AB at the beginning of the operation of the source. The process then repeats, sending out a succession of loops into the slip plane. Eventually these loops meet obstacles such as grain boundaries where they may form a 'pileup'. The dislocations in the pileup exert a stress on the source, called a 'back stress', which eventually reduces the total stress acting on the source to zero. Further operation of the source then requires the applied stress to be increased.

It is instructive to derive an approximate expression for the stress required to operate a Frank–Read source. This is a good example of a 'back-of-an-envelope' estimate that provides insight. Let the length of the pinned segment AB be L. We assume that when the dislocation starts to bow it forms the arc of a circle with a radius R. An element $\mathrm{d}l$ of the arc experiences an outward force due to the applied stress equal to $\tau b \mathrm{d}l$. It also experiences an inward force due to the line tension, given by $T\mathrm{d}l/R$ where $T \approx \mu b^2/2$. At equilibrium $\tau \approx \mu b/(2R)$. As the segment bows further R decreases and τ increases until the critical configuration is reached where R is a minimum. This happens when the bowed segment is a semicircle, with $R = L/2$. Thus, the minimum stress required to operate the source is $\tau_{min} \approx \mu b/L$.

Efforts have been made to improve on this rough estimate. They include replacing the line tension approximation with an evaluation of the dislocation self-interactions when it bows, using anisotropic elasticity and including a friction stress that must be overcome before a dislocation will move. These are all possible but much more difficult than our simple estimate.

The relationship $\tau \approx \mu b/L$ is known as the Orowan flow stress.[26] It provides understanding in a variety of contexts of strengthening mechanisms, and it is one of the most useful in the theory of dislocations. Here are three examples of its usefulness:

- *Precipitation hardening.* One way metals are made stronger is by alloying them with elements that result in a dispersion of second phase particles. Dislocations moving in their slip planes encounter these particles. If the particles are sufficiently large that dislocations cannot cut through them they bow out between them and proceed on the slip plane after leaving a loop around each precipitate. The critical stress required for this process is of order $\mu b/L$, where L is the average separation of the precipitates. This is a key relationship in designing age-hardened alloys.
- *Work hardening.* Another way metals are made stronger is by deforming them plastically. This is called work hardening and it is discussed in Chapter 10. It arises because dislocations moving on one slip plane encounter dislocations moving on inclined slip planes, as a result of which a variety of obstacles may form that impede slip on both slip planes. Further slip then proceeds by dislocations bowing out between these obstacles. If L is the separation between the obstacles then the stress required for further slip is again of order $\mu b/L$. The spacing L is related to the dislocation density ρ, which is defined as the number of dislocations crossing unit area or equivalently the total dislocation length per unit volume: $\rho \approx 1/L^2$. Therefore the stress required for dislocations to glide varies as $\mu b/L \approx \mu b\sqrt{\rho}$. This relationship was first proposed by GI Taylor in his paper of 1934 where he proposed dislocations as the agents of plasticity.
- *Plasticity in nanocrystals.* We saw at the beginning of this chapter that dislocations are the agents of plasticity because the stress required to slide an entire plane of atoms within a crystal over an adjacent plane is far too large. But if the area of the plane where slip occurs is very small, there might be a transition from slip mediated

[26] The flow stress is the stress required to sustain the glide of dislocations when they encounter obstacles.

by dislocations to 'block slip', where an entire plane of atoms *can* slide over another. If a dislocation is introduced into a cubic crystal of side L it will tend to be pinned by the surfaces. The stress required to make it bow out will be of order $\mu b/L$. Thus if $L \approx 10-100b$ the stress required to move the dislocation becomes comparable to that required for block slip.

6.15 Problem set 6

1. In eqn. 4.13 the displacement field of an infinite planar fault was derived:

$$u_i(\mathbf{x}) = -\int_{fault} G_{ij,l'}(\mathbf{x}-\mathbf{x}')\, c_{jlmp}\, t_m\, n_p\, \mathrm{d}^2x', \tag{6.40}$$

where $\mathbf{t}$ is the translation of the medium on the negative side of the fault relative to that on the positive side. In isotropic elasticity we have

$$c_{jlmp}G_{ij,l'}(\mathbf{X}) = \frac{1}{8\pi(1-\nu)}\left[(1-2\nu)\frac{\delta_{mi}X_p + \delta_{pi}X_m - \delta_{mp}X_i}{X^3} + 3\frac{X_m X_p X_i}{X^5}\right], \tag{6.41}$$

where $\mathbf{X} = \mathbf{x}-\mathbf{x}'$. If the fault is in the plane $x_3 = 0$ verify that eqn. 6.40 yields $u_i(\mathbf{x}) = -\frac{1}{2}\mathrm{sgn}(x_3)t_i$.

2. Using Volterra's formula for the displacement field of a dislocation in the form

$$u_i(\mathbf{x}) = -c_{jlmp}\, b_m \int_{S+} G_{ij,l'}(\mathbf{x}-\mathbf{x}')\, n'_p\, \mathrm{d}S',$$

and using eqn. 6.41, show that, with the cut in the half-plane $x_2 = 0, x_1 \geq 0$, the displacement field of a screw dislocation in isotropic elasticity is as follows:

$$u_1(x_1,x_2) = 0$$

$$u_2(x_1,x_2) = 0$$

$$u_3(x_1,x_2) = \frac{b}{2\pi}\left[\tan^{-1}\left(\frac{x_2}{x_1}\right) - \pi\right].$$

3. Using Mura's formula, eqn. 6.32, in isotropic elasticity show that the strain and stress fields of a screw dislocation lying along the x_3-axis with Burgers vector $\mathbf{b} = [0,0,b]$ are as follows:

$$e_{11} = e_{22} = e_{33} = e_{12} = 0$$

$$e_{13} = -\frac{b}{4\pi}\frac{x_2}{x_1^2 + x_2^2}$$

$$e_{23} = \frac{b}{4\pi}\frac{x_1}{x_1^2 + x_2^2}$$

$$\sigma_{11} = \sigma_{22} = \sigma_{33} = \sigma_{12} = 0$$

$$\sigma_{13} = -\frac{\mu b}{2\pi}\frac{x_2}{x_1^2 + x_2^2}$$

$$\sigma_{23} = \frac{\mu b}{2\pi}\frac{x_1}{x_1^2 + x_2^2}. \tag{6.42}$$

Verify that your answer is consistent with the displacement field of the screw dislocation derived in the previous question.

4. Calculate the elastic energy of a screw dislocation in isotropic elasticity two ways:
   (i) by integrating the elastic energy density $\frac{1}{2}\sigma_{ij}e_{ij}$ using the stress and strain tensors derived in the previous question;
   (ii) by using eqn. 6.36.

5. An isotropic elastic crystal contains edge dislocations with Burgers vectors $\pm\mathbf{b}$ gliding on a set of parallel slip planes. Let the dislocations lie along the x_3-axis and let the normal to the slip planes be $[0,1,0]$. An edge dislocation with Burgers vector $\mathbf{b} = [b,0,0]$ is pinned at the origin of the coordinate system. A second edge dislocation with Burgers vector $\mathbf{b} = [b,0,0]$ is gliding on a parallel slip plane. The distance between the slip planes of the two dislocations is D. Show that the position of stable equilibrium of the second dislocation is at $x_1 = 0, x_2 = D$. If now the Burgers vector of the second dislocation is $\mathbf{b} = -[b,0,0]$ show that it has two positions of stable equilibrium at $x_1 = \pm D, x_2 = D$.
6. Stress field of a symmetric tilt boundary.

   This question illustrates how dislocations may organise themselves into configurations where they screen their elastic fields, reducing the overall elastic energy.

   A small angle symmetric tilt grain boundary in the plane $x_1 = 0$ comprises an infinite array of edge dislocations with Burgers vector $[b,0,0]$. The dislocation lines are parallel to the x_3-axis and their positions along the x_2-axis are $x_2 = 0$, $\pm p, \pm 2p, \pm 3p, \ldots, \pm\infty$. Using the components of the stress tensor for an edge dislocation given in eqn. 6.34 show that the non-zero stress components of the grain boundary are

$$\sigma_{11}(X_1,X_2) = -\frac{\mu b}{2\pi(1-\nu)p}\sum_{n=-\infty}^{\infty}(X_2-n)\left(\frac{3X_1^2+(X_2-n)^2}{\left(X_1^2+(X_2-n)^2\right)^2}\right)$$

$$\sigma_{12}(X_1,X_2) = \frac{\mu b}{2\pi(1-\nu)p}\sum_{n=-\infty}^{\infty}X_1\left(\frac{X_1^2-(X_2-n)^2}{\left(X_1^2+(X_2-n)^2\right)^2}\right)$$

$$\sigma_{22}(X_1,X_2) = \frac{\mu b}{2\pi(1-\nu)p}\sum_{n=-\infty}^{\infty}(X_2-n)\left(\frac{X_1^2-(X_2-n)^2}{\left(X_1^2+(X_2-n)^2\right)^2}\right)$$

$$\sigma_{33}(X_1,X_2) = -\frac{\mu b\nu}{\pi(1-\nu)p}\sum_{n=-\infty}^{\infty}\frac{X_2-n}{X_1^2+(X_2-n)^2},$$

where $X_1 = x_1/p$ and $X_2 = x_2/p$.

The sums may be evaluated as contour integrals using the Sommerfeld–Watson transformation. For example,

$$\sum_{n=-\infty}^{\infty}\frac{1}{n+a} = \frac{1}{2\pi i}\oint_C \pi\cot(\pi z)\frac{1}{z+a}\mathrm{d}z = \pi\cot\pi a,$$

where the contour includes all the poles of $\cot(\pi z)$ along the real axis where $z = n$, but excludes the pole at $z = a$ and is closed by a circle at infinity. The result holds for real or complex a. We may use this result to derive all the sums we need to evaluate the stress components for the grain boundary. For example, since

$$\frac{1}{n+X_2+iX_1}+\frac{1}{n+X_2-iX_1} = \frac{2(n+X_2)}{(n+X_2)^2+X_1^2}$$

we deduce that

$$\sum_{n=-\infty}^{\infty}\frac{n+X_2}{(n+X_2)^2+X_1^2} = \frac{\pi\sin(2\pi X_2)}{\cosh(2\pi X_1)-\cos(2\pi X_2)}.$$

Hence show that the non-zero components of the stress tensor associated with the grain boundary are as follows:

$$\sigma_{11} = -\frac{\mu b}{2(1-\nu)p}\frac{\sin(2\pi X_2)[\cosh(2\pi X_1) - \cos(2\pi X_2) + 2\pi X_1 \sinh(2\pi X_1)]}{(\cosh(2\pi X_1) - \cos(2\pi X_2))^2}$$

$$\sigma_{12} = \frac{\pi \mu b}{(1-\nu)p}\frac{X_1[\cosh(2\pi X_1)\cos(2\pi X_2) - 1]}{(\cosh(2\pi X_1) - \cos(2\pi X_2))^2}$$

$$\sigma_{22} = -\frac{\mu b}{2(1-\nu)p}\frac{\sin(2\pi X_2)[\cosh(2\pi X_1) - \cos(2\pi X_2) - 2\pi X_1 \sinh(2\pi X_1)]}{(\cosh(2\pi X_1) - \cos(2\pi X_2))^2}$$

$$\sigma_{33} = -\frac{\mu b \nu}{(1-\nu)p}\frac{\sin(2\pi X_2)}{(\cosh(2\pi X_1) - \cos(2\pi X_2))}.$$

When $|x_1| > p$ show that these stress components are approximately proportional to $e^{-2\pi|x_1|/p}$.

Show that in the limit $p \to \infty$ these stress components become those of an isolated dislocation, given in eqn. 6.34.

7. Having derived the stress component $\sigma_{12}(x_1, x_2)$ for the symmetrical tilt grain boundary in the previous question we may use it to calculate the energy of the boundary. Let the cut associated with each edge dislocation at $x_1 = 0, x_2 = np$ be in the half-plane $x_2 = np, x_1 \geq 0$. Following the same argument using the divergence theorem to derive eqn. 6.36, where the contour C now involves an infinite number of circuits around the cuts and the dislocations in the grain boundary, show that the elastic energy of the grain boundary per unit area is as follows:

$$E_{GB} = \frac{1}{p}\left[\frac{1}{2}\int_{r_c}^{R} \sigma_{12}(x_1, 0)\, b\, \mathrm{d}x_1 + E_c\right]$$

where E_c is the energy per unit length of the material inside the radius r_c of each dislocation that cannot be described with elasticity. It is assumed that E_c does not vary with the misorientation angle θ. This assumption is reasonable provided θ is small. In that case $\theta = b/p$ is also a good approximation.

Given the standard integral

$$\int \frac{u}{\sinh^2 u}\mathrm{d}u = \ln(\sinh u) - u \coth u,$$

show that

$$E_{GB} = \theta\left[\left\{\frac{E_c}{b} + \frac{\mu b}{4\pi(1-\nu)}\ln\left(\frac{eb}{2\pi r_c}\right)\right\} - \frac{\mu b}{4\pi(1-\nu)}\ln\theta\right].$$

This equation has the form $E_{GB} = \theta(A - B\ln\theta)$ where A and B are constants:

$$A = \frac{E_c}{b} + \frac{\mu b}{4\pi(1-\nu)}\ln\left(\frac{eb}{2\pi r_c}\right)$$
$$B = \frac{\mu b}{4\pi(1-\nu)}$$

It is known as the Read–Shockley[27] formula.[28]

8. In this question we repeat the calculation of the energy of a small-angle symmetrical tilt grain boundary in a more heuristic but insightful way. The insight is to exploit the mutual screening of the elastic fields of the edge dislocations it contains. The larger radius R in eqn. 6.37 for the energy of an isolated edge dislocation may be taken as half the spacing p of the dislocations in the boundary. Show that the energy per unit area of the grain boundary is then

$$E_{GB} \approx \frac{1}{p}\left[\frac{\mu b^2}{4\pi(1-\nu)}\ln\left(\frac{p}{2r_c}\right) + E_c\right],$$

where E_c is the energy of the material inside the radius r_c which cannot be described with elasticity. As before, the angle θ of misorientation across the boundary is approximately b/p.

Show that E_{GB} has the form

$$E_{GB} \approx \theta(A - B\ln\theta), \qquad (6.43)$$

where A and B are constants with the dimensions of energy per unit area:

$$A = \frac{E_c}{b} + \frac{\mu b}{4\pi(1-\nu)}\ln\left(\frac{b}{2r_c}\right)$$

$$B = \frac{\mu b}{4\pi(1-\nu)}.$$

This heuristic solution is very close to the more rigorous solution in the previous question.

[27] William Bradford Shockley 1910–89, US Nobel Prize-winning physicist.
[28] Read, WT, and Shockley, W, *Phys. Rev.* **78**, 275 (1950). https://doi.org/10.1103/PhysRev.78.275

9. One application of eqn. 6.25 is to the evolution of prismatic loops in tungsten created by irradiation with high energy neutrons. Tungsten is very close to being elastically isotropic, and the distribution of loops evolves in time driven in part by their elastic interactions. Suppose the prismatic loops are such that $\hat{\mathbf{b}} \cdot \hat{\mathbf{n}} = 1$. Show that eqn. 6.25 simplifies as follows, and discuss its validity:

$$E_{int}^{(AB)} = \frac{\mu b^A b^B (\delta A^A)(\delta A^B)}{4\pi(1-\nu)x^3}\{15(\hat{\mathbf{n}}^A \cdot \hat{\mathbf{x}})^2(\hat{\mathbf{n}}^B \cdot \hat{\mathbf{x}})^2$$
$$-(4\nu-1) - 12\nu(\hat{\mathbf{n}}^A \cdot \hat{\mathbf{n}}^B)(\hat{\mathbf{n}}^A \cdot \hat{\mathbf{x}})(\hat{\mathbf{n}}^B \cdot \hat{\mathbf{x}})$$
$$-(1-2\nu)[3(\hat{\mathbf{n}}^B \cdot \hat{\mathbf{x}})^2 + 3(\hat{\mathbf{n}}^A \cdot \hat{\mathbf{x}})^2 + 2(\hat{\mathbf{n}}^A \cdot \hat{\mathbf{n}}^B)^2]\}. \tag{6.44}$$

Note the angular dependence has been reduced from ten to just three angles between the unit vectors $\hat{\mathbf{x}}, \hat{\mathbf{n}}^A, \hat{\mathbf{n}}^B$ taken in pairs.

10. The elastic interaction energy between a defect D and an applied field A is expressed as a surface integral in eqn. 6.7 where the integration surface encloses D only. In this question we apply this expression to the ellipsoidal inclusion of section 4.7 to derive the interaction energy between an applied elastic field and an ellipsoidal inclusion. The interaction energy was stated without proof in question 6 of Problem set 4, where the various terms used in this question are defined.

Take the surface S of the surface integral in eqn. 6.7 just outside the surface $\mathcal{S}$ of the inclusion. The interaction energy is then

$$E_{int} = \int_{\mathcal{S}} \left(\sigma_{ij}^C u_i^A - \sigma_{ij}^A u_i^C\right) n_j \, \mathrm{d}S.$$

Using the continuity of tractions on the surface of the inclusion, $\sigma_{ij}^C n_j = \sigma_{ij}^I n_j$, and continuity of displacements u_i^C on either side of the surface of the inclusion, and applying the divergence theorem, show that the interaction energy may be expressed as the following volume integral over the interior of the inclusion $\mathcal{R}$:

$$E_{int} = \int_{\mathcal{R}} \left(\sigma_{ij}^I e_{ij}^A - \sigma_{ij}^A e_{ij}^C\right) \mathrm{d}V.$$

Using $\sigma_{ij}^A e_{ij}^C = \sigma_{ij}^C e_{ij}^A$ and $\sigma_{ij}^I = \sigma_{ij}^T + \sigma_{ij}^C$ hence show that

$$E_{int} = -\int_{\mathcal{R}} \sigma_{ij}^T e_{ij}^A \, \mathrm{d}V = -\int_{\mathcal{R}} \sigma_{ij}^A e_{ij}^T \, \mathrm{d}V.$$

7

Hybrid models of dislocations

7.1 Introduction

The treatment of dislocations in an elastic continuum in the previous chapter has a number of faults which are remedied in this chapter by taking into consideration the discrete atomic structure of the crystal. In the continuum theory of dislocations the Burgers vector is arbitrary. But in a crystal to maintain equivalent structures in slipped and unslipped regions of a slip plane the Burgers vector must be a crystal lattice translation vector. The stress associated with a Volterra dislocation becomes infinite at its centre because the dislocation is treated as a line singularity separating slipped and unslipped regions. Infinite stresses cannot be sustained. In reality the relative displacement across the slip plane by the Burgers vector is accumulated over a finite number of interatomic distances in the dislocation core, giving the core a finite width within which the Burgers vector is distributed. By including the energy of distorted bonds between atoms on either side of the slip plane the Frenkel–Kontorova and Peierls–Nabarro models predict a finite core width, and stresses remain finite in the dislocation core. The Peierls–Nabarro model also shows there is a finite stress required to move a dislocation, which is a direct result of its discrete atomic structure. This leads to a mechanism of dislocation motion involving kinks on dislocations, which is thermally activated, and which provides an explanation for brittle to ductile transitions in some pure crystals, as well as the strain rate dependence of the stress at which dislocations move in pure crystals.

A key concept in this chapter is the γ-surface introduced in section 7.3. The γ-surface provides the restoring force tending to keep the dislocation core as narrow as possible. The γ-surface also indicates the possible existence of metastable planar faults in the crystal which may enable dislocations with Burgers vectors equal to lattice translation vectors to split into two or more partial dislocations separated by the metastable fault(s), as discussed in section 7.4.

Physics of elasticity and crystal defects. Adrian P. Sutton, Oxford University Press (2020). © Adrian P. Sutton.
DOI: 10.1093/oso/9780198860785.001.0001

7.2 Frenkel–Kontorova model

The Frenkel–Kontorova model[1] is a development of a model introduced[2] ten years earlier by Dehlinger.[3] The model considers a chain of atoms connected by identical harmonic springs, with spring constant κ and natural length a, placed on a rigid substrate characterised by a sinusoidally varying potential $V\sin^2(\pi x/a)$. In the ground state the position of each atom along the x-axis is $x = na$, where n is an integer, and since all springs have their natural length a and all atoms sit in the minima of the substrate potential the total energy is zero. If u_n is the displacement of atom n from its ground state position $x = na$ the total energy of the system is as follows:

$$E = \sum_n \frac{1}{2}\kappa(u_n - u_{n-1})^2 + V\sin^2(\pi u_n/a). \tag{7.1}$$

Now suppose the chain is stretched by a. If the chain were not in the presence of the substrate potential the stretch would be taken up equally by all the springs in the chain. But in the presence of the substrate potential this is unlikely because the constant strain would put too many atoms at and near the maximum in the potential. The substrate potential tends to localise the strain to a region where the strain is correspondingly large, so that the stretch displaces a smaller number of atoms from the minima in the potential. Minimisation of the total energy in eqn. 7.1 leads to a balance between the elastic energy, tending to delocalise and minimise the strain in a large region, and the substrate potential, tending to localise the strain in a small region where it is large.

The relationship of this one-dimensional model to an edge dislocation is illustrated in Fig. 7.1. It is assumed the dislocation core is planar lying in the slip plane, and corresponds to the atoms shown in red in Fig. 7.1. But rather than there being one chain of atoms separated by harmonic springs sitting in a rigid periodic potential there are two harmonic chains, which interact with each other. Far from the location of the extra half plane of the dislocation each harmonic chain provides a periodic potential for the other, as in the original Frenkel–Kontorova model. But in the dislocation core both chains distort, with one being compressed by $a/2$ and the other stretched by $a/2$, to accommodate the Burgers vector with magnitude equal to a, as illustrated in the lower sketch in Fig. 7.1.

It is straightforward to modify the original Frenkel–Kontorova model to have two interacting harmonic chains. If v_n describes the deviation of atom n in the second harmonic chain we may express the total energy of both chains as follows:

$$E = \sum_n \frac{1}{2}\kappa(u_n - u_{n-1})^2 + \frac{1}{2}\kappa(v_n - v_{n-1})^2 + V\sin^2\left(\frac{\pi(u_n - v_n)}{a}\right). \tag{7.2}$$

[1] Frenkel, J, and Kontorova, T, *J. Phys. Acad. Sci. USSR* **1**, 137–49 (1939).

[2] Dehlinger, U, *Ann. Phys.* **394** 749–93 (1929). https://doi.org/10.1002/andp.19293940702

[3] Ulrich Dehlinger 1901–81, theoretical physicist at the Technische Hochschule Stuttgart (now University of Stuttgart).

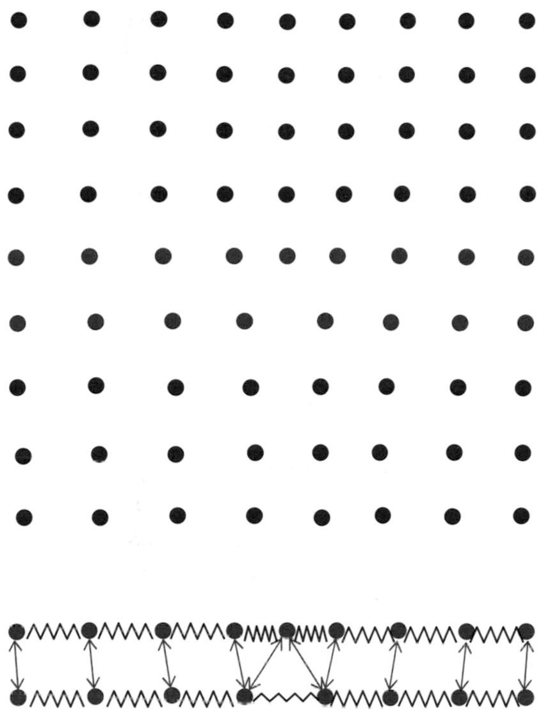

Figure 7.1 *A sketch to illustrate the relationship between the Frenkel–Kontorova model and an edge dislocation in a two-dimensional square lattice. The upper figure is a sketch of the atomic structure of an edge dislocation. The atoms coloured red are on either side of the horizontal slip plane. These atoms are reproduced in the lower figure, which is a sketch of the Frenkel–Kontorova model. In each chain of atoms nearest neighbours are linked together by harmonic springs. The disregistry between nearest neighbour atoms in either chain is shown by blue arrows. When there is no disregistry the arrows are vertical. The disregistry is seen to increase to a maximum where the extra half plane of the dislocation terminates.*

The first two terms are the elastic energies of the harmonic chains. The third term is the energy associated with the disregistry $u_n - v_n$ between atoms n in the two chains.

Provided the displacement fields u_n and v_n are slowly varying we may replace the discrete sums in eqn. 7.2 with integrals. In that limit $\sum_n \to \int \frac{dx}{a}$, $u_n \to u(x)$, $u_{n+1} \to u(x+a) \to u(x) + a\mathrm{d}u/\mathrm{d}x$ and similarly for v. We can then write eqn. 7.2 as follows:

$$E = \int_{-\infty}^{\infty} \left\{ \frac{\kappa a^2}{2} \left[\left(\frac{\mathrm{d}u}{\mathrm{d}x} \right)^2 + \left(\frac{\mathrm{d}v}{\mathrm{d}x} \right)^2 \right] + V \sin^2 \left(\frac{\pi(u-v)}{a} \right) \right\} \frac{\mathrm{d}x}{a}. \tag{7.3}$$

Minimisation of the integral in eqn. 7.3 leads to the following coupled Euler–Lagrange equations:

$$\kappa a \frac{\mathrm{d}^2 u}{\mathrm{d}x^2} = \frac{\pi V}{a^2} \sin\left(\frac{2\pi(u-v)}{a}\right) \tag{7.4}$$

$$\kappa a \frac{\mathrm{d}^2 v}{\mathrm{d}x^2} = -\frac{\pi V}{a^2} \sin\left(\frac{2\pi(u-v)}{a}\right). \tag{7.5}$$

By subtracting and adding eqns. 7.4 and 7.5 we obtain

$$\frac{\mathrm{d}^2(u-v)}{\mathrm{d}x^2} = \frac{2\pi V}{\kappa a^3} \sin\left(\frac{2\pi(u-v)}{a}\right) \tag{7.6}$$

$$\frac{\mathrm{d}^2(u+v)}{\mathrm{d}x^2} = 0. \tag{7.7}$$

To model a dislocation with a Burgers vector of magnitude a we impose the boundary conditions $u(-\infty) = v(-\infty) = 0$ and $u(+\infty) = a/2, v(+\infty) = -a/2$. It follows from these boundary conditions and eqn. 7.7 that $u(x) = -v(x)$ for all x. It then follows from eqn. 7.6 that $u(x)$ is determined by the sine-Gordon equation:

$$\frac{\mathrm{d}^2 u}{\mathrm{d}x^2} = \frac{\pi V}{\kappa a^3} \sin\left(\frac{4\pi u}{a}\right). \tag{7.8}$$

The solution to this equation satisfying the boundary conditions for $u(x)$ at $\pm\infty$ is as follows:

$$u = \frac{a}{\pi} \tan^{-1}\left[\exp\left\{\sqrt{\frac{V}{\kappa a^2}} \frac{2\pi x}{a}\right\}\right]. \tag{7.9}$$

The strains in the two chains are equal and opposite:

$$\frac{\mathrm{d}u}{\mathrm{d}x} = -\frac{\mathrm{d}v}{\mathrm{d}x} = \sqrt{\frac{V}{\kappa a^2}} \operatorname{sech}\left(\sqrt{\frac{V}{\kappa a^2}} \frac{2\pi x}{a}\right). \tag{7.10}$$

These strains are plotted in Fig. 7.2, where it is seen they are localised around the origin $x = 0$. The width of the strain profiles may be defined by

$$W = \frac{a}{2\pi}\sqrt{\frac{\kappa a^2}{V}}. \tag{7.11}$$

The physics of the Frenkel–Kontorova model is clear. The disregistry associated with the Burgers vector between atoms on either side of the slip plane is accommodated in a finite width characterised by W. The magnitude of W increases with increasing spring stiffness κ and decreasing amplitude V of the interaction potential between the chains. When $V \to \infty$ the curves in Fig. 7.2 become two delta functions at $x = 0$ of opposite sign, and the relative displacement of the chains by the Burgers vector becomes a step

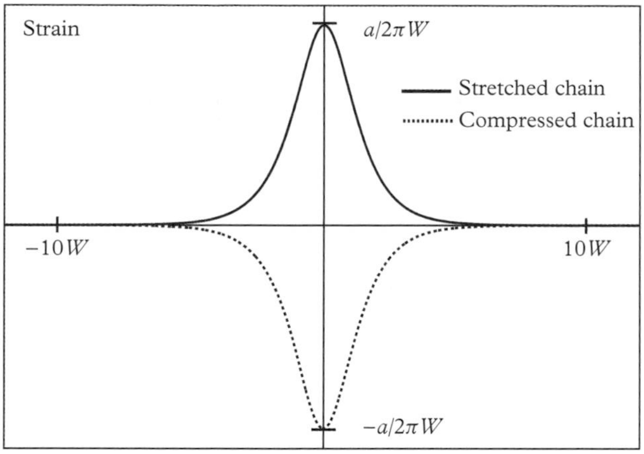

Figure 7.2 *Plot of the strains in a pair of interacting chains stretched (solid line) and compressed (broken line) by $a/2$. Reproduced from Hammad, A, Swinburne, TD, Hasan, H, Del Rosso, S, Iannucci, L and Sutton, AP,* Proc. R. Soc. A **471** *20150171 (2015). http://dx.doi.org/10.1098/rspa.2015.0171. This is an open access paper.*

function at the origin: the core of a Volterra dislocation has no width. In the next section the Peierls–Nabarro model also predicts a finite core width, and goes further by treating the crystals above and below the red atoms in Fig. 7.1.

Exercise 7.1

Derive the solution, eqn. 7.9, to the sine-Gordon equation, eqn. 7.8, with the boundary conditions stated above.

Hint: Multiply both sides of eqn. 7.8 by $\mathrm{d}u/\mathrm{d}x$ and integrate. Then use an arctangent substitution.

Exercise 7.2

Show that in the ground state there is equipartition of the total energy between the elastic energy of the harmonic springs (i.e. the first two terms in eqn. 7.3 and the misfit energy (the third term in eqn. 7.3 and that each is equal to $2VW/a = \kappa a^3/(2\pi^2 W)$.

(In the problems at the end of this chapter it is shown that this is an example of the equipartition of the total energy between the elastic energy and the misfit energy in the Frenkel–Kontorova model for *any* functional form of the misfit energy.)

7.3 Peierls–Nabarro model

Although the physics of the Frenkel–Kontorova model is appealing it treats the interaction between only two linear chains of atoms. It does not address the elastic energy stored in the semi-infinite crystals on either side of the slip plane. The Peierls[4]–Nabarro model treats both the elastic energy of the dislocation and the misfit energy in the slip plane, albeit in a simplified manner. We will develop the model for an edge dislocation, following Eshelby's approach[5] of treating the dislocation as a continuous distribution of edge dislocations along the slip plane with infinitesimal Burgers vectors.[6]

We regard the edge dislocation shown in Fig. 7.1 as a projection of its atomic structure viewed along the dislocation line. The atoms coloured red are in (010) planes, seen edge on, on either side of the geometrical slip plane, which is midway between them. As in the Frenkel–Kontorova model the disregistry between these atoms will be treated, but as the relative translation along x between columns of atoms on either side of the slip plane rather than between single atoms in linear chains.

Vitek[7] introduced the γ-surface in 1968.[8] There is a γ-surface for every plane within a crystal. Each γ-surface describes the energy γ associated with the imposition of a relative translation by $\mathbf{t}$ of one crystal half with respect to the other on either side of the plane. The relative translation vector is in the plane. The planar defect formed as a result of the relative translation is called a 'generalised fault'. The γ-surface is an atomic-scale property of the crystal, and it is calculated increasingly using electronic density functional theory. An example is shown if Fig. 7.3. The relative translation $\mathbf{t}$ parallel to the chosen crystal plane is imposed as a step function on atoms on either side of the plane. Atoms are allowed to move only along the plane normal to positions where they experience no normal force. $\gamma(\mathbf{t})$ is the excess energy per unit area of the whole crystal containing the fault in this partially relaxed and constrained state. When the plane contains two non-parallel sets of lattice vectors $\gamma(\mathbf{t})$ is periodic in two dimensions. If $\mathbf{t} = 0$ coincides with the perfect crystal configuration there is a global energy minimum at $\mathbf{t} = 0$ and at all translation vectors equal to lattice vectors within the plane. If there are n atoms associated with each lattice site of the crystal then there are up to n γ-surfaces on each plane. We will return to the γ surface in section 7.4. The purpose of introducing it here is to put the Peierls–Nabarro model, in particular eqn. 7.12, in the context of current thinking.

In the Frenkel–Kontorova and Peierls–Nabarro models we consider relative translations of the two crystal halves on either side of the slip plane, where the translations are parallel to the Burgers vector of the dislocation. The energy associated with these relative translations is a section through the γ-surface on the slip plane. Equation 6.1 is a general expression for the energy per unit area of the generalised faults created by

[4] Sir Rudolf Ernst Peierls FRS 1907–95, German born British physicist.

[5] Eshelby, JD, *Phil. Mag.* **40**, 903–12 (1949). https://doi.org/10.1080/14786444908561420

[6] We have chosen to present the model for an edge dislocation because it is most easily visualised. The model is equally applicable to screw and mixed edge and screw dislocations, although in all cases the core is usually assumed to be planar.

[7] Vaclav Vitek 1940–, Czech born British and US materials physicist.

[8] Vitek, V, *Phil. Mag.* **18**, 773–86 (1968). https://doi.org/10.1080/14786436808227500

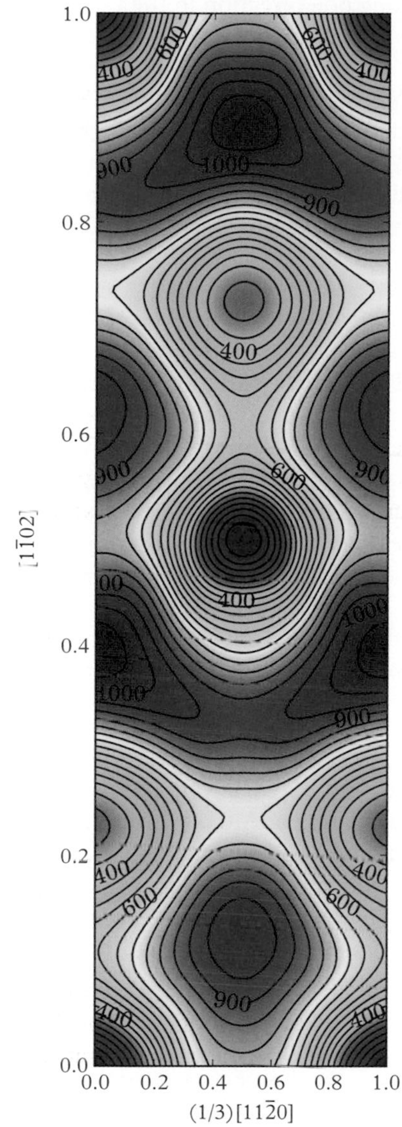

Figure 7.3 *The γ-surface for widely spaced* $(1\bar{1}0\bar{1})$ *atomic planes in titanium computed by density functional theory. The energy surface is viewed from above like a map showing hills in red and valleys in blue with contours separated by* 50 mJm^{-2}. *The periodic cell contains two lattice sites - one at the four corners and one in the centre. There is a local minimum at [0.00, 0.22] in fractional coordinates which is seen most clearly at [0.5, 0.72], where the energy is* 279 mJm^{-2}. *Reproduced with permission by Taylor & Francis (www.tandfonline.com) from Ready, AJ, Haynes, PD, Rugg, D, and Sutton, AP,* Phil. Mag. **97**, *1129–43 (2017). http://dx.doi.org/10.1080/14786435.2017.1292059*

these relative translations along one direction in the γ-surface. The Peierls–Nabarro and Frenkel–Kontorova models simplify eqn. 6.1 by taking just the first harmonic of the Fourier expansion.

To illustrate the Peierls–Nabarro model consider an edge dislocation in a simple cubic crystal, with just one atom at each lattice site. The slip plane is assumed to be (010) and the Burgers vector $\mathbf{b} = [a, 0, 0]$, where a is the lattice constant of the simple cubic crystal. Let w be the relative translation in the [100] direction of the crystal halves on either side of the (010) slip plane. As with the Frenkel–Kontorova model $\gamma(w)$ is approximated as follows:

$$\gamma(w) = V \sin^2\left(\frac{\pi w}{a}\right), \tag{7.12}$$

where V is the amplitude of the energy variation with relative translation w.

The dislocation creates a relative displacement parallel to [100] between atomic columns along [001] that were directly opposite each other on either side of the slip plane (010) before the dislocation was introduced. This relative displacement is the disregistry, w, between the atomic columns, and inside the dislocation core it varies with position: $w = w(x)$. Provided the disregistry does not vary too rapidly with position we may define the misfit energy, E_c, of the dislocation per unit length of the dislocation line by the following integral:

$$E_c = \int_{-\infty}^{\infty} \gamma(w(x))\, \mathrm{d}x. \tag{7.13}$$

This is the energy associated with the misfitting atomic bonds across the slip plane. Far from the dislocation core $\gamma(w)$ is zero. Although the limits of the integral extend from minus infinity to plus infinity the dominant contribution to E_c is confined to the region where $w(x)$ is changing, which is in the dislocation core. It is worth noting that Peierls and Nabarro did not allow any displacements of atoms normal to the slip plane. This has two consequences. The first is that their $\gamma(w)$ differs from a section of the γ-surface as defined above, where relaxations normal to the plane are allowed. Secondly, the misfit energy of eqn. 7.13 would be reduced if normal relaxations were allowed.

So far the Peierls–Nabarro model does not differ much from the Frenkel–Kontorova model. It is the treatment of the elastic energy where they diverge, because it is non-local in the Peierls–Nabarro model. The dislocation is now viewed as a continuous distribution of dislocations with infinitesimal Burgers vectors. In the case of a Volterra dislocation of Burgers vector a at the origin the distribution is $a\delta(x)$. The δ-function is smeared out by writing the distribution of the Burgers vector as $af(x)$ where the integral $\int_{-\infty}^{+\infty} f(x)\mathrm{d}x = 1$. Then $af(x)\mathrm{d}x$ is the infinitesimal Burgers vector of a dislocation between x and $x + \mathrm{d}x$. Note that $f(x)$ has the dimensions of one over length. It is clear that when $f(x) \neq 0$ the disregistry $w(x)$ is changing. More precisely, $af(x)\mathrm{d}x = \mathrm{d}w$, which leads to the following relation:

$$\frac{\mathrm{d}w}{\mathrm{d}x} = af(x). \tag{7.14}$$

The Peach–Koehler force on an edge dislocation with Burgers vector b_1 at x due to another edge dislocation with Burgers vector b_2 at x_2 in isotropic elasticity is as follows:

$$F = \frac{\mu b_1 b_2}{2\pi(1-\nu)} \frac{1}{(x-x_2)}.$$

Integrating $-F$ between $x = x_2 + r_c$, where r_c is the radius of the core of each dislocation,[9] and $x = x_1$ we obtain their interaction energy:

$$E_{int} = -\frac{\mu b_1 b_2}{2\pi(1-\nu)} \ln\left(\frac{|x_1 - x_2|}{r_c}\right). \tag{7.15}$$

It follows that for the continuous distribution of dislocations comprising the dislocation with Burgers vector a the elastic energy is as follows:

$$E_{el} = -\frac{1}{2}\frac{\mu a^2}{2\pi(1-\nu)} \int_{-\infty}^{\infty} dx_1 \int_{-\infty}^{\infty} dx_2\, f(x_1)f(x_2) \ln\left(\frac{|x_1 - x_2|}{r_c}\right), \tag{7.16}$$

where the factor of one half is to correct for the double counting in the double integral. Combining eqns. 7.13, 7.14 and 7.16, Christian[10] and Vitek obtain[11] the following expression for the total energy of the dislocation per unit length in the Peierls–Nabarro model:

$$E_{PN} = \int_{-\infty}^{\infty} V \sin^2\left(\frac{\pi w(x)}{a}\right) dx - \frac{\mu}{4\pi(1-\nu)} \int_{-\infty}^{\infty} dx_1 \int_{-\infty}^{\infty} dx_2 \left(\frac{dw}{dx}\right)_{x=x_1} \left(\frac{dw}{dx}\right)_{x=x_2} \ln\left(\frac{|x_1 - x_2|}{r_c}\right). \tag{7.17}$$

This equation is the analogue of eqn. 7.3 in the Frenkel–Kontorova model, where the disregistry is $u(x) - v(x)$. Minimising the integral in eqn. 7.17 with respect to variations $\delta w(x)$ we obtain the following Euler–Lagrange equation:

$$\frac{\mu}{2\pi(1-\nu)} P\int_{-\infty}^{\infty} dx_1 \frac{(dw/dx)_{x=x_1}}{x - x_1} = -\frac{d\gamma}{dw(x)} = -\frac{V\pi}{a} \sin\left(\frac{2\pi w(x)}{a}\right). \tag{7.18}$$

Although mathematically this is a nonlinear, singular, integro-differential equation, physically it is just a balance of shear stresses at each point on the slip plane: the left hand side is the elastic shear stress at x arising from the continuous distribution of dislocations, and the right hand side the shear stress at x arising from the $\gamma(w)$. The P in front of

[9] It has to be non-zero because the force becomes infinite when the dislocations are coincident. It is not present in the final equation, so its value is unimportant.

[10] John Wyrill Christian FRS 1926–2001, British materials scientist.

[11] Christian, JW and Vitek, V, *Rep. Prog. Phys.* **33**, 307 (1970). https://doi.org/10.1088/0034-4885/33/1/307

the integral signifies it is a Cauchy principal value integral because each dislocation in the distribution interacts with all the other dislocations in the distribution, but not with itself. As with the Frenkel–Kontorova model, the elastic energy is tending to make the distribution of dislocations as wide as possible, so that the gradient of the disregistry is minimised. But $\gamma(w)$ is tending to make the distribution of dislocations as narrow as possible. The actual distribution is determined by a balance between these two opposing tendencies.

The boundary conditions on the disregistry are that $w(-\infty) = 0$ and $w(+\infty) = a$. Peierls[12] guessed the following solution and showed that it solved eqn. 7.18:

$$w(x) = (a/\pi)\tan^{-1}(x/\zeta) + a/2 \tag{7.19}$$

since it satisfies the boundary conditions. The derivative of $w(x)$ shows that the distribution $f(x)$ of the Burgers vector is a Lorentzian:

$$f(x) = \frac{1}{\pi}\frac{\zeta}{x^2 + \zeta^2}. \tag{7.20}$$

The full width at half maximum of this distribution is 2ζ. Therefore, we may define 2ζ as the width of the dislocation core.

Exercise 7.3

Verify that eqn. 7.19 solves eqn. 7.18 provided

$$\zeta = \frac{\mu a^2}{4\pi^2 V(1-\nu)}. \tag{7.21}$$

Hint: Differentiate eqn. 7.19 and substitute the derivative dw/dx into the principal value integral, which may be evaluated using contour integration. When the expression for $w(x)$ is substituted into the right hand-side of eqn. 7.18 it is found the equation is satisfied provided eqn. 7.21 holds. It is customary to invoke the Frenkel argument of section 6.2 to express the amplitude V as $V = \mu a^2/(2\pi^2 d)$, where d is the spacing of atomic planes parallel to the slip plane. Then we obtain

$$\zeta = \frac{d}{2(1-\nu)}.$$

Show that the misfit energy in eqn. 7.13 is given by

$$E_c = \pi V\zeta = \frac{\mu a^2}{4\pi(1-\nu)}. \tag{7.22}$$

[12] Peierls, R, *Proc. Phys. Soc.* **52**, 34 (1940). https://doi.org/10.1088/0959-5309/52/1/305

Having replaced a Volterra edge dislocation with its singular core by a more realistic dislocation with a finite core width, the stress and strain fields of the dislocation no longer become infinite at the centre of the dislocation. This may be seen by convolving the components of the stress tensor for a Volterra edge dislocation in eqn. 6.34 with the Burgers vector distribution function $af(x)$ given by eqn. 7.20:

$$\sigma_{11}(x_1,x_2) = -\frac{\mu a}{2\pi(1-\nu)} \int_{-\infty}^{\infty} dx_1' \frac{x_2\left(3(x_1-x_1')^2+x_2^2\right)}{\left((x_1-x_1')^2+x_2^2\right)^2} \frac{\zeta}{\pi\left(\zeta^2+(x_1')^2\right)}$$

$$\sigma_{22}(x_1,x_2) = \frac{\mu a}{2\pi(1-\nu)} \int_{-\infty}^{\infty} dx_1' \frac{x_2\left((x_1-x_1')^2+x_2^2\right)}{\left((x_1-x_1')^2+x_2^2\right)^2} \frac{\zeta}{\pi\left(\zeta^2+(x_1')^2\right)}$$

$$\sigma_{12}(x_1,x_2) = \frac{\mu a}{2\pi(1-\nu)} \int_{-\infty}^{\infty} dx_1' \frac{(x_1-x_1')\left((x_1-x_1')^2-x_2^2\right)}{\left((x_1-x_1')^2+x_2^2\right)^2} \frac{\zeta}{\pi\left(\zeta^2+(x_1')^2\right)}.$$

These integrals may be evaluated using contour integration to yield the following stress components:

$$\sigma_{11}(x_1,x_2) = \frac{\mu a}{2\pi(1-\nu)}\left(\frac{3x_2+2\zeta}{x_1^2+(x_2+\zeta)^2} - \frac{2x_2(x_2+\zeta)^2}{\left(x_1^2+(x_2+\zeta)^2\right)^2}\right)$$

$$\sigma_{22}(x_1,x_2) = \frac{\mu a}{2\pi(1-\nu)}\left(\frac{x_2}{x_1^2+(x_2+\zeta)^2} - \frac{2x_1^2x_2}{\left(x_1^2+(x_2+\zeta)^2\right)^2}\right)$$

$$\sigma_{12}(x_1,x_2) = -\frac{\mu a}{2\pi(1-\nu)}\left(\frac{x_1}{x_1^2+(x_2+\zeta)^2} - \frac{2x_1x_2(x_2+\zeta)}{\left(x_1^2+(x_2+\zeta)^2\right)^2}\right)$$

$$\sigma_{33}(x_1,x_2) = \nu(\sigma_{11}(x_1,x_2)+\sigma_{22}(x_1,x_2)). \tag{7.23}$$

The presence of the finite value of ζ in the denominators of these expressions removes the singularity, which is present in the stress field of a Volterra dislocation as $\sqrt{x_1^2+x_2^2} \to 0$.

7.3.1 Comments on the Peierls–Nabarro model

Like its predecessor the Frenkel–Kontorova model, the Peierls–Nabarro model captures the physics of the balance between the misfit energy tending to make the dislocation as narrow as possible and the elastic energy tending to make it as wide as possible. In both cases this is achieved by introducing new physics into the Volterra treatment of a dislocation, namely atomic interactions across the slip plane, and a new length scale, namely the atomic spacing. Whereas the Frenkel–Kontorova model involves just two interacting linear chains of atoms, the Peierls–Nabarro model treats the dislocation in a three-dimensional medium. In both models the misfit energy arising from the distortion of bonds across the slip plane is approximated by a simple sinusoidal form. This is undoubtedly a gross simplification, as atomistic calculations of γ-surfaces have shown for a variety of slip planes in different crystal structures. The small value of the calculated width 2ζ of the Burgers vector distribution leads to strains near the dislocation centre that are far too large to be described by linear elasticity. It also raises serious doubts about the neglect of gradient terms in eqn. 7.13 for the misfit energy. But the most serious weakness of both the Frenkel–Kontorova and Peierls-Nabarro models is the presumption that the dislocation core is *planar*. As Vitek[13] has stressed, in any given crystal there may be some orientations of the dislocation line where the cores are planar, but there may be other orientations where the cores are non-planar. Orientations of dislocations with non-planar cores govern plastic deformation because they require higher stresses to make them glide. For example, in bcc metals screw dislocations have non-planar cores, whereas edge dislocations have planar cores; plastic deformation in these metals is limited by the motion of screw dislocations.[14]

The genesis of the Peierls–Nabarro model has been described by Peierls.[15] He explained that the key idea of embedding a slab on the slip plane, where the misfit is treated atomistically, between elastic continua, was conceived by neither Peierls nor Nabarro, but by Orowan. Orowan approached Peierls for help with the mathematical formulation of the model. Peierls derived the integral equation, eqn. 7.18, but not by the method followed here. Seeing its nonlinear form he thought it was probably insoluble. He guessed the arctan solution, eqn. 7.19, and when he inserted it into the integral equation he was amazed to discover it was *the* solution. He introduced an error of a factor of two in the algebra associated with the stress required to move the dislocation, which appears in a large exponent and therefore had dramatic consequences. The error was corrected by Nabarro seven years later,[16] and thereafter it was known as the Peierls–Nabarro model. Peierls had tried to persuade Orowan to publish the 1940 paper under Orowan's name alone, or as a joint paper, because Orowan had conceived the model. But Orowan refused. Peierls' concern about the authorship of the 1940 paper grew as its fame increased, and he felt he should have insisted Orowan was at least a coauthor, if not the sole author.

[13] Vitek, V, private communication (2019).

[14] Vitek, V, *Prog. Mater. Sci.* **56**, 577–85 (2011). https://doi.org/10.1016/j.pmatsci.2011.01.002

[15] Peierls, RE, *Proc. R. Soc. A* **371**, 28–38 (1980). https://doi.org/10.1098/rspa.1980.0053. A slightly more complete account by Peierls is given on pages xiii–xiv of *Dislocation dynamics*, ed. AR Rosenfield, GT Hahn, AL Bement Jr and RI Jaffee. McGraw-Hill: New York (1968). Peierls' original derivation of the integral equation and his solution is transcribed on pages xvii–xx of this book.

[16] Nabarro, FRN, *Proc. Phys. Soc.* **59**, 256 (1947). https://doi.org/10.1088/0959-5309/59/2/309

7.4 Stacking faults and partial dislocations

Sometimes there are local minima in the γ-surface in addition to global minima. Suppose there is a local minimum at $\mathbf{t} = \mathbf{t}_0$ relative to the nearest global energy minimum. A metastable planar fault in the crystal may arise at which there is a relative translation of the crystals on either side of the fault equal to $\mathbf{t}_0$. It sometimes happens that the fault may also be created through a local change in the normal sequence of the stacking of atomic planes parallel to the fault. It is then called a stacking fault. When the energy associated with these faults is relatively small they may enable dislocations to reduce their elastic energy by dissociating into two or more partial dislocations separated by one or more faults.

We will illustrate these ideas with the example of the dissociation of crystal lattice dislocations into partial dislocations in face-centred cubic (fcc) lattices. Slip takes place in these lattices on $\{111\}$ planes with Burgers vectors $\frac{1}{2}\langle 110\rangle$, where we have set the lattice constant to 1. In the γ-surface of the (111) plane there are global minima at the three lattice vectors $\frac{1}{2}[1\bar{1}0]$, $\frac{1}{2}[10\bar{1}]$ and $\frac{1}{2}[01\bar{1}]$ and integer combinations of these vectors, where there is no misfit and the energy γ may be set to zero. Relative to each of these lattice vectors there are three local minima at $\mathbf{t} = \frac{1}{6}[2\bar{1}\bar{1}]$, $\frac{1}{6}[\bar{1}2\bar{1}]$ and $\frac{1}{6}[\bar{1}\bar{1}2]$, where the energy $\gamma(\mathbf{t}) = \gamma_{SF} > 0$. Therefore, faults may exist in these lattices with $\frac{1}{6}\langle 211\rangle$ fault vectors on $\{111\}$ planes. In addition to being created by these relative translations they may also be created by altering the stacking sequence of $\{111\}$ planes by removing a plane, so that the perfect crystal sequence...*ABCABCABC*...becomes ...*ABCBCABCA*.... These particular faults are therefore 'stacking faults'.

Consider a dislocation with Burgers vector $\frac{1}{2}[1\bar{1}0]$ on the (111) slip plane. This dislocation may dissociate into two partial dislocations, called Shockley partials, as follows:

$$\frac{1}{2}[1\bar{1}0] \rightarrow \frac{1}{6}[2\bar{1}\bar{1}] + \frac{1}{6}[1\bar{2}1]. \tag{7.24}$$

Thus, a dislocation with Burgers vector $\frac{1}{2}[1\bar{1}0]$ becomes two dislocations with Burgers vectors $\frac{1}{6}[2\bar{1}\bar{1}]$ and $\frac{1}{6}[1\bar{2}1]$ separated by a ribbon of stacking fault. The driving force for the dissociation is the reduction of the elastic energy of the dislocation which depends on the square of the magnitude of the Burgers vector. The separation of the partial dislocations is determined by a balance between the Peach–Koehler force of repulsion between them and the magnitude of the stacking fault energy γ_{SF}. In fcc crystals with large stacking fault energies, such as aluminium, the separation of the partials, if it exists, may be too small to resolve experimentally. In silver the stacking fault energy is relatively small and the partial dislocations are clearly resolved in the transmission electron microscope.

In isotropic elasticity the force per unit length acting between two parallel, straight dislocations with Burgers vectors $\mathbf{b}_1$ and $\mathbf{b}_2$ and line direction $\hat{\xi}$, with a separation S, is as follows:[17]

$$F_{el} = \frac{\mu}{2\pi S}(\mathbf{b}_1 \cdot \hat{\xi})(\mathbf{b}_2 \cdot \hat{\xi}) + \frac{\mu}{2\pi(1-\nu)S}\left[(\mathbf{b}_1 \times \hat{\xi}) \cdot (\mathbf{b}_2 \times \hat{\xi})\right]. \tag{7.25}$$

If the separation increases by dS the change in the energy per unit length of the elastic interaction between the dislocations is $-F_{el}\mathrm{d}S$. At the same time the change in the energy of the stacking fault, per unit length of dislocation line, is $\gamma_{SF}\mathrm{d}S$. At equilibrium these two changes cancel exactly, so that $\gamma_{SF} = F_{el}$. Thus, a measurement of S enables the stacking fault energy to be determined. The tails of the Burgers vector distributions associated with each partial dislocation extend into the stacking fault slightly, which introduces an error that diminishes with increasing S/ζ.

The existence of partial dislocations separated by stacking faults reduces the ability of screw dislocations to cross slip because the partials have to be forced to form a constriction before they can cross-slip.[18] The ease of cross-slip has consequences for macroscopic plasticity because it enables dislocations to overcome obstacles and it may also lead to new slip systems being activated. It is a remarkable fact that the origin of these macroscopic properties can be traced to local minima in the γ-surfaces of slip planes, which in turn are determined by the quantum mechanical behaviour of electrons in the distorted bonding environments sampled at each point of the γ-surface.

7.5 The static friction stress on a dislocation

Peierls and Nabarro calculated the minimum stress required to initiate glide of an edge dislocation on its slip plane. This is a static friction stress originating from the discrete atomic structure of the crystal, and it is often called a Peierls stress or lattice friction stress. In section 10.4 we will consider sources of dynamic friction a dislocation has to overcome to keep gliding once it has overcome the Peierls stress.

Consider the glide of the edge dislocation shown in Fig. 7.1. As the dislocation centre moves from one atomic row to the next along the slip plane its elastic energy does not change if we assume the disregistry $w(x)$ is rigidly translated with the dislocation with no change to its functional form. But the misfit energy E_c changes as bonds are stretched across the slip plane, switch from one neighbour to the next, and return to their original length. The maximum slope of this periodic function is the stress required to initiate dislocation glide. This was a significant advance because it had been presumed since 1934, when dislocations were proposed as the agents of plastic deformation, that the stress required to move dislocations in metals was several orders of magnitude less

[17] Hirth, JP and Lothe, J, *Theory of dislocations*, 2nd edn., Krieger: Malabar, FL (1982), p.117. ISBN 0-89464-617-6. John Price Hirth 1930–, US theoretical materials engineer. Jens Lothe 1931–2016, Norweigan physicist.

[18] See section 2.4 of Kubin, LP, *Dislocations, mesoscale simulations and plastic flow*, Oxford University Press: Oxford (2013). ISBN 978-0-19-852501-1. Ladislas P Kubin, French materials scientist.

than the shear modulus. But it was not demonstrated theoretically until Peierls calculated it. To evaluate the Peierls stress we have to evaluate the misfit energy E_c as a discrete sum rather than a continuous integral, as the dislocation centre moves along the slip plane.

To evaluate the misfit energy E_c as a sum of discrete interactions between atoms we assume eqn. 7.19 for the disregistry $w(x)$ remains valid as the dislocation moves. The calculation proceeds by evaluating the energy per unit length of each row of atoms on either side of the slip plane. The misfit energy $\gamma(w)$ per unit area given by eqn. 7.12 is shared by all atoms in the two atomic planes on either side of the slip plane. The energy per unit length per row of atoms on either side of the slip plane is then given by

$$\begin{aligned}
E_{row} &= \frac{aV}{2}\sin^2\left(\frac{\pi w(x)}{a}\right) \\
&= \frac{\mu a^3}{4\pi^2 d}\sin^2\left(\frac{\pi w(x)}{a}\right) \\
&= \frac{\mu a^3}{8\pi^2 d}\left[1-\cos\left(\frac{2\pi}{a}\left(\frac{a}{\pi}\tan^{-1}\left(\frac{x}{\zeta}\right)+\frac{a}{2}\right)\right)\right] \\
&= \frac{\mu a^3}{8\pi^2 d}\left(1+\cos\left(2\tan^{-1}\left(\frac{x}{\zeta}\right)\right)\right) \\
&= \frac{\mu a^3}{4\pi^2 d}\frac{\zeta^2}{\zeta^2+x^2}.
\end{aligned} \tag{7.26}$$

Before the dislocation is introduced the positions of atomic rows on say the lower side of the slip plane are $x_n = na$ and on the upper side they are $x_n = (n+\frac{1}{2})a$. Assuming the disregistry $w(x)$ is accommodated equally by the two crystal halves, the positions of atomic rows on the lower side of the slip plane become $x_n = na + \beta a/2$ and on the upper side $x_n = (n+\frac{1}{2})a - \beta a/2$, where $0 \le \beta \le 1$. The misfit energy then becomes

$$\begin{aligned}
E_c &= \frac{\mu a^3}{4\pi^2 d}\frac{\zeta^2}{a^2}\left(\sum_{n=-\infty}^{\infty}\frac{1}{\frac{\zeta^2}{a^2}+\left(n+\frac{1}{2}-\frac{\beta}{2}\right)^2}+\frac{1}{\frac{\zeta^2}{a^2}+\left(n+\frac{\beta}{2}\right)^2}\right) \\
&= \frac{\mu a^3}{4\pi^2 d}\frac{\zeta^2}{a^2}\frac{\pi a}{\zeta}\left(\frac{\sinh(2\pi\zeta/a)}{\cosh(2\pi\zeta/a)+\cos(\pi\beta)}+\frac{\sinh(2\pi\zeta/a)}{\cosh(2\pi\zeta/a)-\cos(\pi\beta)}\right) \\
&= \frac{\mu a^2}{2\pi(1-\nu)}\frac{\sinh(4\pi\zeta/a)}{\cosh(4\pi\zeta/a)-\cos(2\pi\beta)} \\
&\approx \frac{\mu a^2}{2\pi(1-\nu)}(1+2\cos(2\pi\beta)\exp(-4\pi\zeta/a)).
\end{aligned} \tag{7.27}$$

As expected the energy is a periodic function of β with the periodicity of the crystal lattice. The periodic variations in the energy is called the Peierls potential and its amplitude is equal to $[\mu a^2/(\pi(1-\nu))]e^{-4\pi\zeta/a}$. The stress required to initiate glide is determined by the maximum value of $(1/a^2)\partial E_c/\partial\beta$, which is called the Peierls stress, σ_P:

$$\sigma_P = \frac{2\mu}{(1-\nu)}e^{-4\pi\zeta/a}. \tag{7.28}$$

Provided a dislocation has a planar core eqn. 7.28 explains why the resolved shear stress required to initiate glide in single crystals of pure metals is so much less than the shear modulus. The Peierls stress is smaller for dislocations with smaller Burgers vectors in slip planes with larger interplanar spacings. These predictions are broadly consistent with the selection of slip systems found experimentally in a wide range of crystals, but they are unreliable when the core is non-planar.

The details of the Peierls analysis may be criticised on many fronts, but the central ideas have been seminal. The existence of a minimum stress required to initiate glide over the periodic energy surface, E_c in eqn. 7.27, is a key distinction between dislocation motion in a crystal lattice as compared to a continuum. The maxima and minima in this energy surface are often referred to as Peierls barriers and Peierls valleys respectively. The undulations in this energy surface reflect the changes in bonding between atoms on either side of the slip plane as the dislocation glides.[19] When the core is very narrow a small number of atomic rows undergo a short sequence of relatively large shear displacements on either side of the slip plane. With a wider core a larger number of atomic rows undergo a longer sequence of smaller shear displacements, which require a smaller stress σ_P to bring about.

In some pure fcc metals, such as copper, the resistance to dislocation motion is extremely small, and plastic deformation can occur at cryogenic temperatures. But in many body-centred cubic metals, such as iron and tungsten, the resistance to the motion of screw dislocations is greater because their cores are non-planar. Consequently, at low temperatures these metals become brittle because cracks can grow before there is significant dislocation motion to reduce their exposure to an applied stress. However, at higher temperatures they undergo a brittle-to-ductile transition, and they deform plastically more readily. In pure single crystals of these metals the brittle to ductile transition is associated with the temperature dependence of the formation and mobility of kinks on the dislocations (see the next section).

In crystals with strong bonding the Peierls stress of dislocations even with planar cores can be very high. For example, diamond has an fcc lattice with the same slip systems as copper, but dislocation motion in diamond at room temperature is much more difficult because the Peierls barriers are much larger owing to the strong covalent bonding. In this

[19] Note that unlike the γ-surface the Peierls energy surface is not a property of the slip plane alone because it also depends on the orientation of the dislocation line: there are different Peierls barriers and valleys for dislocations with different line directions in the same slip plane.

case also dislocation motion can occur only through a kink mechanism, which requires thermal activation.

Vitek and Paidar[20] have shown[21] that non-planar dislocation cores are common in crystalline materials, even when atomic bonding is not directional. The modern approach to Peierls barriers is to use atomistic modelling of dislocation motion with a variety of models of atomic interactions from empirical potentials to density functional theory. Since bond breaking and making are always involved in dislocation motion methods that treat the quantum mechanical nature of bonding explicitly should be more reliable, *provided* the size of the system modelled is sufficient for the calculation to be credible.

7.6 Dislocation motion by a kink mechanism

Dislocations are the agents of slip because they localise in the dislocation core the far more extensive bond breaking and making that would occur if one plane of atoms were to slide *en masse* over another. But when the Peierls barrier is too large for the dislocation line itself to move *en masse* over the barrier the bond breaking and making is further localised to a small region where the dislocation crosses the Peierls barrier, called a kink. The motion of the dislocation line over the Peierls barrier is then effected by the sideways propagation of the kink along the Peierls barrier, as illustrated in Fig. 7.4(a). The energy of the kink along the line is also periodic and the maxima are called secondary Peierls barriers. The further localisation of the bond breaking and making associated with the motion of a kink reduces the barrier to dislocation motion significantly. Movement of the

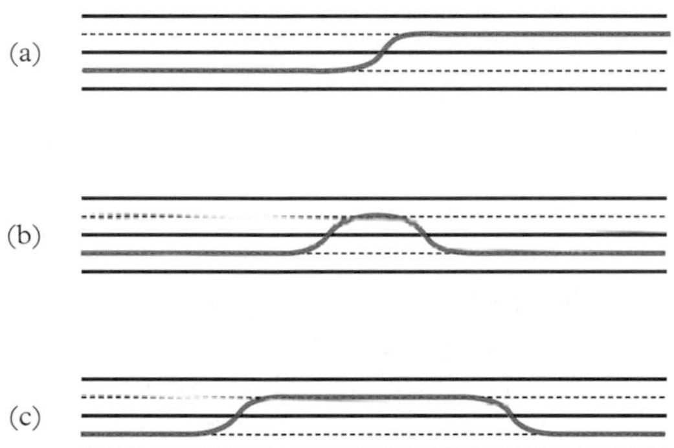

Figure 7.4 *Schematic illustrations of kinks. Black solid lines are Peierls barriers. Black broken lines are Peierls valleys. The dislocation line is shown in red. If the single kink in (a) moves to the left(right) the dislocation line moves up(down). In (b) a kink pair is nucleated and separates in (c) enabling a segment of the dislocation line to move up.*

[20] Vaclav Paidar 1946–, Czech materials physicist.

[21] Vitek, V and Paidar, V, *Dislocations in solids*, ed. FRN Nabarro, Vol. 14, Elsevier: Amsterdam (2008), pp.439–514. ISBN 9780444531667.

kink over the secondary Peierls barriers may be thermally activated, with an activation Gibbs free energy of migration, G_m.

Kinks are geometrically necessary when a dislocation line is inclined to the Peierls valleys. Changes in direction of the dislocation line are effected through variations in the number and sense of kinks per unit length of dislocation line. Kink pairs may be introduced into a straight dislocation by a nucleation process as illustrated in Fig. 7.4(b) and (c).

As an example of a calculation using current computational techniques, Fig. 7.5 shows the relaxed structure of a kink on a Shockley partial dislocation in silicon calculated with density functional theory. The bonds within the kink are fully reconstructed and comparable in strength to those in the crystal and along the dislocation line. For the kink to move one crystal period to the left the bond between atoms 1 and 2 has to be rotated into the orientation of the bond between atoms 3 and 4. When this rotation was applied in nine equal steps, using a constrained energy minimisation procedure, the activation energy for migration of the kink was found to be 1.1 eV, in agreement with experiment.

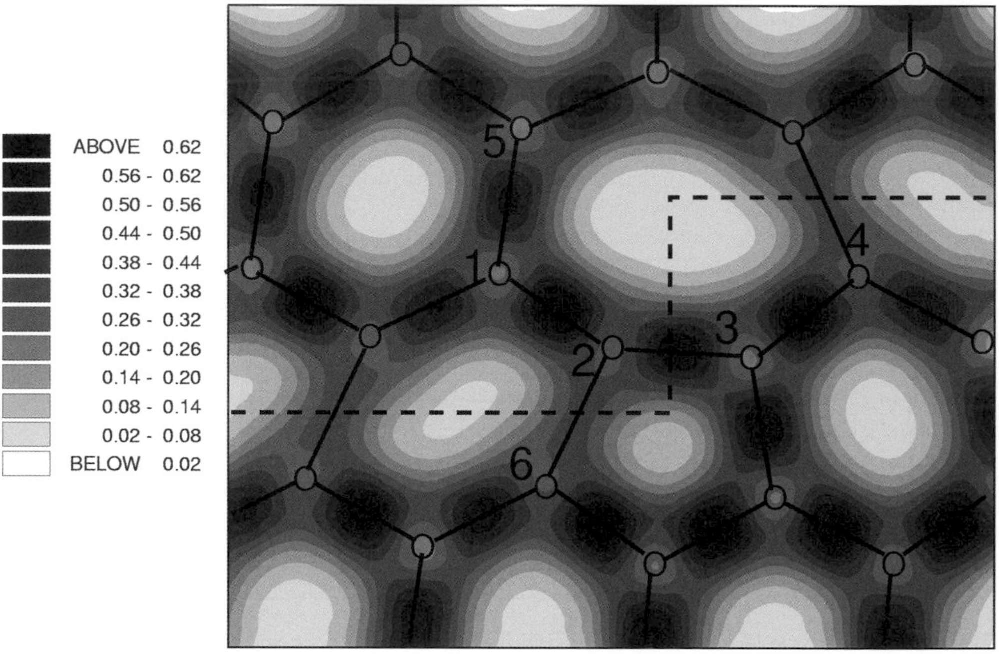

Figure 7.5 *Valence electronic charge density (in electrons* $Å^{-3}$*), viewed in the slip plane, of a relaxed kink on a Shockley partial edge dislocation in silicon calculated quantum mechanically with density functional theory. The broken line is the centre of the dislocation. Atoms (circles) lie either slightly above or below the plane of the figure. Solid lines signify bonds between atoms. Reprinted with permission from Valladares, A, White, JA and Sutton, AP,* Phys. Rev. Lett., **81**, *4903–6 (1998). https://doi.org/10.1103/PhysRevLett.81.4903. Copyright 1997 by the American Physical Society.*

Since kinks are defects on linear objects they are point defects, and the thermodynamics of points defects applies to them. The lowest energy state of a dislocation at absolute zero is achieved when it lies in a Peierls valley along its entire length. At a finite temperature kink pairs may be nucleated through thermal fluctuations along the dislocation. Let the Gibbs free energy of formation of a kink pair be $G^f_{kp} = H^f_{kp} - TS^f_{kp}$, where H^f_{kp} is the enthalpy of formation and S^f_{kp} is the vibrational entropy of formation of a kink pair. Then if λ is the period of the atomic structure along the dislocation line in the Peierls valley there are $1/\lambda$ possible sites for the nucleation of a kink pair per unit length of dislocation line. In each kink pair we may designate one kink as positive and the other as negative. Let c_+ and c_- be the number of positive and negative kinks per unit length of dislocation line. In thermal equilibrium the rate at which positive and negative kinks annihilate each other is equal to the rate at which they are generated. This leads to the following law of mass action:[22]

$$c_+ c_- = \frac{1}{\lambda^2} \exp(-G_{kp}/k_B T). \tag{7.29}$$

Once the separation between the positive and negative kinks of a pair exceeds a few atomic spacings they escape their mutual attraction and they perform a random walk along the dislocation line. There is an activation free energy for migration of the kinks, which in general is not the same for positive and negative kinks. In the presence of a resolved shear stress on the slip plane in the slip direction the nucleation and migration of the kinks is biased and the dislocation acquires a drift velocity, which is proportional to the resolved shear stress.[23] Since both the formation and migration of kinks are thermally activated plasticity becomes easier with increasing temperature, and the transition from brittle to ductile behaviour may be quite abrupt owing to the exponential dependence of the rates of these processes on temperature.

The macroscopic strain rate achieved by dislocation motion depends on the average velocity of mobile dislocations, which varies linearly in the kink mechanism with stress. Conversely if a strain rate is imposed on the material the stress required to maintain the strain rate will increase as it increases. Here we have an explanation for the dependence of the stress required to achieve plastic deformation on the rate at which it is applied.

7.7 Problem set 7

1. In this question we show how the energy for the two-chain Frenkel–Kontorova model in eqn. 7.2 arises from a description of atomic interactions. To be specific we assume atoms interact through Lennard-Jones pair potentials:

[22] It may be helpful to compare this with the concentrations, n and p, of electrons and holes in an intrinsic semiconductor with a band gap of E_g, where $np = \text{constant} \times \exp(-E_g/k_B T)$. In an n-type semiconductor this relation still holds but $n \gg p$. The analogue of doping in the case of kinks is to tilt the dislocation line at some angle to the Peierls valley. If it is tilted in a positive sense then $c_+ \gg c_-$

[23] For a detailed discussion see Hirth, JP and Lothe, J, *Theory of dislocations*, 2nd edn., Krieger: Malabar, FL (1982), section 15-2. ISBN 0-89464-617-6.

$$E_{LJ} = \frac{1}{2}\sum_i \sum_{j \neq i} \varepsilon \left[\left(\frac{r_0}{r_{ij}} \right)^{12} - 2\left(\frac{r_0}{r_{ij}} \right)^{6} \right], \tag{7.30}$$

where ε and r_0 are parameters with the dimensions of energy and length, i and j label atoms in the two chains and r_{ij} is the distance between atoms i and j. Since the Frenkel–Kontorova model assumes harmonic springs exist only between neighbouring atoms in either chain we also assume that atoms in the same chain interact only with their nearest neighbours. In that case show that r_0 must equal a, and $\kappa = 72\varepsilon/a^2$.

Let the separation of the chains be ρ. Let atoms in the first chain be at $x = na$, where n is an integer. Let the atoms in the second chain be at $x = na + \Delta$ where Δ is a constant. The distance between the atom at $x = 0$ in the first chain and atom n in the second chain is then $\left(\rho^2 + (na+\Delta)^2\right)^{1/2}$. The energy of interaction between the atom at $x = 0$ in the first chain and all atoms in the second chain is then

$$E_{int} = \varepsilon \sum_{n=-\infty}^{\infty} \left\{ \left(\frac{a^{12}}{[\rho^2 + (na+\Delta)^2]^6} \right) - 2\left(\frac{a^6}{[\rho^2 + (na+\Delta)^2]^3} \right) \right\}. \tag{7.31}$$

The sum in this equation is taken over all atoms n in the second chain to ensure that the interaction energy is a smooth, periodic function of Δ. These sums may be evaluated exactly using the Sommerfeld–Watson transformation. A simpler approach is to note that the interaction energy is an even periodic function of Δ, and therefore it can be expanded as a Fourier cosine series:

$$E_{int} = c_0 + \sum_{m=1}^{\infty} c_m \cos\left(\frac{2m\pi\Delta}{a} \right). \tag{7.32}$$

The constant c_0 is of no significance. Show that for $m \neq 0$, c_m is given by the Fourier integral:

$$c_m = 2\varepsilon \int_{-\infty}^{\infty} \left[\frac{1}{(\zeta^2 + z^2)^6} - \frac{2}{(\zeta^2 + z^2)^3} \right] e^{i2m\pi z}\, dz, \tag{7.33}$$

where $\zeta = \rho/a \approx 1$. There are third and sixth order poles at $z = i\zeta$. Without evaluating the integral in detail it is clear[24] it is going to be a polynomial multiplied

[24] For example,

$$I_6 = \int_{-\infty}^{\infty} \frac{e^{i2m\pi z}}{(z^2+\zeta^2)^3} dz = \frac{4\pi^2 m^2 \zeta^2 + 6\pi m\zeta + 3}{8\zeta^5} \pi\, e^{-2m\pi\zeta}$$

and

$$I_{12} = \int_{-\infty}^{\infty} \frac{e^{i2m\pi z}}{(z^2+\zeta^2)^6} dz = -\frac{1}{60} \frac{d^3 I_6}{d(\zeta^2)^3}.$$

by $\exp(-2m\pi\zeta)$. Therefore, if we assume that terms with $m > 1$ in eqn. 7.32 are negligible then E_{int} may be expressed approximately as $V\sin^2(\pi\Delta/a)$, ignoring the unimportant term which is independent of Δ. Equation 7.2 then follows if we make a local approximation for Δ, that is, if we ignore the contributions to the interaction energy arising from gradients in Δ.

2. In the Frenkel–Kontorova model let $\gamma(w(x))$ be the misfit energy associated with the disregistry at x, where $w(x) = u(x) - v(x)$. In the continuum approximation the total energy of the two interacting linear chains in eqn. 7.3 becomes

$$E = \int_{-\infty}^{\infty} \left\{ \frac{\kappa a^2}{2} \left[\left(\frac{\mathrm{d}u}{\mathrm{d}x} \right)^2 + \left(\frac{\mathrm{d}v}{\mathrm{d}x} \right)^2 \right] + \gamma(w(x)) \right\} \frac{\mathrm{d}x}{a}.$$

By minimising the integral with respect to variations in $u(x)$ and $v(x)$ and applying the same boundary conditions as in section 7.2 show that

$$\frac{\kappa a^2}{2} \frac{\mathrm{d}^2 w}{\mathrm{d}x^2} = \frac{\mathrm{d}\gamma}{\mathrm{d}w}. \tag{7.34}$$

Show that the misfit energy is equal to the elastic energy stored in the chains, that is,

$$\int_{-\infty}^{\infty} \gamma(w(x)) \frac{\mathrm{d}x}{a} = \int_{-\infty}^{\infty} \frac{\kappa a^2}{2} \left[\left(\frac{\mathrm{d}u}{\mathrm{d}x} \right)^2 + \left(\frac{\mathrm{d}v}{\mathrm{d}x} \right)^2 \right] \frac{\mathrm{d}x}{a}.$$

Hint: Integrate the left hand side by parts and use eqn. 7.34.

3. Prove that for *any* function $\gamma(w)$ satisfying the integral equation in eqn. 7.18, the misfit energy of eqn. 7.13 is equal to $\mu a^2/[4\pi(1-\nu)]$. This surprising result, due originally to AJE Foreman,[25] demonstrates that the total misfit energy is independent of the functional form of the γ-surface in the Peierls–Nabarro model.

[25] See Nabarro, FRN, *Adv. Phys.* **1**, 269–394 (1952), p.360. https://doi.org/10.1080/00018735200101211

8

The force on a defect

8.1 Introduction

We have already encountered the concept of the force on a defect with the Peach–Koehler force on a dislocation. In this chapter we will generalise the concept to other defects including point defects and interfaces. In the following chapter we will consider the force on a crack tip. The forces we are concerned with are often called configurational forces. They arise from changes in the total potential energy of the system when a defect is displaced. The total potential energy comprises the elastic energy of the body in which the defects reside and the potential energy of any external mechanism applying tractions to the surface of the body. The configurational force on the defect is then defined as minus the gradient of the total potential energy of the system with respect to the position of the defect. It follows that it is always a conservative force.[1]

This chapter is based on a series of papers by Eshelby,[2] who first introduced the concept of the force on an elastic singularity in a paper published in 1951.[3] He defined an elastic singularity in an infinite homogeneous elastic medium in the following way. Draw a closed surface S inside the medium. There is an elastic singularity inside S if the stresses inside S could not be produced by body forces outside S or tractions on S. The stresses inside S are not necessarily singular in the mathematical sense, e.g. the stress field of a dislocation with a finite core width does not display a mathematical singularity. Nevertheless, a dislocation with a finite core width is an elastic singularity in the sense defined by Eshelby.

[1] There are other forces on defects arising from an exchange of momentum with another field, such as an electric current flowing through the medium. Such current-induced forces may be regarded as body forces and it has been shown that they are generally non-conservative—see Todorov, TN, Dundas, D, Lü, J-T, Brandbyge M, and Hedegård, P, *Eur. J. Phys.* **35**, 065004 (2014). https://doi.org/10.1088/0143-0807/35/6/065004.

[2] Eshelby, JD, *Phil. Trans. R. Soc. A* **244**, 87–111 (1951) https://doi.org/10.1098/rsta.1951.0016; *Solid State Phys.* **3**, 79–144 (1956) https://doi.org/10.1016/S0081-1947(08)60132-0; In *Inelastic behaviour of solids*, ed. MF Kanninen, WF Adler, AR Rosenfield and RI Jaffee, McGraw-Hill: New York (1970), pp.77–115; *J. Elast.* **5**, 321–35 (1975) https://doi.org/10.1007/BF00126994; In *Continuum models of discrete systems*, ed. E Kröner and K-H Anthony, University of Waterloo Press: Waterloo, p.651–665, (1980). All these papers are available in *Collected works of J. D. Eshelby*, ed. X Markenscoff and A Gupta, Springer: Dordrecht (2006). ISBN 9781402044168 hard back, ISBN 9789401776448 soft cover.

[3] Published one year after his PhD in Physics at the University of Bristol.

Physics of elasticity and crystal defects. Adrian P. Sutton, Oxford University Press (2020). © Adrian P. Sutton.
DOI: 10.1093/oso/9780198860785.001.0001

8.2 The electrostatic force on a charge

To introduce the idea of the force on a defect we will first review the use of the Maxwell electrostatic stress tensor to calculate the force on a charge. Suppose there is some charge density $\rho(\mathbf{x})$ in a vacuum and an associated electric field $\mathbf{E}(\mathbf{x})$. The two are related by $\nabla \cdot \mathbf{E} = \rho(\mathbf{x})/\varepsilon_0$, where ε_0 is the permittivity of free space. The charge density may consist of a set of point charges or a continuous distribution. Consider a closed region $\mathcal{R}$ of the distribution of charges, and let $\mathcal{S}$ be the surface of $\mathcal{R}$. We can write down the force $\mathbf{F}$ acting on the charges inside $\mathcal{R}$ as the following volume integral over $\mathcal{R}$:

$$\mathbf{F} = \int_{\mathcal{R}} \rho(\mathbf{x})\, \mathbf{E}(\mathbf{x})\, \mathrm{d}V. \tag{8.1}$$

In this expression $\mathbf{E}(\mathbf{x})$ includes contributions to the electric field from both inside and outside $\mathcal{R}$. If the integrand could be expressed as the divergence of a second rank tensor $M_{ij}(\mathbf{x})$, that is, $\rho(\mathbf{x})E_i(\mathbf{x}) = M_{ij,j}$, it would be possible to use the divergence theorem to express the force as a surface integral over $\mathcal{S}$:

$$F_i = \int_{\mathcal{R}} \rho(\mathbf{x})\, E_i(\mathbf{x})\, \mathrm{d}V = \int_{\mathcal{R}} M_{ij,j}\, \mathrm{d}V = \int_{\mathcal{S}} M_{ij}\, n_j\, \mathrm{d}S. \tag{8.2}$$

As shown in texts on electromagnetism,[4] the tensor M_{ij} does exist and it is the Maxwell electrostatic stress tensor:

$$M_{ij} = \varepsilon_0 \left\{ E_i E_j - \frac{E^2}{2} \delta_{ij} \right\} \tag{8.3}$$

Exercise 8.1

Prove that $M_{ij,j} = \rho E_i$.

Imagine the charge distribution is a set of point charges and we wish to calculate the force on one of them. Suppose the surface $\mathcal{S}$ surrounding our chosen point charge is deformed into another surface $\mathcal{S}'$ in such a way that no other point charges are included within $\mathcal{S}'$. Since there are no charges between $\mathcal{S}$ and $\mathcal{S}'$ the divergence $M_{ij,j}$ is zero everywhere between the surfaces, and therefore the surface integrals $\int_{\mathcal{S}} M_{ij} n_j \mathrm{d}S$ and $\int_{\mathcal{S}'} M_{ij} n_j \mathrm{d}S$ are equal.

In the following sections we will derive a tensor whose divergence can replace the integrand in a volume integral representing the force on a defect. Then, using

[4] For example, Griffiths, DJ, *Introduction to electrodynamics*, international edn., Prentice-Hall (1999), p.351. ISBN 978-9332550445.

the divergence theorem this volume integral can be converted into a surface integral enclosing the defect, which can be deformed arbitrarily so long as it does not include any other defects, to give an expression for the configurational force on the defect.

8.3 The pressure on an interface

In this section we derive the force per unit area on an interface separating two regions A and B of a body. The interface may enclose a region or it may extend through the body and terminate at its surface. The forces we are concerned with arise from differences in the elastic energy densities W^A and W^B, for example stemming from different elastic constants in A and B when A and B are elastically deformed, or differences in the densities of dislocations and point defects which drive recrystallisation. The interface may also be an agent of deformation, as in mechanical twinning and martensitic transformations, where movement of the interface is accompanied by deformation of the region through which it sweeps. There are other sources of forces on interfaces, which we will not be concerned with, arising from curvature of the interface, and chemical free energy differences on either side of the interface during a phase change. In the following we assume the interface is coherent in the sense that the elastic displacement field and tractions are continuous across the interface.

Let the local normal vector to the interface be directed from B to A, with its tail on the interface. At each point on the interface $\mathcal{I}$ we imagine it migrates by the infinitesimal vector $\delta\xi n_i$ to a new position $\mathcal{I}'$. With $\delta\xi > 0$ this results in the consumption of A and the growth of B. Let δE_{tot} be the change in the total energy of the system resulting from this migration.

Cut out and remove the sliver of A between $\mathcal{I}$ and $\mathcal{I}'$ before the interface has migrated, applying suitable tractions to the freshly created surfaces to prevent relaxation. The change in the energy of A accompanying this first step is

$$\delta E^{(1)} = -\int_{\mathcal{I}} W^A \delta\xi \, \mathrm{d}S. \tag{8.4}$$

The elastic displacements on the newly created surface of A are $u_i^A + \delta\xi n_k u_{i,k}^A$, and the tractions are $\sigma_{ij}^A n_j + O(\delta\xi)$. The elastic displacements on the surface of A will relax to final values u_i^{FA}, where the difference $u_i^{FA} - \left(u_i^A + \delta\xi n_k u_{i,k}^A\right)$ is of order $\delta\xi$. Therefore, the surface tractions on the surface of A do work, which gives a second contribution to the change of the energy:

$$\delta E^{(2)} = -\int_{\mathcal{I}} \sigma_{ij}^A n_j \left[u_i^{FA} - \left(u_i^A + \delta\xi n_k u_{i,k}^A\right)\right] \mathrm{d}S \quad + \quad O(\delta\xi^2). \tag{8.5}$$

Now consider changes on the B-side. First there is the elastic energy arising from the creation of the sliver B between $\mathcal{I}$ and $\mathcal{I}'$:

$$\delta E^{(3)} = \int_{\mathcal{J}} W^B \delta\xi \mathrm{d}S. \tag{8.6}$$

The elastic displacements on the newly created surface of B are $u_i^B + \delta\xi n_k u_{i,k}^B$ and they relax in the final state to u_i^{FB}. The tractions on the surface of B are $\sigma_{ij}^B n_j + O(\delta\xi)$. The work done by them during the relaxation to the final state is

$$\delta E^{(4)} = +\int_{\mathcal{J}} \sigma_{ij}^B n_j \left(u_i^{FB} - (u_i^B + \delta\xi n_k u_{i,k}^B)\right) \mathrm{d}S \ + \ O(\delta\xi^2). \tag{8.7}$$

The minus sign in front of the integral in eqn. 8.5 is a plus sign in eqn. 8.7 because the sense of the normal is reversed.

In the final configuration the tractions and elastic displacements are continuous across the interface. Therefore, $\sigma_{ij}^A n_j = \sigma_{ij}^B n_j = \sigma_{ij} n_j$ and $u_i^{FA} = u_i^{FB} = u_i$. The change in the total energy is then

$$\begin{aligned}
\delta E_{tot} &= \delta E^{(1)} + \delta E^{(2)} + \delta E^{(3)} + \delta E^{(4)} \\
&= -\int_{\mathcal{J}} n_j \left\{ (W^A \delta_{jk} - \sigma_{ij} u_{i,k}^A) - (W^B \delta_{jk} - \sigma_{ij} u_{i,k}^B) \right\} \delta\xi n_k \mathrm{d}S \\
&= -\int_{\mathcal{J}} n_j \left\{ P_{jk}^A - P_{jk}^B \right\} \delta\xi n_k \mathrm{d}S,
\end{aligned} \tag{8.8}$$

where

$$P_{jk} = W\delta_{jk} - \sigma_{ij} u_{i,k} \tag{8.9}$$

is called the static energy-momentum tensor of the elastic field. A formal derivation of the static energy-momentum tensor is given in section 8.5. This is the first time we meet it and it plays a central role in the configurational force on any defect, not just an interface. The pressure acting on an element of area of the interface is then

$$p = -\frac{\delta E_{tot}}{\delta\xi \ \mathrm{d}S} = n_j \left(P_{jk}^A - P_{jk}^B\right) n_k. \tag{8.10}$$

Exercise 8.2

Consider a flat interface with unit normal $\hat{\mathbf{n}}$ between two semi-infinite media with elastic constants c_{ijkl}^A and c_{ijkl}^B with reference to a common coordinate system. Suppose a uniform distortion $u_{i,j}$ is applied to both media, such that continuity of tractions and displacements at the interface is maintained. Show that the pressure generated on the interface is equal to the difference in the elastic energy densities.

8.4 The force on a static defect

Consider a body occupying a region $\mathcal{R}$ containing a defect D at (ξ_1, ξ_2, ξ_3). For simplicity we will assume the elastic constants are the same throughout the body. The external surface $\mathcal{S}$ of the body may be loaded with tractions by some external mechanism. If F_k is the force on the defect and if it undergoes a small displacement $\delta\xi_m$ then the change in the total energy of the body and the external loading mechanism, is

$$\delta E_{tot} = -F_m \delta\xi_m, \tag{8.11}$$

where

$$F_m = \int_{\mathcal{S}} P_{mj} n_j \mathrm{d}S \tag{8.12}$$

and P_{jk} is the static energy-momentum tensor given by eqn. 8.9.

To prove eqn. 8.11 we proceed by first rigidly displacing the entire displacement field of the defect, thus $u_i^D(\mathbf{x}) \to u_i^D(\mathbf{x} - \delta\boldsymbol{\xi})$. This will then need to be adjusted at the surface of the body to satisfy the boundary conditions on $\mathcal{S}$. To calculate the change in the total energy we consider changes in both the elastic energy of the body and the potential energy of the loading mechansim.

In the first step the change in elastic energy of the body when the displacement field of D is rigidly translated is given by

$$\begin{aligned}
\delta E_{el}^{(1)} &= \int_{\mathcal{R}} W(\mathbf{x} - \delta\boldsymbol{\xi}) - W(\mathbf{x}) \mathrm{d}V \\
&= -\delta\xi_m \int_{\mathcal{R}} \frac{\partial W}{\partial x_m} \mathrm{d}V + O(\delta\xi^2) \\
&= -\delta\xi_m \int_{\mathcal{S}} W n_m \mathrm{d}S + O(\delta\xi^2).
\end{aligned} \tag{8.13}$$

The derivative of the elastic energy in the integrand of the volume integral may diverge, and this raises the question of whether it is integrable. But by converting the volume integral into a surface integral in the last line only a thin sliver around the surface of the body contributes to the integral, which is far from the elastic singularity at D and therefore the surface integral always converges.

In the second step we make a correction to satisfy the boundary conditions on the surface $\mathcal{S}$ of the body. Following the rigid displacement of the elastic field of D in the first step the surface tractions on $\mathcal{S}$ are $\{\sigma_{ij}(\mathbf{x}) - \delta\xi_m \sigma_{ij,m}(\mathbf{x})\} n_j + O(\delta\xi^2)$. Following the correction these surface tractions are returned to $\sigma_{ij}(\mathbf{x})n_j$ plus a term of order $\delta\xi$ depending on the 'hardness' of the mechanism applying the surface tractions. If the mechanism is soft the loading will be 'dead' and the traction will be returned to $\sigma_{ij}(\mathbf{x})n_j$. If the loading is rigid (infinitely hard) the difference in the surface traction from $\sigma_{ij}(\mathbf{x})n_j$ will still be of order $\delta\xi$. The surface displacements after the rigid translation of the field of D are $u_i(\mathbf{x}) - \delta\xi_m u_{i,m}(\mathbf{x})$, where $\mathbf{x}$ is on $\mathcal{S}$. After the adjustment they become $u_i^F(\mathbf{x})$, which can be determined only by detailed calculation. Therefore the change in the elastic energy of the body resulting from the restoration of the original surface tractions on the surface $\mathcal{S}$ is

$$\delta E_{el}^{(2)} = \int_{\mathcal{S}} \sigma_{ij}\{u_i^F(\mathbf{x}) - u_i(\mathbf{x}) + \delta\xi_m u_{i,m}(\mathbf{x})\} n_j \, \mathrm{d}S + O(\delta\xi^2). \tag{8.14}$$

We see that this expression is independent of the hardness of the loading mechanism because variations in $\delta E_{el}^{(2)}$ arising from the hardness are second order in $\delta\xi$. The work done by the external loading mechanism at the end of both steps is also independent of its hardness to first order in $\delta\xi$:

$$\delta w_{ext} = \int_{\mathcal{S}} \sigma_{ij}\{u_i^F(\mathbf{x}) - u_i(\mathbf{x})\} n_j \mathrm{d}S + O(\delta\xi^2). \tag{8.15}$$

The change in the potential energy of the loading mechanism is $-\delta w_{ext}$. Adding $\delta E_{el}^{(1)}$, $\delta E_{el}^{(2)}$ and $-\delta w_{ext}$ we obtain the change in the total energy of the body and loading mechanism:

$$\delta E_{tot} = -\delta\xi_m \int_{\mathcal{S}} (W\delta_{mj} - \sigma_{ij} u_{i,m}) n_j \mathrm{d}S + O(\delta\xi^2), \tag{8.16}$$

from which eqn. 8.12 follows.

So far we have considered the somewhat unrealistic case of a body containing just one defect. The extension to the more realistic case of the force on D arising from the elastic fields of other defects within the body, in addition to or instead of loads applied to the surface of the body, is straightforward. Construct an internal surface S surrounding D and no other defects. The sum of the elastic energy and the potential energy of the external loading mechanism remains the same if we redefine the elastic energy to be the elastic energy of the region R within S, and redefine the potential energy to be the sum of the elastic energy between S and the external surface $\mathcal{S}$ and the potential energy of the external loading mechanism. The above steps may then be repeated regarding R as the body loaded by surface tractions on S arising from other defects between S and $\mathcal{S}$ and from external loads applied to $\mathcal{S}$. Thus, we arrive at the more general result that the force on D is the following integral:

$$F_m = \int_S (W\delta_{mj} - \sigma_{ij}u_{i,m})n_j \mathrm{d}S, \tag{8.17}$$

where S is any surface enclosing only D inside the body.

The invariance of the force in eqn. 8.17 if the surface S is changed to S', provided no defects exist between S and S', follows because the divergence of the integrand is then zero. We will now prove this for a general elastic energy density W that depends on the elastic displacement field and the displacement gradients:[5] $W = W(u_i, u_{i,j})$.

The partial derivative of W with respect to x_m is as follows:

$$\begin{aligned}\frac{\partial W}{\partial x_m} &= \frac{\partial W}{\partial u_p}\frac{\partial u_p}{\partial x_m} + \frac{\partial W}{\partial u_{p,q}}\frac{\partial u_{p,q}}{\partial x_m} \\ &= -f_p\, u_{p,m} + \sigma_{pq}\, u_{p,qm},\end{aligned} \tag{8.18}$$

where $-\partial W/\partial u_p = f_p$ is a body force and $\partial W/\partial u_{p,q} = \sigma_{pq}$ is a component of the stress tensor. Therefore the divergence of the integrand in eqn. 8.17 is as follows:

$$\begin{aligned}P_{mj,j} &= (W\delta_{mj} - \sigma_{ij}u_{i,m})_{,j} \\ &= -f_p\, u_{p,m} + \sigma_{pq}\, u_{p,qm} - \sigma_{ij,j}u_{i,m} - \sigma_{ij}u_{i,mj}.\end{aligned} \tag{8.19}$$

We see that $P_{mj,j} = 0$ because the equilibrium condition ensures $\sigma_{ij,j} + f_i = 0$. It follows that S may be deformed into S' and the integral in eqn. 8.17 is invariant.

The divergence $P_{mj,j}$ is non-zero when there is a defect because the elastic energy density then has an explicit dependence on position: $W = W(u_i, u_{i,j}, x_i)$. For then the gradient of W with respect to position becomes

$$\frac{\partial W}{\partial x_m} = -f_p\, u_{p,m} + \sigma_{pq}\, u_{p,qm} + \left(\frac{\partial W}{\partial x_m}\right)_{exp}. \tag{8.20}$$

The explicit partial derivative $(\partial W/\partial x_m)_{exp}$ is evaluated by displacing the point x_m by a virtual amount δx_m while keeping fixed all other points in the body and keeping the elastic displacements and displacement gradients fixed. If there is a defect at $\mathbf{X}$ then when $\mathbf{x} = \mathbf{X}$ the partial derivative $(\partial W/\partial x_m)_{exp}$ is nonzero because the defect undergoes a virtual displacement.[6] The divergence $P_{mj,j}$ is then $(\partial W/\partial x_m)_{exp}$ and the force F_m becomes

$$F_m = \int_R \left(\frac{\partial W}{\partial x_m}\right)_{exp} \mathrm{d}V. \tag{8.21}$$

If there is just one defect in the region R at $\mathbf{X}$ then $(\partial W/\partial x_m)_{exp} = (\partial W/\partial X_m)_{exp}\,\delta(\mathbf{x} - \mathbf{X})$.

[5] Note that the relation between stress and strain may be nonlinear

[6] Alternatively, the explicit partial derivative is non-zero in regions where the elastic constants depend on position.

It is emphasised that although a virtual displacement of the defect will induce a change in the displacement gradients of the same order throughout the body they must not be taken into account in the evaluation of the force. For example, for a point defect at **X** with an elastic dipole tensor ρ_{ij} the elastic interaction energy with a displacement gradient $u_{i,j}(\mathbf{x})$ is $-\rho_{ij}u_{i,j}(\mathbf{X})$. This is the part of the elastic energy density that depends explicitly on **X**. Therefore the force on the point defect is

$$F_m = -\rho_{ij}u_{i,jm}(\mathbf{X}), \tag{8.22}$$

where $u_{i,j}(\mathbf{X})$ is the *frozen* displacement gradient at **X** arising from other defects and from the boundary conditions on the surface of the body. This is reminiscent of the force theorem in electronic density functional theory[7] (DFT) for the force on a nucleus, where the changes in the self-consistent electronic charge density caused by the virtual change in the position of the nucleus must not be taken into account in the evaluation of the force.

Exercise 8.3

For the elastic energy density $W = W(u_i, u_{i,j})$ show that the elastic energy of a body is minimised when:

$$\frac{\partial W}{\partial u_i} - \frac{\partial}{\partial x_j}\left(\frac{\partial W}{\partial u_{i,j}}\right) = 0. \tag{8.23}$$

What is the physical meaning of this equation?

8.5 Relationship to the static energy-momentum tensor

In this section we will use the formal methods of classical field theory to derive the static energy-momentum tensor. We will show that it is equal to the integrand of the expression in eqn. 8.17 for the force on a defect.

We begin with the time-independent Lagrangian density, which we write as $L = L(u_i, u_{i,j}, x_m)$. As usual u_i and $u_{i,j}$ are the displacements and displacement gradients. The presence of inhomogeneities such as defects introduces an explicit dependence of

[7] The force theorem is due to DG Pettifor and was published in *Commun. Phys.* **1**, 141–6 (1976), which is unavailable online. David Godfrey Pettifor FRS 1945–2017, British theoretical materials physicist. An online account of it may be found in Heine, V, *Solid State Phys.* **35**, 114–20 (1980). https://doi.org/10.1016/S0081-1947(08)60503-2. Volker Heine FRS 1930–, German born British theoretical condensed matter physicist.

the Lagrangian density on position, as in eqn. 8.20. Since we are ignoring kinetic energy the Lagrangian density is just $-W$, where W is the elastic energy density.

The integral of the Lagrangian density over the volume of the body is minimised with respect to variations δu_i of the displacement field, where $\delta u_i = 0$ on the surface of the body, when the following Euler–Lagrange equation is satisfied:

$$\frac{\partial}{\partial x_j}\left(\frac{\partial L}{\partial u_{i,j}}\right) - \frac{\partial L}{\partial u_i} = 0. \tag{8.24}$$

With $L = -W$ this is the equation of mechanical equilibrium.

Consider the partial derivative of the Lagrangian density with respect to x_l. To be clear, it is defined as

$$\frac{\partial L}{\partial x_l} = \lim_{\delta x_l \to 0} \frac{L(x_l + \delta x_l) - L(x_l)}{\delta x_l}. \tag{8.25}$$

It has both implicit contributions and an explicit contribution:

$$\begin{aligned}\frac{\partial L}{\partial x_l} &= \frac{\partial L}{\partial u_i} u_{i,l} + \frac{\partial L}{\partial u_{i,j}} u_{i,jl} + \left(\frac{\partial L}{\partial x_l}\right)_{exp} \\ &= \left[\frac{\partial L}{\partial u_i} u_{i,l} - \frac{\partial}{\partial x_j}\left(\frac{\partial L}{\partial u_{i,j}}\right) u_{i,l}\right] + \frac{\partial}{\partial x_j}\left(\frac{\partial L}{\partial u_{i,j}}\right) u_{i,l} + \frac{\partial L}{\partial u_{i,j}} u_{i,jl} + \left(\frac{\partial L}{\partial x_l}\right)_{exp} \\ &= \frac{\partial}{\partial x_j}\left\{\left(\frac{\partial L}{\partial u_{i,j}}\right) u_{i,l}\right\} + \left(\frac{\partial L}{\partial x_l}\right)_{exp}.\end{aligned} \tag{8.26}$$

The term in square brackets is zero, owing to the Euler–Lagrange equation eqn. 8.24. Equation 8.26 can be rewritten as follows:

$$\left(\frac{\partial L}{\partial x_l}\right)_{exp} = -\frac{\partial}{\partial x_j} P_{jl}, \tag{8.27}$$

where

$$P_{jl} = \frac{\partial L}{\partial u_{i,j}} u_{i,l} - L\delta_{jl} \tag{8.28}$$

is the static energy-momentum tensor. Setting $L = -W$, and using $\partial W/\partial u_{i,j} = \sigma_{ij}$, we see that eqn. 8.28 is the same as eqn. 8.9. Note that in eqn. 8.27 the explicit partial derivative of the Lagrangian is equated to the divergence of the static energy-momentum tensor, which is analogous to the divergence of the Maxwell electrostatic stress tensor in eqn. 8.3 as an expression of the electrostatic force on a point charge in Exercise 8.1.

8.6 The force due to an applied stress and image forces

The elastic energy density W, the stress tensor σ_{ij} and the displacement gradient $u_{i,m}$ in eqn. 8.17 include contributions from all elastic fields present in the body. In addition to the elastic fields $\sigma_{ij}^{D\infty}, u_i^{D\infty}$ of the defect in an infinite medium there are contributions from other defects and from forces applied to the surface of the body, which we collectively label σ_{ij}^{A}, u_i^{A}, and from satisfying the boundary conditions on the surface of the body σ_{ij}^{I}, u_i^{I} which are often called image fields. In linear elasticity we have

$$\sigma_{ij} = \sigma_{ij}^{D\infty} + \sigma_{ij}^{A} + \sigma_{ij}^{I}$$

$$u_i = u_i^{D\infty} + u_i^{A} + u_i^{I}$$

When these sums are inserted in eqn. 8.17 the force becomes

$$F_m = \int_\Sigma (D\infty, D\infty) + (A,A) + (I,I) + (D\infty, A) + (A,I) + (I, D\infty)\,\mathrm{d}S. \tag{8.29}$$

The diagonal terms are given by

$$(X,X) = \left(\frac{1}{2}\sigma_{pq}^{X} u_{p,q}^{X}\delta_{mj} - \sigma_{ij}^{X} u_{i,m}^{X}\right) n_j \tag{8.30}$$

and the off-diagonal terms by

$$(X,Y) = \left(\frac{1}{2}\left(\sigma_{pq}^{X} u_{p,q}^{Y} + \sigma_{pq}^{Y} u_{p,q}^{X}\right)\delta_{mj} - \left(\sigma_{ij}^{X} u_{i,m}^{Y} + \sigma_{ij}^{Y} u_{i,m}^{X}\right)\right) n_j. \tag{8.31}$$

To proceed we need a version of Stokes' theorem:

$$\oint_C \varepsilon_{jli} w_i \mathrm{d}x_j = \int_\Sigma \left(w_{m,m}\delta_{jl} - w_{j,l}\right) n_j \mathrm{d}S, \tag{8.32}$$

where w_i is a differentiable function, and the left hand side is a line integral taken around the closed contour C and Σ is an *open* surface bounded by C.

Exercise 8.4

Derive eqn. 8.32 by applying the familiar form of Stokes' theorem to $A_{jl} = \varepsilon_{jli} w_i$. Hence, for a *closed* surface show that

$$\int_{\Sigma} w_{m,m} \delta_{jl} n_j \mathrm{d}S = \int_{\Sigma} w_{j,l} n_j \mathrm{d}S. \tag{8.33}$$

Another way to derive this formula is to consider the change in the integral of $\mathbf{w} \cdot \hat{\mathbf{n}}$, where $\hat{\mathbf{n}}$ is the local unit normal, taken over a closed surface Σ when the surface is displaced by an infinitesimal amount ε along the x_l-axis. The change in the surface integral is

$$\varepsilon \int_{\Sigma} \frac{\partial}{\partial x_l} \mathbf{w} \cdot \hat{\mathbf{n}} \, \mathrm{d}S.$$

This integral is equal to the integral on the right hand side of eqn. 8.33 multiplied by ε. But this is also equal to

$$\int_{\Sigma'} w_j n_j \mathrm{d}S - \int_{\Sigma} w_j n_j \mathrm{d}S,$$

where Σ' is the surface following the displacement of Σ by ε along x_l. By the divergence theorem this difference in surface integrals is equal to the volume integral of the divergence of $\mathbf{w}$ taken over the thin sliver between Σ and Σ'. A volume element of this sliver is $\varepsilon \hat{\mathbf{e}}_l \cdot \hat{\mathbf{n}} \, \mathrm{d}S$, where $\hat{\mathbf{e}}_l$ is a unit vector along the x_l-axis. Therefore,

$$\varepsilon \int_{\Sigma} \nabla \cdot \mathbf{w} \, (\hat{\mathbf{e}}_l \cdot \hat{\mathbf{n}}) \, \mathrm{d}S = \varepsilon \int_{\Sigma} \frac{\partial}{\partial x_l} \mathbf{w} \cdot \hat{\mathbf{n}} \, \mathrm{d}S.$$

Equation 8.33 then follows.

Using eqn. 8.33 with $w_q = \sigma^X_{pq} u^Y_p + \sigma^Y_{pq} u^X_p$ we obtain

$$\begin{aligned}\int_{\Sigma} (\sigma^X_{pq} u^Y_{p,q} + \sigma^Y_{pq} u^X_{p,q}) n_m \mathrm{d}S &= \int_{\Sigma} \left(\sigma^X_{pj} u^Y_p + \sigma^Y_{pj} u^X_p\right)_{,m} n_j \mathrm{d}S \\ &= \int_{\Sigma} \left(\sigma^X_{pj,m} u^Y_p + \sigma^X_{pj} u^Y_{p,m} + \sigma^Y_{pj,m} u^X_p + \sigma^Y_{pj} u^X_{p,m}\right) n_j \mathrm{d}S,\end{aligned} \tag{8.34}$$

where the equilibrium condition $\sigma_{pq,q} = 0$ has been used. The contribution $F^{(XY)}_m$ to the force F_m on the defect arising from the (XY) term in eqn. 8.31 is then

$$F_m^{(XY)} = \frac{1}{2}\int_\Sigma \left(\sigma_{pj,m}^X u_p^Y + \sigma_{pj,m}^Y u_p^X - \sigma_{pj}^X u_{p,m}^Y - \sigma_{pj}^Y u_{p,m}^X\right) n_j \,\mathrm{d}S. \tag{8.35}$$

Using the divergence theorem we may convert this to a volume integral over the volume V enclosed by Σ:

$$F_m^{(XY)} = \frac{1}{2}\int \left(\sigma_{pj,m}^X u_{p,j}^Y + \sigma_{pj,m}^Y u_{p,j}^X - \sigma_{pj}^X u_{p,mj}^Y - \sigma_{pj}^Y u_{p,mj}^X\right) \mathrm{d}V, \tag{8.36}$$

where we have again used the equilibrium condition $\sigma_{ij,j} = 0$. If $\sigma_{pj}^X = c_{pjkl}u_{k,l}^X$ and $\sigma_{pj}^Y = c_{pjkl}u_{k,l}^Y$ then it is easy to show that this volume integral is zero remembering that $c_{ijkl} = c_{klij}$. But if X and/or Y is $D\infty$ this is no longer true because $e_{pj}^{D\infty} \neq (1/2)(u_{p,j}^{D\infty} + u_{j,p}^{D\infty})$ throughout V. Thus all the off-diagonal terms in eqn. 8.31 are zero except those involving $D\infty$. The diagonal terms (A,A) and (I,I) are zero by the same argument. The diagonal term $(D\infty, D\infty)$ is zero because Σ may be taken at infinity where the integral tends to zero provided σ_{ij} decays at least as rapidly as $1/r^2$ in three dimensions and at least as rapidly as $1/r$ in two dimensions.

The conclusion is that only the two off-diagonal terms involving $D\infty$ are non-zero: $(D\infty, A)$ and $(D\infty, I)$. These are the forces arising from the applied stress and the image field. The force arising from an applied stress is

$$F_m^A = \int_\Sigma \left(\sigma_{pj,m}^{D\infty} u_p^A - \sigma_{pj}^A u_{p,m}^{D\infty}\right) n_j \mathrm{d}S \tag{8.37}$$

and the force arising from the image field is

$$F_m^I = \int_\Sigma \left(\sigma_{pj,m}^{D\infty} u_p^I - \sigma_{pj}^I u_{p,m}^{D\infty}\right) n_j \mathrm{d}S. \tag{8.38}$$

In eqn. 8.37 we see that the force on a defect in a finite body due to other defects or tractions applied to the external surface of the body involves the field of the defect in an *infinite* medium, not those in the *finite* body. This is a very significant simplification.

8.7 The Peach–Koehler force revisited

In this section we apply the general expression for the force on a defect in eqn. 8.17 to rederive the Peach–Koehler formula for the force per unit length on a dislocation due to an applied stress σ_{ij}^A, which was first obtained in section 6.6.

To apply eqn. 8.37 to a dislocation consider a straight Volterra dislocation with its positive line sense along the positive x_3-axis of a Cartesian coordinate system with an arbitrary Burgers vector b_i. The surface S of the integral in eqn. 8.37 has to enclose only the dislocation, but is otherwise arbitrary. It is defined by the line integral (see eqn. 6.2)

$$b_i = \oint_C \frac{\partial u_i^{D\infty}}{\partial x_k} \mathrm{d}x_k, \tag{8.39}$$

where the circuit C is any clockwise circuit taken around the positive sense of the dislocation line. The absence of a resultant body force associated with the dislocation ensures that the first term in the integrand in eqn. 8.37 integrates to zero. We are left with

$$F_m^A = -\int_S \sigma_{ij}^A u_{i,m}^{D\infty} n_j \mathrm{d}S. \tag{8.40}$$

Define axes x_1 and x_2 in the plane normal to the dislocation line. Let the cut plane be the half-plane $x_2 = 0, x_1 > 0$. We choose S to consist of the surfaces S^+ and S^- extending from $x_1 = -\infty$ to $x_1 = +\infty$, infinitesimally above and below the x_1-axis, as shown in Fig. 8.1, and closed by infinitesimal segments along x_2 at $x_1 = \pm\infty$. On S^+ and S^- we have $\sigma_{ij}^A(x_1, 0^+, x_3)n_j = \sigma_{ij}^A(x_1, 0^-, x_3)n_j$. We also have $u_i^{D\infty}(x_1, 0^+, x_3) - u_i^{D\infty}(x_1, 0^-, x_3) = -b_i$ for all $x_1 > 0$ and for all x_3, and $u_i^{D\infty}(x_1, 0^+, x_3) - u_i^{D\infty}(x_1, 0^-, x_3) = 0$ for all $x_1 < 0$ and for all x_3. Therefore, $u_{i,1}^{D\infty}(x_1, 0^+, x_3) - u_{i,1}^{D\infty}(x_1, 0^-, x_3) = -b_i\delta(x_1)$, which is consistent with eqn. 8.39. Following section 6.6 the positive normal to the cut half-plane is along $-x_2$, that is $n_j = -\delta_{j2}$. Carrying out the integration around S, in the sense indicated in Fig. 8.1, we obtain the force per unit length of dislocation line, $F_1^A = \sigma_{i2}^A b_i$, where σ_{i2}^A is evaluated at the origin. This agrees with eqn. 6.11 with $t_p = \delta_{p3}$.

To find F_2^A we choose the cut plane to be the half-plane $x_1 = 0, x_2 > 0$ and we choose S^+ and S^- to extend from $x_2 = -\infty$ to $x_2 = +\infty$ with $x_1 = \pm\varepsilon$, where ε is a

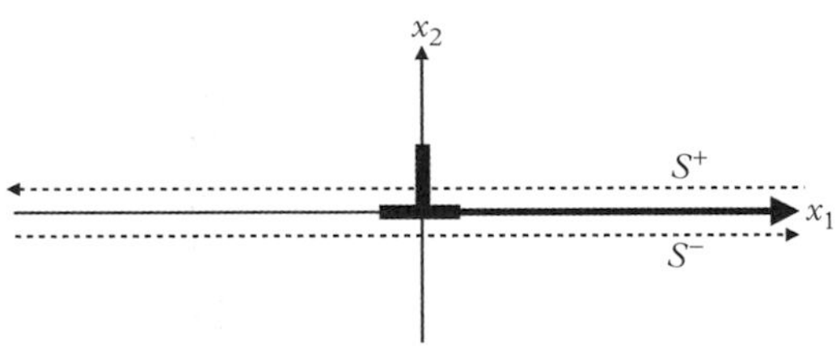

Figure 8.1 *To illustrate the evaluation of the Peach–Koehler force. A straight dislocation lies along the x_3-axis, with its positive line sense along the positive x_3 direction. The cut is in the plane $x_2 = 0$, along $x_1 \geq 0$. The surface integral in eqn. 8.40 is taken in the clockwise sense along the positive direction of the dislocation line sense, which appears anticlockwise in the figure because we are looking along $-x_3$. S^+ and S^- are surfaces infinitesimally above and below the x_1-axis.*

positive infinitesimal number. Repeating the argument with $n_j = \delta_{j1}$ and $u^{D\infty}_{i,2}(0^+, x_2, x_3) - u^{D\infty}_{i,2}(0^-, x_2, x_3) = b_i\delta(x_2)$, which is consistent with eqn. 8.39, we obtain $F^A_2 = -\sigma_{i1} b_i$. This also agrees with eqn. 6.11.

8.8 Problem set 8

1. In an invariant plane strain deformation an interface separates deformed and undeformed regions of a medium. The interface is an invariant plane of the deformation, so that it is neither deformed nor rotated. Consider a flat interface between semi-infinite regions A and B. The positive sense of the normal $\hat{\mathbf{n}}$ to the interface points from region B to region A. Region A has not undergone the invariant plane strain, but region B has. If $\hat{\mathbf{e}}$ is a unit vector perpendicular to the unit normal $\hat{\mathbf{n}}$ show that the distortion tensor of a general invariant plane strain is as follows:

$$u^B_{i,k} = s\hat{e}_i\hat{n}_k + \lambda\hat{n}_i\hat{n}_k,$$

and state the meaning of s and λ. Verify that any vector lying in the interface is unchanged by this deformation, and hence the interface is invariant. In deformation twinning $\lambda = 0$, and the twin crystal and the parent crystal are related by a simple shear.

If only *linear* elastic forces were at play then any movement of the interface induced by an applied stress would be reversed when the stress is removed. In both deformation twinning and martensitic phase changes the elastic energy density W becomes a *nonlinear* function of the distortion tensor. That is because after the crystal has been twinned or transformed martensitically it corresponds to a local minimum of the elastic energy density W in the space of distortions. The elastic energy density W then describes the change in the energy of the crystal along a path between the initial and final states of the crystal in this space. The variation of the elastic energy density along such a path can be calculated using modern DFT methods. When the distortion from a local minimum in this space is sufficiently small the change in the energy is described adequately by linear elasticity. But linear elasticity fails to capture the existence of other local minima.

In the absence of an applied stress we write $W^A = W^{A0}$ where W^{A0} is the energy density of region A at the local minimum. Similarly, $W^B = W^{B0}$ is the energy density of region B at its local minimum. Show that the pressure on the interface is then $W^{A0} - W^{B0}$. This difference in the energy densities may depend on other variables such as temperature and applied magnetic or electric fields. The pressure $W^{A0} - W^{B0}$ may the regarded as a *chemical* pressure because it arises from atomic interactions in undistorted crystals A and B. In mechanical twinning $W^{A0} = W^{B0}$ because the crystal structures of states A and B are related by a rotation.

When there is an applied stress the displacement field in region A is just the elastic displacement u_i^A from the atomic configuration at the local energy minimum. The displacement field in region B is $u_i^{B0} + u_i^B$ where u_i^{B0} is the displacement due to the invariant plane strain and u_{iB} is the additional elastic displacement caused by the applied stress. Show that the pressure on the interface in the presence of an applied stress becomes:

$$p = n_j\left[\left(W^{A0} + W_{el}^A - W^{B0} - W_{el}^B\right)\delta_{jk} + \sigma_{ij}\left(u_{i,k}^{B0} + u_{i,k}^B - u_{i,k}^A\right)\right]n_k$$

In general the components of the distortion tensor of the invariant plane strain are much larger than those due to elastic distortions of regions A and B. In that case we may ignore the elastic distortions $u_{i,k}^A$ and $u_{i,k}^B$ in comparison to $u_{i,k}^{B0}$. Show that the pressure on the interface then becomes:

$$p = \left(W^{A0} - W^{B0} + W_{el}^{A0} - W_{el}^{B0}\right) + \left[\{n_i\sigma_{ij}e_j\}s + \{n_i\sigma_{ij}n_j\}\lambda\right].$$

In a martensitic transformation the term in round brackets may be influenced by other fields such as thermal, magnetic or electric. The term in square brackets shows that resolved components of the applied stress may drive the transformation, converting region A into region B, by promoting the simple shear and the tensile strain normal to the interface. In shape memory alloys these two terms are played off against each other. Even though W^{B0} may be slightly greater than W^{A0} a suitably applied stress may drive the transformation and convert region A into the martensite phase B: this is 'stress-induced martensite'. Then by adjusting the temperature or some other field $W^{B0} - W^{A0}$ may become sufficiently negative to drive the transformation in the reverse direction. This reversibility of the martensitic phase change is the basis of the 'shape memory effect'.

2. In the derivation of the pressure on an interface in eqn. 8.10 it was assumed there is continuity of displacements and tractions across the interface. These assumptions apply when the interface is 'coherent'. At high temperatures, or when there is a large misfit strain between the crystal lattices on either side of the interface, some interfaces become incoherent. In that case only the normal displacements and normal tractions are continuous across the interface. The tractions parallel to the interface are assumed to be relaxed to zero through sliding of one crystal with respect to the other as if the interface were greased. In general there are then discontinuous changes in the displacement vector parallel to the interface. Show that the pressure on an incoherent interface is $p = n_i\left(Q_{ij}^A - Q_{ij}^B\right)n_j$, where $Q_{ij} = W\delta_{ij} - \sigma_{ij}\left(n_k u_{m,k} n_m\right)$.

9

Cracks

9.1 Introduction

The nucleation and propagation of cracks leads to fracture. For this reason cracks have been studied for about a century. There are many shapes and sizes of cracks, and a range of theoretical and computational methods has been developed to solve their elastic fields. In this chapter we will focus on just one theoretical technique in which cracks are represented by continuous distributions of dislocations. There are two reasons for choosing this technique. Cracks in all but the most brittle crystalline materials are associated with some degree of plasticity in their vicinity, called plastic zones. Plastic zones contain crystal dislocations which interact with the elastic field of the crack. The representation of the crack itself in terms of dislocations enables these interactions to be described conveniently using dislocation theory. Conversely, the elastic fields of cracks are intimately related to the elastic fields of dislocation pileups, where dislocations pile up against barriers such as grain boundaries and precipitates during plastic deformation. Dislocation pileups may also nucleate cracks. There are thus physical and mathematical close relationships between cracks and dislocations.

Various criteria for crack propagation have been proposed in the literature. The most fundamental is the Griffith criterion, which states that for a crack to grow the total energy of the system must decrease, where the 'system' includes the external loading mechanism. The Griffith criterion is discussed in section 9.5. A recurring feature of the theory of fracture is its incompleteness—the lack of knowledge about atomic mechanisms at the root of a crack and how they are influenced by temperature, strain rate, local stress, local chemical composition and microstructure. In 1921 Griffith wrote in the seminal paper containing his criterion for fracture that no criterion for strength or fracture could be considered complete without taking into consideration interatomic forces. This incompleteness will emerge several times in this chapter. We will see in section 9.6 that dislocations ahead of a crack may either reduce or increase the stress acting on the crack tip. Forces acting on atoms at the crack tip are thus strongly influenced by groups of dislocations, and other defects, that may be further away. The multi-scale nature of fracture is also one of its key features. Screening of applied stresses acting at crack tips by dislocations is another central theme of this chapter.

Physics of elasticity and crystal defects. Adrian P. Sutton, Oxford University Press (2020). © Adrian P. Sutton.
DOI: 10.1093/oso/9780198860785.001.0001

9.2 Mathematical preliminaries

In this chapter we will encounter singular integral equations of the Cauchy type. They have the following form:

$$P\int_{-1}^{1} \frac{F(u)}{v-u}\mathrm{d}u = H(v), \tag{9.1}$$

where v lies between -1 and 1. The P in front of the integral signifies it is a principal value type. The standard way to deal with these equations is to use complex analysis following the work[1] in this area by Muskhelishvili.[2] We shall follow a simpler approach using Chebyshev[3] polynomials.[4] This will limit our use of complex analysis to evaluating contour integrals, of which there will be many in this chapter.

9.2.1 Chebyshev polynomials

Chebyshev polynomials of the first kind are defined as follows:

$$T_n(\cos\theta) = \cos n\theta, \quad \text{where } n \geq 0. \tag{9.2}$$

Using the addition formula for cosines it is easy to show

$$T_{n+1}(\cos\theta) = 2\cos\theta\, T_n(\cos\theta) - T_{n-1}(\cos\theta),$$

where $n \geq 1$. Replacing $\cos\theta$ by x we have the recurrence relation:

$$T_{n+1}(x) = 2xT_n(x) - T_{n-1}(x), \tag{9.3}$$

where $T_n(x) = \cos(n\cos^{-1} x)$, and $T_0(x) = 1,\ T_1(x) = x$.

Chebyshev polynomials of the second kind are defined as follows:

$$U_n(\cos\theta) = \frac{\sin(n+1)\theta}{\sin\theta}, \quad \text{where } n \geq 0. \tag{9.4}$$

[1] Muskhelishvili, NI, *Singular integral equations*, Dover Publications: New York (2008), ISBN 9780486462424.

[2] Nikolay Ivanovich Muskhelishvili 1891–1976, Georgian mathematician, physicist and engineer.

[3] Pafnuty Lvovich Chebyshev 1821–94, Russian mathematician.

[4] This is the approach taken by Bilby, BA, and Eshelby, JD, Dislocations and the theory of fracture, in *Fracture*, ed. H Liebowitz, volume 1, Academic Press: New York (1968), Chapter 2. Although these authors did not explicitly mention Chebyshev polynomials the mathematics they used is the mathematics of Chebyshev polynomials.

Using the addition formula for sines it is easy to show

$$U_{n+1}(\cos\theta) = 2\cos\theta U_n(\cos\theta) - U_{n-1}(\cos\theta),$$

where $n \geq 1$. Replacing $\cos\theta$ by x we have the recurrence relation:

$$U_{n+1}(x) = 2xU_n(x) - U_{n-1}(x), \tag{9.5}$$

where $U_n(x) = \sin((n+1)\cos^{-1} x)/\sin(\cos^{-1} x)$, and $U_0(x) = 1$, $U_1(x) = 2x$.

Exercise 9.1

Using the recurrence relations determine the Chebyshev polynomials of the first and second kind up to $n = 6$.

Prove the following orthogonality relations on the interval $-1 \leq x \leq 1$:

$$\int_{-1}^{1} \frac{T_m(x)T_n(x)}{\sqrt{1-x^2}}\,\mathrm{d}x = \begin{cases} 0, & n \neq m \\ \pi, & n = m = 0 \\ \pi/2 & n = m \neq 0, \end{cases} \tag{9.6}$$

and

$$\int_{-1}^{1} U_m(x)U_n(x)\sqrt{1-x^2}\,\mathrm{d}x = \begin{cases} 0, & n \neq m \\ \pi/2 & n = m. \end{cases} \tag{9.7}$$

Hint: Make the substitution $x = \cos\theta$.

As a result of the orthogonality relations, eqn. 9.6, any integrable function $F(x)$ on the interval $-1 \leq x \leq 1$ may be expanded in terms of Chebyshev polynomials of the first kind:

$$F(x) = \sum_{n=0}^{\infty} a_n T_n(x), \tag{9.8}$$

where

$$a_0 = \frac{1}{\pi}\int_{-1}^{1} \frac{F(x)}{\sqrt{1-x^2}}\,\mathrm{d}x$$

$$a_m = \frac{2}{\pi} \int_{-1}^{1} \frac{F(x) T_m(x)}{\sqrt{1-x^2}} \, \mathrm{d}x. \tag{9.9}$$

Chebyshev polynomials of the first and second kind are related by the following principal value integrals:

$$P \int_{-1}^{1} \frac{T_n(y)}{(y-x)\sqrt{1-y^2}} \, \mathrm{d}y = \pi U_{n-1}(x), \; |x| \leq 1 \tag{9.10}$$

$$P \int_{-1}^{1} \frac{\sqrt{1-y^2}\, U_{n-1}(y)}{(y-x)} \, \mathrm{d}y = -\pi T_n(x), \; |x| \leq 1. \tag{9.11}$$

Exercise 9.2

Prove eqns. 9.10 and 9.11.

Hint: Making the substitutions $y = \cos\theta$ and $x = \cos\phi$ show that eqn. 9.10 becomes

$$P \int_{0}^{\pi} \frac{\cos n\theta}{\cos\theta - \cos\phi} \, \mathrm{d}\theta = \frac{\pi \sin n\phi}{\sin\phi}$$

and eqn. 9.11 becomes

$$-P \int_{0}^{\pi} \frac{\sin\theta \sin n\theta}{\cos\theta - \cos\phi} \, \mathrm{d}\theta = \pi \cos n\phi.$$

These integrals are even in θ. They may be transformed to Cauchy principal value integrals around the unit circle in the complex plane by making the substitution $z = e^{i\theta}$. There are simple poles on the contour at $z = e^{\pm i\phi}$, for which the residues cancel in both integrals. This leaves poles of order n, $n-1$ and $n+1$ at $z = 0$, for which the residues may be evaluated by splitting the integrands into partial fractions. These contour integrals take a bit of effort to evaluate but the method outlined here is straightforward if somewhat tedious. The results of the above two integrals are stated without proof in Appendix C of the paper by Bilby and Eshelby (1968).

9.2.2 Solution of the Cauchy type integral equation

We now have the tools we need to solve the integral equation, eqn. 9.1. Let $F(u) = L(u)/\sqrt{1-u^2}$, where $L(u)$ is an integrable function on $[-1, 1]$ to be determined by solving the integral equation. We may expand $L(u)$ in terms of Chebyshev polynomials of the first kind:

$$L(u) = a_0 + \sum_{n=0}^{\infty} a_n\, T_n(u). \tag{9.12}$$

Inserting this expansion into eqn. 9.1 we obtain

$$P\int_{-1}^{1} \left(a_0 + \sum_{n=0}^{\infty} a_n T_n(u)\right) \frac{1}{\sqrt{(1-u^2)}} \frac{1}{(v-u)}\, \mathrm{d}u = H(v). \tag{9.13}$$

Using eqn. 9.10 to evaluate the integral this equation becomes

$$\sum_{n=1}^{\infty} a_n \pi U_{n-1}(v) = -H(v). \tag{9.14}$$

Note that a_0 is undetermined because

$$P\int_{-1}^{1} \frac{1}{\sqrt{1-u^2}} \frac{1}{(v-u)}\, \mathrm{d}u = 0, \tag{9.15}$$

as may be verified by contour integration. The constant a_0 is the solution of the homogeneous equation, where $H(v)$ is zero, and we shall see it is determined by the boundary conditions at $u = \pm 1$.

If we now multiply both sides of eqn. 9.14 by $\sqrt{1-v^2}/(v-u)$ and integrate over v from $v = -1$ to $v = 1$, making use of eqn. 9.11, we obtain

$$\pi^2 \sum_{n=1}^{\infty} a_n T_n(u) = P\int_{-1}^{1} \frac{H(v)\sqrt{1-v^2}}{v-u}\, \mathrm{d}v. \tag{9.16}$$

Using eqn. 9.12 and the definition $L(u) = \sqrt{1-u^2}\, F(u)$ we reach the solution of the integral equation:

$$F(u) = \frac{a_0}{\sqrt{1-u^2}} + \frac{1}{\pi^2} P\int_{-1}^{1} \frac{H(v)}{(v-u)} \sqrt{\frac{1-v^2}{1-u^2}}\, \mathrm{d}v. \tag{9.17}$$

Quite often we are confronted with a different range of integration for the integral equation, such as

$$P\int_a^b \frac{D(y)}{x-y}\,\mathrm{d}y = Q(x), \tag{9.18}$$

where $a \leq x \leq b$. But if we let $u = (2y-a-b)/(b-a)$ and $v = (2x-a-b)/(b-a)$ this integral equation becomes identical to eqn. 9.1, and the corresponding solution is readily obtained from eqn. 9.17:

$$D(y) = \frac{1}{\pi^2\sqrt{(y-a)(b-y)}}\left[C + P\int_a^b \frac{Q(x)}{x-y}\sqrt{(x-a)(b-x)}\,\mathrm{d}x\right], \tag{9.19}$$

where C is again an arbitrary constant determined by the boundary conditions on $D(y)$ at $y = a, b$. The solution in eqn. 9.19 diverges at $y = a, b$. If a solution is sought where $D(y)$ is bounded at $y = a$ then the term in square brackets in eqn. 9.19 is set to zero at $y = a$, which yields

$$C = -P\int_a^b \frac{Q(x)}{x-a}\sqrt{(x-a)(b-x)}\,\mathrm{d}x.$$

When this is substituted back into eqn. 9.19 we obtain

$$D(y) = \frac{1}{\pi^2}\sqrt{\frac{(y-a)}{(b-y)}}\,P\int_a^b Q(x)\sqrt{\frac{(b-x)}{(x-a)}}\,\frac{1}{(x-y)}\,\mathrm{d}x. \tag{9.20}$$

Similarly, a solution bounded at $y = b$ is as follows:

$$D(y) = \frac{1}{\pi^2}\sqrt{\frac{(b-y)}{(y-a)}}\,P\int_a^b Q(x)\sqrt{\frac{(x-a)}{(b-x)}}\,\frac{1}{(x-y)}\,\mathrm{d}x. \tag{9.21}$$

If this solution is also required to be bounded at $y = a$ then the integral in eqn. 9.21 has to be zero at $y = a$, or equivalently the integral in eqn. 9.20 has to be zero at $y = b$. Both are satisfied if the following relationship holds:

$$\int_a^b \frac{Q(x)}{\sqrt{(x-a)(b-x)}}\,dx = 0. \tag{9.22}$$

The solution of the integral equation then becomes

$$D(y) = \frac{1}{\pi^2}\sqrt{\frac{(b-y)}{(y-a)}}\left\{P\int_a^b Q(x)\sqrt{\frac{(x-a)}{(b-x)}}\,\frac{1}{(x-y)}\,dx - \int_a^b \frac{Q(x)}{\sqrt{(x-a)(b-x)}}\,dx\right\}$$

$$= \frac{1}{\pi^2}\sqrt{(y-a)(b-y)}\,P\int_a^b \frac{Q(x)}{\sqrt{(x-a)(b-x)}}\,\frac{1}{(x-y)}\,dx. \tag{9.23}$$

9.2.3 Some contour integrals

In this chapter we use contour integration extensively. There are some contour integrals that involve a careful analysis of the contribution from the closure of the contour at infinity. For example, consider the following familiar integral:

$$I = \int_{-a}^{a} \sqrt{a^2 - x^2}\,dx.$$

It is trivial to evaluate this integral by making the substitution $x = a\sin\theta$ or $x = a\cos\theta$ to obtain $I = \pi a^2/2$. Although there are no poles it may also be evaluated in the complex plane by considering the following contour integral:

$$J_1 = \oint_C \sqrt{z^2 - a^2}\,dz, \tag{9.24}$$

where the contour C is illustrated in Fig. 9.1. On the upper surface of the branch cut between $-a$ and $+a$ we find $\sqrt{z^2-a^2} = i\sqrt{a^2-x^2}$, and on the lower surface $\sqrt{z^2-a^2} = -i\sqrt{a^2-x^2}$. Therefore the contribution to the contour integral from the loop around the branch cut is $2iI$. The contributions from the small circles at $z = \pm a$ are zero. The contour is closed at infinity by a large circle $z = Re^{i\theta}$, where $0 \leq \theta \leq 2\pi$ and we take the limit[5] $R \to \infty$. The contribution from this large circle is evaluated as follows:

[5] The process of taking limits may be illustrated by the following example attributed to Konrad Zacharias Lorenz 1903–89, Nobel Prize-winning Austrian zoologist: Philosophers are people who know less and less about more and more, until they know nothing about everything. Scientists are people who know more and more about less and less, until they know everything about nothing.

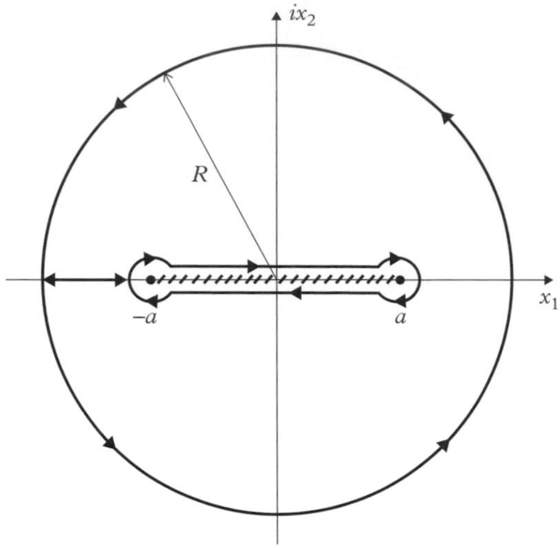

Figure 9.1 *Contour to evaluate the integral in eqn. 9.24. The limit $R \to \infty$ is taken.*

$$
\begin{aligned}
\lim_{R\to\infty} \int_0^{2\pi} \left(R^2 e^{2i\theta} - a^2\right)^{\frac{1}{2}} iRe^{i\theta}\,\mathrm{d}\theta &= \lim_{R\to\infty} \int_0^{2\pi} Re^{i\theta}\left(1 - \frac{a^2}{R^2 e^{2i\theta}}\right)^{\frac{1}{2}} iRe^{i\theta}\,\mathrm{d}\theta \\
&= \lim_{R\to\infty} \int_0^{2\pi} Re^{i\theta}\left(1 - \frac{a^2}{2R^2 e^{2i\theta}} + \ldots\right) iRe^{i\theta}\,\mathrm{d}\theta \\
&= -\pi i a^2.
\end{aligned}
$$

Since there are no poles inside the contour Cauchy's theorem tells us that $2iI - \pi i a^2 = 0$, and therefore $I = \pi a^2/2$, as before.

We will also encounter principal value integrals of the form

$$
I = P\int_{-a}^{a} \frac{\sqrt{a^2 - x^2}}{x - b}\,\mathrm{d}x, \quad -a \le b \le a.
$$

It may also be evaluated by considering a contour integral:

$$
J_2 = \oint_C \frac{\sqrt{z^2 - a^2}}{z - b}\,\mathrm{d}z, \tag{9.25}
$$

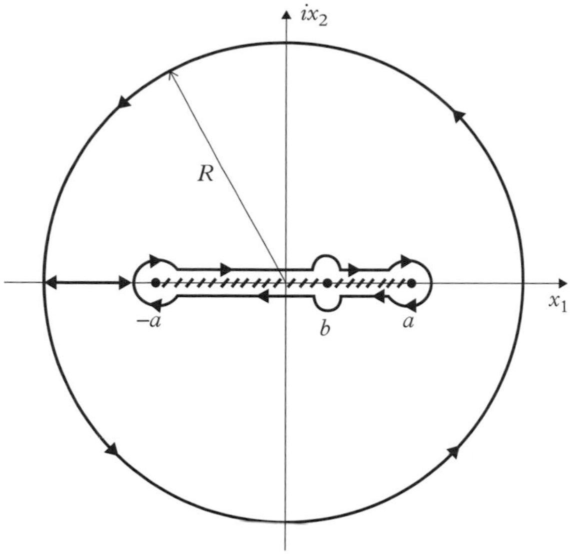

Figure 9.2 *Contour to evaluate the integral in eqn. 9.25. The limit $R \to \infty$ is taken.*

where C is the contour shown in Fig. 9.2. There is a simple pole on the branch cut at $z = b$. However, the contributions from the semi circles above and below the pole are equal and opposite because the square root changes sign on either side of the branch cut. Therefore the contribution from the loop around the branch cut is $2iI$, including the semicircles around the pole. The contour is again closed by a large circle of radius R, giving the following contribution:

$$\lim_{R\to\infty} \int_0^{2\pi} \frac{\left(R^2 e^{2i\theta} - a^2\right)^{\frac{1}{2}}}{\left(Re^{i\theta} - b\right)} iRe^{i\theta}\, \mathrm{d}\theta = \lim_{R\to\infty} \int_0^{2\pi} \frac{Re^{i\theta}\left(1 - \frac{a^2}{R^2 e^{2i\theta}}\right)^{\frac{1}{2}}}{Re^{i\theta}\left(1 - \frac{b}{Re^{i\theta}}\right)} iRe^{i\theta}\, \mathrm{d}\theta$$

$$= \lim_{R\to\infty} \int_0^{2\pi} \left(1 - \frac{a^2}{2R^2 e^{2i\theta}}\right)\left(1 + \frac{b}{Re^{i\theta}}\right) iRe^{i\theta}\, \mathrm{d}\theta$$

$$= 2\pi ib.$$

Since there are no poles inside the contour C Cauchy's theorem tells us that $2iI + 2\pi ib = 0$, or $I = -\pi b$.

Exercise 9.3

Prove the following integrals using contour integration:

$$\int_{-a}^{a} \frac{1}{\sqrt{a^2 - x^2}} \, \mathrm{d}x = \pi$$

$$P\int_{-a}^{a} \frac{1}{\sqrt{a^2 - x^2}} \frac{1}{x - b} \, \mathrm{d}x = 0, \quad -a \leq b \leq a$$

$$\int_{-a}^{a} \frac{1}{\sqrt{a^2 - x^2}} \frac{1}{x - b} \, \mathrm{d}x = -\frac{\pi}{\sqrt{b^2 - a^2}}, \quad b > a.$$

9.3 Representation of loaded cracks as distributions of dislocations

Consider a slit crack in the plane $x_2 = 0$ between $x_1 = a$ and $x_1 = b$, with $a < b$, and infinitely long along the x_3-axis. A slit crack is a rectangular cut in an infinite elastic continuum, which is infinitely long in one direction and finite in the perpendicular direction. The faces of the cut undergo a relative displacement in response to an applied load. Suppose a uniform tensile stress $\sigma_{22} = \sigma$ is applied to the medium at $x_2 = \pm\infty$. The faces of the crack separate slightly under the action of the applied stress. Let $s(x_1)$ denote the x_2-coordinate of the upper crack face at $a \leq x_1 \leq b$ minus the x_2-coordinate of the lower crack face at x_1, where $s(a) = s(b) = 0$. The function $s(x_1)$ is called the crack opening displacement.

The variation of $s(x_1)$ with x_1 may be modelled mathematically as the result of a continuous distribution of edge dislocations in $a \leq x_1 \leq b$, with their lines parallel to the x_3-axis and their Burgers vectors along the x_2-axis. Let the positive sense of the dislocation lines be along the positive direction of the x_3-axis. Let $f(x_1)\mathrm{d}x_1$ be the Burgers vector of dislocations between x_1 and $x_1 + \mathrm{d}x_1$. Then the relationship between the crack opening displacement and $f(x_1)$ is as follows:

$$s(x_1) = -\int_{a}^{x_1} f(x_1') \, \mathrm{d}x_1' \tag{9.26}$$

where the negative sign in front of the integral is consistent with the FS/RH convention and the definition of $s(x_1)$. The distribution of dislocations in $a \leq x_1 \leq b$ representing the

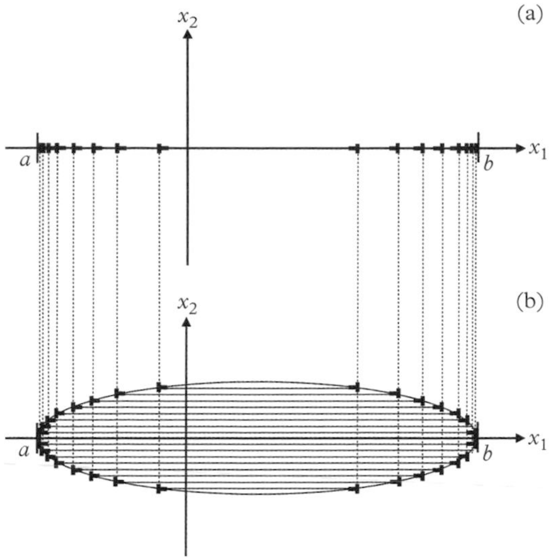

Figure 9.3 *To illustrate the representation of a slit crack under a normal tensile load by a distribution of dislocations. (a) A slit crack occupies $a \leq x_1 \leq b$ and is subjected to a tensile stress along x_2. The crack faces separate in (b) forming an ellipse. The elliptical crack faces may also be thought of as the interface surrounding an inclusion comprising a stack of interstitial loops shown in (b). When the locations of the edge dislocations bounding these interstitial loops are projected by the vertical broken lines onto the slit crack in (a) we obtain the representation of the loaded slit crack as a distribution of edge dislocations. As the spacing between the horizontal planes on which the interstitial loops lie becomes infinitesimal the distribution of dislocations representing the crack becomes continuous, and their Burgers vectors become infinitesimal. The normal tractions created on the crack faces by the dislocations cancel the tractions arising from the applied normal stress so that the crack faces are free of tractions.*

crack opening is illustrated in Fig. 9.3(a).[6] In Fig. 9.3(b) it is shown that the distribution in (a) may be viewed as a distribution of interstitial loops collapsed onto the plane $x_2 = 0$. The loops transform the slit crack into the ellipse shown in (b). Since these loops are interstitial in character they create a compressive stress field inside the ellipse which generates tractions on the crack faces that exactly cancel the tractions arising from the

[6] Historical note. There is a drawing similar to Fig. 9.3 on p.215 of the book *Les dislocations* by Friedel (Jacques Friedel ForMemRS 1921–2014, French materials physicist and President of the French Academy of Sciences 1992–94.), where he called the dislocations representing the crack *les dislocations de clivage*—cleavage dislocations. The first suggestion that cracks and slip bands may both be represented by arrays of dislocations appears to be due to Zener (Clarence Melvin Zener 1905–93, US physicist) in 1948 in *Fracturing of metals*, Symposium, American Society for Metals: Cleveland, OH, p.3. The close relationship between a crack and a pileup of dislocations was also discussed in Eshelby, JD, Frank, FC, and Nabarro, FRN, *Phil. Mag.* **42**, 351–64 (1951), https://doi.org/10.1080/14786445108561060. The mathematics for the replacement of arrays of discrete dislocations in pileups by continuous distributions of infinitesimal dislocations was developed by Zener, Leibfried (Leibfried, G, *Z. Phys.* **130**, 244 (1951) https://doi.org/10.1007/BF01337695) and by Head and Louat (Head, AK, and Louat, N, *Aust. J. Phys.* **8**, 1 (1955). https://doi.org/10.1071/PH550001). Alan Kenneth Head FRS 1925–2010, Australian physicist. Norman P Louat 1920–2010, Australian and US physicist.

applied load. *The requirement that there are no net tractions on the faces of the loaded slit crack is a boundary condition that must be satisfied by* $f(x_1)$.

So far we have discussed the loading of a slit crack with normal along x_2 by a normal stress σ_{22}. This is called mode I loading. The crack may also be loaded by a shear stress σ_{12} and this is called mode II loading. A slit crack loaded in mode II may be represented by a distribution of edge dislocations with Burgers vectors parallel to x_1. If the slit crack is loaded by a shear stress σ_{23} this is called mode III loading. A slit crack loaded in mode III may be represented by a distribution of screw dislocations along the x_3-axis. More complex loadings on a slit crack may be modelled by combining these dislocation representations.

9.4 Elastic field of a mode I slit crack and the stress intensity factor

In this section we will set up and solve the integral equation for the Burgers vector density $f(x_1)$ representing the mode I slit crack loaded by an applied normal tensile stress σ of the previous section. Far from the crack the stress field is just the applied normal stress. But this field is perturbed very significantly near the crack, and we shall see it becomes singular at the crack tips at $x_1 = a, b$. In this section we assume the response of the medium is purely elastic so that no plasticity takes place.

The applied normal stress σ creates tractions $-\sigma\hat{\mathbf{e}}_2$ on the crack faces. These tractions have to be cancelled by tractions created by the edge dislocations representing the crack. An edge dislocation at $(x'_1, 0)$ with Burgers vector $\mathbf{b} = [0, b, 0]$ creates a stress field at (x_1, x_2) with the following components:

$$\sigma_{11}(x_1,x_2) = \frac{\mu b}{2\pi(1-\nu)} \frac{(x_1 - x'_1)((x_1 - x'_1)^2 - x_2^2)}{((x_1 - x'_1)^2 + x_2^2)^2}$$

$$\sigma_{22}(x_1,x_2) = \frac{\mu b}{2\pi(1-\nu)} \frac{(x_1 - x'_1)(3x_2^2 + (x_1 - x'_1)^2)}{((x_1 - x'_1)^2 + x_2^2)^2}$$

$$\sigma_{12}(x_1,x_2) = \sigma_{21}(x_1,x_2) = \frac{\mu b}{2\pi(1-\nu)} \frac{x_2((x_1 - x'_1)^2 - x_2^2)}{((x_1 - x'_1)^2 + x_2^2)^2}$$

$$\sigma_{33}(x_1,x_2) = \nu(\sigma_{11}(x_1,x_2) + \sigma_{22}(x_1,x_2))$$

$$\sigma_{13}(x_1,x_2) = \sigma_{31}(x_1,x_2) = \sigma_{23}(x_1,x_2) = \sigma_{32}(x_1,x_2) = 0. \tag{9.27}$$

In the plane of the slit crack $x_2 = 0$ and the only component of the stress field of a dislocation at $(x'_1, 0)$ contributing to tractions on the crack faces is $\sigma_{22}(x_1, 0)$:

$$\sigma_{22}(x_1,0) = \frac{\mu b}{2\pi(1-\nu)} \frac{1}{(x_1 - x_1')}. \tag{9.28}$$

The total normal stress at $(x_1, 0)$ arising from the distribution of edge dislocations in the crack is

$$\Sigma_{22}(x_1,0) = \frac{\mu}{2\pi(1-\nu)} P \int_a^b \frac{f(x_1')}{(x_1 - x_1')} \, dx_1'. \tag{9.29}$$

The integral is a principal value integral because the edge dislocation at x_1 does not exert a stress upon itself. The requirement of zero tractions on the crack faces leads to the following integral equation for the Burgers vector density:

$$P \int_a^b \frac{f(x_1')}{(x_1 - x_1')} \, dx_1' + \frac{2\pi(1-\nu)}{\mu} \sigma = 0. \tag{9.30}$$

The solution to this equation follows from eqn. 9.19. The arbitrary constant C in eqn. 9.19 is determined by the condition that the total Burgers vector content of the distribution is zero:

$$f(x_1) = \frac{2(1-\nu)\sigma}{\mu} \frac{1}{\sqrt{(x_1 - a)(b - x_1)}} \left(x_1 - \left(\frac{a+b}{2} \right) \right). \tag{9.31}$$

This solution is plotted in Fig. 9.4.

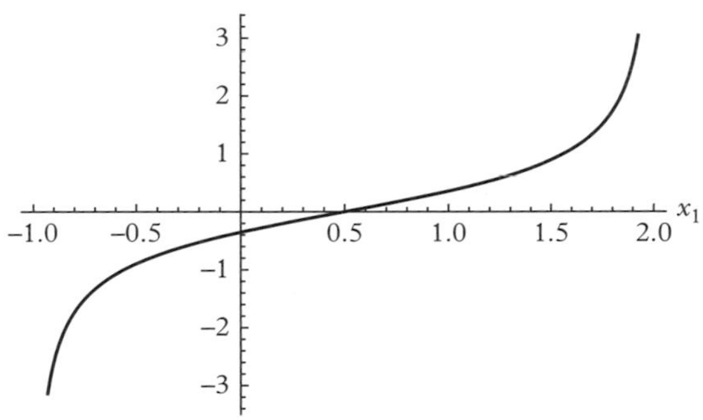

Figure 9.4 *Plot of the Burgers vector density for an elastic mode I crack, given by eqn. 9.31. In this example the crack tips are at $x_1 = -1$ and $x_1 = 2$, where the Burgers vector density diverges. The vertical axis is in units of $2(1-\nu)\sigma/\mu$.*

Exercise 9.4

Derive eqn. 9.31 from eqn. 9.19 by using contour integration to evaluate the integrals.[a]

Compare the continuous distribution $f(x_1)$ with the distribution of discrete dislocations sketched in Fig. 9.3.

Show that under the action of the applied stress the crack opening displacement $s(x)$ attains a maximum value of $(1-\nu)(\sigma/\mu)(b-a)$ in the middle of the crack, and that the crack shape is an ellipse with the following equation:

$$\left(x_1 - \left(\frac{a+b}{2}\right)\right)^2 + \left(\frac{\mu x_2}{(1-\nu)\sigma}\right)^2 = \left(\frac{b-a}{2}\right)^2. \tag{9.32}$$

[a] The following integral may be evaluated by contour integration paying attention to the contribution from the contour at infinity, as discussed in section 9.2.3:

$$P\int_a^b \frac{\sqrt{(x_1'-a)(b-x_1')}}{x_1'-x_1}\,dx_1' = -\pi(x_1-(a+b)/2).$$

Along the x_1-axis at $x_1 > b$ the total stress $\sigma_{22}^T(x_1,0)$ is given by:[7]

$$\begin{aligned}\sigma_{22}^T(x_1,0) &= \sigma + \frac{\sigma}{\pi}\int_a^b \frac{1}{(x_1-x_1')}\frac{x_1'-(a+b)/2}{\sqrt{(x_1'-a)(b-x_1')}}\,dx_1' \\ &= \sigma\left(\frac{x_1-(a+b)/2}{\sqrt{(x_1-a)(x_1-b)}}\right).\end{aligned} \tag{9.33}$$

The crack generates stress singularities at its tips. In eqn. 9.33 let $x_1 = b+\Delta$ where $\Delta \ll (b-a)$. We obtain

$$\sigma_{22}^T(b+\Delta,0) = \frac{\sigma}{2}\sqrt{\frac{(b-a)}{\Delta}} + O\left(\sqrt{\frac{\Delta}{(b-a)}}\right). \tag{9.34}$$

Thus, the applied stress field is intensified near the crack tips, displaying an inverse square root singularity. The *stress intensity factor* is defined by

$$K_I = \lim_{\Delta\to 0}\sqrt{\Delta}\,\sigma_{22}^T(b+\Delta,0) = \sigma\frac{\sqrt{(b-a)}}{2} = \sigma\sqrt{c/2}, \tag{9.35}$$

[7] Equation 9.33 was derived by Zener (1948), using the method followed here, but for a mode II slit crack.

where $2c = (b - a)$ is the crack length. The units of the stress intensity factor are $\mathrm{Pa\,m}^{1/2}$ or equivalently $\mathrm{Nm}^{-3/2}$.

Expressions for the stress field of the loaded crack throughout the medium may be obtained using the Burgers vector density in eqn. 9.31 and the stress components of an individual edge dislocation, eqn. 9.27. For example,

$$\sigma_{12}^{T}(x_1, x_2) = \frac{\sigma x_2}{\pi} \int_a^b \frac{((x_1 - x_1')^2 - x_2^2)}{((x_1 - x_1')^2 + x_2^2)^2} \frac{(x_1' - (a+b)/2)}{\sqrt{(x_1' - a)(b - x_1')}} \mathrm{d}x_1', \tag{9.36}$$

which can be evaluated by contour integration. See the problems at the end of this chapter.

9.5 Energy considerations and Griffith's fracture criterion

It is often said that when a crack grows the elastic energy stored in the body is reduced. Thus, one often sees in the literature the expression 'the elastic energy release rate', which is supposed to drive the growth of a crack. Following Bilby and Eshelby (1968) we will show that the elastic energy of the body *increases* as the crack grows at constant applied load,[8] but the *sum* of the elastic energy and the potential energy of the external loading mechanism decreases when the crack grows. It is this sum which drives crack growth and leads to Griffith's thermodynamic criterion for fracture. Since the elastic energy stored in the body is a form of potential energy we may call the rate of decrease of the sum of the elastic energy stored in the body and the potential energy of the external loading mechanism the *potential energy release rate* rather than the elastic energy release rate. We will show that the potential energy release rate is identical to what is commonly called the elastic energy release rate.

Consider a body subjected to constant forces on its external surface. The body contains a crack which can grow. Let the stress and displacement fields in the body when the crack has a length L be σ_{ij} and u_i. The elastic energy stored in the body is the usual expression:

$$\begin{aligned} E_{el} &= \frac{1}{2} \int_V \sigma_{ij} u_{i,j} \mathrm{d}V \\ &= \frac{1}{2} \int_{S_0} \sigma_{ij} u_i n_j \mathrm{d}S. \end{aligned} \tag{9.37}$$

[8] If you are unconvinced consider a catapult. For a given applied force the elastic energy stored in the catapult bands increases as the compliance of the bands increases. As a crack grows the body it is in becomes more deformable, that is, more compliant. Therefore, for the same forces applied to the surface of the body the elastic energy increases as the crack grows.

The integral over the volume of the body in the first line is transformed in the second line using the divergence theorem into a surface integral. The surface comprises the external surface $\mathcal{S}_0$ of the body and the faces of the crack. But the tractions on the surface of the crack are zero, and therefore the crack faces do not contribute to the surface integral, which may therefore be taken over the external surface only. The equilibrium condition $\sigma_{ij,j} = 0$ has been used in the second line.

When the crack grows by δL let the stress and displacement fields in the body become σ'_{ij} and u'_i. The elastic energy of the body changes by

$$\begin{aligned}\delta E_{el} &= \frac{1}{2}\int_{\mathcal{S}_0} \left(\sigma'_{ij}u'_i - \sigma_{ij}u_i\right) n_j \mathrm{d}S \\ &= \frac{1}{2}\int_{\mathcal{S}_0} \sigma_{ij}\left(u'_i - u_i\right) n_j \mathrm{d}S,\end{aligned} \tag{9.38}$$

where use has been made of $\sigma_{ij}n_j = \sigma'_{ij}n_j$ because the loading is constant. At the same time the potential energy of the external loading mechanism is reduced because it does work on the body:

$$\delta U_{ext} = -\int_{\mathcal{S}_0} \sigma_{ij}\left(u'_i - u_i\right) n_j \mathrm{d}S. \tag{9.39}$$

Therefore the sum δU of the changes in the elastic energy of the body and the potential energy of the external loading mechanism is as follows:

$$\delta U = \delta E_{el} + \delta U_{ext} = -\delta E_{el} = \frac{1}{2}\delta U_{ext} = -\frac{1}{2}\int_{\mathcal{S}_0} \sigma_{ij}\left(u'_i - u_i\right) n_j \mathrm{d}S. \tag{9.40}$$

Thus, the potential energy release rate is equal in magnitude to the change of the elastic energy but they have opposite signs. As anticipated at the beginning of this section the elastic energy of the body *increases* as the crack grows under a constant load.

To calculate the change in the total potential energy of the body and the loading mechanism when the crack is introduced we may extend the same argument. Let σ^A_{ij} and u^A_i be the stress and displacement fields in the loaded body before the crack is introduced. Let σ'_{ij} and u'_i be the stress and displacement fields in the body after the crack has been introduced, and where the surface of the body is subjected to the same surface tractions. The change in elastic energy of the body associated with the introduction of the crack is then

$$\begin{aligned}\Delta E_{el} &= \frac{1}{2}\int_{V} \sigma'_{ij}u'_{i,j} - \sigma^A_{ij}u^A_{i,j}\,\mathrm{d}V \\ &= \frac{1}{2}\int_{\mathcal{S}_0} \left(\sigma'_{ij}u'_i - \sigma^A_{ij}u^A_i\right) n_j \mathrm{d}S \\ &= \frac{1}{2}\int_{\mathcal{S}_0} \sigma^A_{ij}\left(u'_i - u^A_i\right) n_j \mathrm{d}S \\ &= \frac{1}{2}\int_{\mathcal{S}_0} \sigma'_{ij}\left(u'_i - u^A_i\right) n_j \mathrm{d}S. \end{aligned} \tag{9.41}$$

The surface $\mathcal{S}_0$ includes the faces of the crack. But in the last line $\sigma'_{ij}n_j = 0$ on the crack faces, so the contributions from the crack faces are zero. Therefore, $\mathcal{S}_0$ may be taken as the external surface of the body.

The change in the potential energy of the external loading mechanism following the introduction of the crack is

$$\Delta U_{ext} = -\int_{\mathcal{S}_0} \sigma^A_{ij}\left(u'_i - u^A_i\right) n_j \mathrm{d}S. \tag{9.42}$$

The change in the total potential energy following the introduction of the crack is

$$\Delta U = \Delta E_{el} + \Delta U_{ext} = -\frac{1}{2}\int_{\mathcal{S}_0} \sigma^A_{ij}\left(u'_i - u^A_i\right) n_j \mathrm{d}S = -\frac{1}{2}\int_{\mathcal{S}_0} \left(\sigma^A_{ij}u'_i - \sigma'_{ij}u^A_i\right) n_j \mathrm{d}S. \tag{9.43}$$

If there are no other defects present in the body the divergence of $\left(\sigma^A_{ij}u'_i - \sigma'_{ij}u^A_i\right)$ is zero throughout the body except at the crack. The surface $\mathcal{S}_0$ may therefore be deformed to a closed surface $\mathcal{S}$ just outside and infinitesimally close to the crack faces. Equation 9.43 then becomes

$$\Delta U = -\frac{1}{2}\int_{\mathcal{S}} \left(\sigma^A_{ij}u'_i - \sigma'_{ij}u^A_i\right) n_j \mathrm{d}S = -\frac{1}{2}\int_{\mathcal{S}} \sigma^A_{ij}u'_i n_j \mathrm{d}S, \tag{9.44}$$

where use has been made of $\sigma'_{ij}n_j = 0$ on the crack faces. Since u^A_i and σ^A_{ij} are continuous at the crack we obtain

$$\Delta U = -\frac{1}{2}\int_{\mathcal{S}} \sigma^A_{ij}u^C_i n_j \mathrm{d}S, \tag{9.45}$$

where u^C_i is the displacement of the crack faces due to the applied load.

When this equation is applied to the slit crack loaded in mode I of the previous section it becomes

$$\Delta U = -\frac{1}{2}\int_a^b \sigma\, s(x_1)\,\mathrm{d}x_1$$

$$= -\frac{1}{2}\frac{(1-\nu)\sigma^2}{\mu}\int_a^b \sqrt{(b-a)^2-(2x_1-(a+b))^2}\,\mathrm{d}x_1$$

$$= -\frac{\pi(1-\nu)\sigma^2 c^2}{2\mu}, \qquad (9.46)$$

where ΔU is the change of potential energy per unit length of the slit crack and σ is the applied normal stress far from the crack.

The total energy of the body per unit length of the slit crack is ΔU plus the energy of the crack surfaces. If γ is the energy per unit area of the crack surfaces then since there are two surfaces, each of length $2c$, the crack surface energy is $4\gamma c$. Griffith[9] argued[10] that for the crack to grow the total energy of the system must decrease with increasing crack length. The critical crack length c^*, above which the total energy of the system decreases with increasing crack length (and below which it increases with increasing crack length), is thus

$$c^* = \frac{4\gamma\mu}{\pi(1-\nu)\sigma^2}. \qquad (9.47)$$

This equation is equivalent to saying that the stress intensity factor, $K_I = \sigma\sqrt{c/2}$, has to reach a critical value K_{Ic} for the crack to grow:

$$K_{Ic} = \sqrt{\frac{2\gamma\mu}{\pi(1-\nu)}} = \sqrt{\frac{\gamma Y}{\pi(1-\nu^2)}}, \qquad (9.48)$$

where Y is the Young's modulus.

This is a very important result because K_{Ic} is a property of the material. For $\gamma \approx 1$ Jm^{-2}, $Y \approx 10^{11}$ Pa, $\nu \approx 1/3$, we find $K_{Ic} \approx 0.2$ MPam$^{1/2}$. The presence of the surface energy in K_{Ic} reflects the work that has to be done to break bonds if the crack is to grow. The presence of the elastic constants reflects the increase in the elastic energy of the entire system if the crack were to grow under the influence of a constant external load. It is evident that the energetics of crack growth involves a wide range of length scales from atomic bonds to the size of the component.

[9] Alan Arnold Griffith FRS 1893–1963, British engineer.

[10] Griffith, AA, *Phil. Trans. R. Soc. A* **221**, 163–98 (1921). https://doi.org/10.1098/rsta.1921.0006

The analysis of this section has explicitly excluded plasticity, and other forms of irreversibility such as the excitation of crystal lattice vibrations. Griffith's analysis applies to a crack in a perfectly elastic medium where the crack may grow or shrink quasistatically and hence reversibly. In practice there are very few cases where this applies. The stress singularity ahead of a crack in metallic systems and many other crystalline materials invariably leads to some dislocation generation even in so-called 'brittle' materials. The work of fracture in such materials is typically an order of magnitude larger than twice the surface energy, and in more ductile materials it may be several orders of magnitude larger. It is therefore essential to include some degree of plasticity in a more realistic model of a crack in a crystal. Orowan[11] proposed that Griffith's criterion can be generalised to include plasticity by reinterpreting the γ in eqn. 9.47 as the sum of the plastic work of fracture and the surface energy of the crack.

Although the plastic work of fracture usually vastly exceeds the surface energy of the crack faces, the surface energy remains crucially important. That is because the degree of plasticity is determined by the maximum stress that can be sustained by bonds at the crack tip, which is directly related to the surface energy.[12] A greater ease of breaking bonds at the crack tip is one mechanism by which crystalline materials may be embrittled by impurities along the crack path, for example in intergranular embrittlement where cracks propagate along grain boundaries containing segregated impurities.

The stress intensity factor for any particular crack in a specimen depends on the size and geometry of both the crack and the sample in which it is located. Calculating the stress intensity factors in different geometries is the field of linear elastic fracture mechanics. It has enabled standardised sample geometries to be introduced to measure critical stress intensity factors. Although the critical stress intensity factor is a material property it does depend on temperature and the strain rate of the loading mechanism. It also depends sensitively on the microstructure of the material, which is determined by the history of its thermal and mechanical treatment and its impurity content. As a material parameter the critical stress intensity factor is similar in this respect to the yield stress.

9.6 The interaction between a dislocation and a slit crack

The stresses created by defects such as dislocations may act as loads on cracks. The principal difference compared to the previous sections is that the loading on the crack faces is no longer uniform. But the boundary condition that there can be no tractions on the crack faces remains the same. Such interactions between other defects and cracks may decrease or increase the local loading on a crack by externally applied forces on the body. In this section we consider the case of an edge dislocation with Burgers vector $\mathbf{b} = [0, b, 0]$ with its line parallel to the x_3-axis at $(D, 0)$. It is placed ahead of a slit crack

[11] Orowan, E, *Trans. Inst. Eng. Shipbuilders, Scotland* **89**, 165 (1945). Unavailable online. See also Orowan, E, *Rep. Prog. Phys.* **12**, 185–232 (1949). https://doi.org/10.1088/0034-4885/12/1/309

[12] This insight was due to Jokl, ML, Vitek, V and McMahon Jr., CJ, *Acta Metall.* **28**, 1479–88 (1980). https://doi.org/10.1016/0001-6160(80)90048-6. Charles J McMahon Jr. 1933–, US metallurgist.

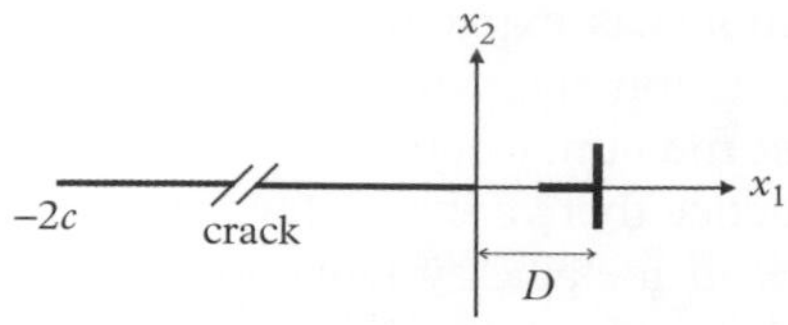

Figure 9.5 *To illustrate the geometry for the calculation of the Burgers vector density induced by an edge dislocation at a distance D ahead of a long crack of length 2c, where D ≪ 2c. The crack is not subjected to any loading apart from that created by the edge dislocation at D.*

with $a = -2c$ and $b = 0$, as shown in Fig. 9.5. No external forces are applied to the body, so that the only tractions on the crack faces are those arising from the dislocation at $(D, 0)$.

The traction created on the crack faces at x_1 by the dislocation at $(D, 0)$ arises from its stress component $\sigma_{22}^D(x_1, 0)$:

$$\sigma_{22}^D(x_1, 0) = \frac{\mu b}{2\pi(1-\nu)} \frac{1}{x_1 - D}. \tag{9.49}$$

The crack surfaces deform in such a way as to annihilate these tractions. The induced deformation may be modelled by a continuous distribution of dislocations in the crack with Burgers vector density $f(x_1')$. As before, $f(x_1')\mathrm{d}x_1'$ is the infinitesimal Burgers vector of induced edge dislocations between x_1' and $x_1' + \mathrm{d}x_1'$ with their lines parallel to the x_3-axis. The requirement that there are no resultant tractions on the crack faces then leads to the following Cauchy principal value integral equation:

$$\frac{\mu}{2\pi(1-\nu)} P\int_{-2c}^{0} \frac{f(x_1')}{x_1 - x_1'} \mathrm{d}x_1' = -\frac{\mu}{2\pi(1-\nu)} \frac{b}{x_1 - D}, \tag{9.50}$$

where $-2c \leq x_1 \leq 0$. Using eqn. 9.19 we can write down the solution:

$$f(x_1) = \frac{b}{\pi^2\sqrt{(x_1+2c)(-x_1)}} \left[C - P\int_{-2c}^{0} \frac{1}{x_1' - D} \frac{\sqrt{(x_1'+2c)(-x_1')}}{x_1' - x_1} \mathrm{d}x_1' \right]. \tag{9.51}$$

Evaluating the integral by contour integration we obtain

$$f(x_1) = \frac{b}{\pi\sqrt{(x_1+2c)(-x_1)}} \left[C' - \frac{\sqrt{D(D+2c)}}{D - x_1} \right], \tag{9.52}$$

where the constant C' is determined by the condition that the integral of $f(x_1)$ over the length of the crack must be zero. This condition follows from the conservation of the total

Burgers vector, and it yields $C' = 1$. The induced Burgers vector density is therefore as follows:

$$f(x_1) = \frac{b}{\pi\sqrt{(x_1+2c)(-x_1)}}\left[1 - \frac{\sqrt{D(D+2c)}}{D-x_1}\right]. \tag{9.53}$$

Exercise 9.5

Verify that the Burgers vector density of eqn. 9.53 satisfies eqn. 9.50.

9.6.1 Shielding and anti-shielding of cracks by dislocations

It is interesting to see how the stress field $\sigma_{22}(x_1, 0)$ of the dislocation is modified by the presence of the crack. For this purpose we evaluate $\sigma_{22}(x_1, 0)$ with $x_1 > 0$:

$$\sigma_{22}(x_1,0) = \frac{\mu b}{2\pi(1-\nu)}\frac{1}{(x_1-D)} + \frac{\mu}{2\pi(1-\nu)}\int_{-2c}^{0}\frac{f(x_1')}{(x_1-x_1')}\,\mathrm{d}x_1'. \tag{9.54}$$

Inserting the Burgers vector density in eqn. 9.53 into the integral we obtain

$$\sigma_{22}(x_1,0) = \frac{\mu b}{2\pi(1-\nu)}\frac{1}{\sqrt{x_1(x_1+2c)}}\left(\frac{\sqrt{D(2c+D)}}{(x_1-D)}+1\right). \tag{9.55}$$

This is plotted in Fig. 9.6.

The presence of the crack introduces the square root pre-factor into the stress field of the dislocation. This can be made clearer by considering the case where the crack length $2c$ is much larger than D and x_1. In that case:

$$\sigma_{22}(x_1,0) \approx \frac{\mu b}{2\pi(1-\nu)}\sqrt{\frac{D}{x_1}}\frac{1}{(x_1-D)}. \tag{9.56}$$

For x_1 close to D the stress field is very close to that of the isolated dislocation. But as x_1 approaches the crack tip at $x_1 = 0$ the stress field has an inverse square root singularity. The stress intensity factor K_I^D at the crack tip due to the dislocation is as follows:

$$K_I^D \approx \lim_{x_1\to 0}\sqrt{x_1}\frac{\mu b}{2\pi(1-\nu)}\sqrt{\frac{D}{x_1}}\frac{1}{(x_1-D)} \approx -\frac{b}{2\pi(1-\nu)D}\mu\sqrt{D}. \tag{9.57}$$

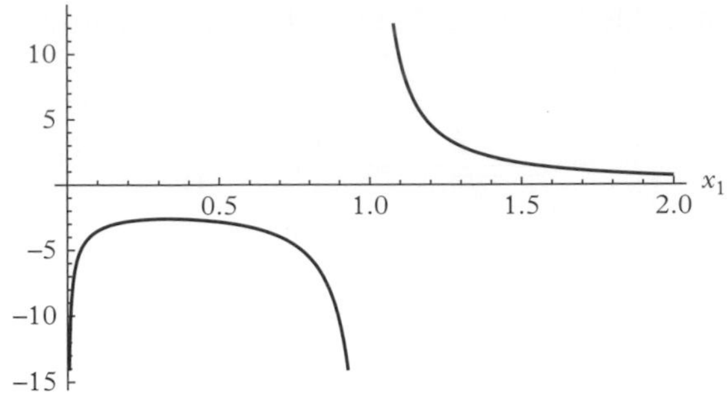

Figure 9.6 *Plot of the normal stress component $\sigma_{22}(x_1, 0)$, as given by eqn. 9.55, for an edge dislocation with Burgers vector $[0, b, 0]$ located at $x_1 = 1, x_2 = 0$ in front of an otherwise unloaded slit crack between $x_1 = -100$ and $x_1 = 0$ on $x_2 = 0$. The vertical axis is in units of $\mu b/[2\pi(1-\nu)]$. In addition to the normal $1/x_1$ divergence at the dislocation core there is a $1/\sqrt{x_1}$ divergence in the field near the crack tip.*

If the crack tip is subjected to an applied stress intensity factor K_I, dislocations with positive Burgers vectors will reduce K_I. Such dislocations are described as 'shielding' because they screen the crack tip from the stress field that is creating K_I. Dislocations with negative Burgers vectors are described as 'anti-shielding' because they increase the stress intensity factor K_I. The resultant stress intensity factor comprising the applied stress intensity factor and the intensity factors contributed by dislocations outside the crack is called the local stress intensity factor. The local stress intensity factor is the resultant stress intensity factor acting on the crack tip. At the other side of the crack, at $x_1 = -2c$, the Burgers vectors of shielding and anti-shielding dislocations are negative and positive respectively.

The induced Burgers vector density in eqn. 9.53 generates a Peach–Koehler force along the x_1-axis per unit length on the dislocation at $(D > 0, 0)$ given by

$$
\begin{aligned}
F_1^{ind}(D,0) &= \frac{\mu b}{2\pi(1-\nu)} \int_{-2c}^{0} \frac{f(x_1)}{(D-x_1)} \mathrm{d}x_1 \\
&= \frac{\mu b^2}{2\pi^2(1-\nu)} \int_{-2c}^{0} \frac{\mathrm{d}x_1}{(D-x_1)} \frac{1}{\sqrt{(-x_1)(x_1+2c)}} \left[1 - \frac{\sqrt{D(D+2c)}}{(D-x_1)}\right] \\
&= -\frac{\mu b^2}{2\pi(1-\nu)} \left[\frac{(D+c)}{D(D+2c)} - \frac{1}{\sqrt{D(D+2c)}}\right].
\end{aligned}
\tag{9.58}
$$

This is often called an image force and it always attracts the dislocation towards the crack tip. If $c \gg D$ then the image force is the force of attraction between two edge dislocations with Burgers vectors $[0,b,0]$ and $[0,-b,0]$ spaced $2D$ apart along the x_1-axis. As with image charges in a metal surface the 'image dislocation' has the opposite sign Burgers vector and it is located at D on the other side of the crack tip from the real dislocation.

Using eqn. 9.33 we can write down the stress $\sigma_{22}(D,0)$ arising from the applied normal stress σ and the distribution of dislocations in the crack required to annihilate the surface tractions on the crack faces:

$$\sigma_{22}(D,0) = \sigma \frac{(D+c)}{\sqrt{D(D+2c)}}. \tag{9.59}$$

Combining the last two equations, the total force per unit length acting on the edge dislocation at $(D>0,0)$ when there is an applied normal stress σ is therefore as follows:

$$F_1^T(D,0) = \left[\sigma \frac{(D+c)}{\sqrt{D(D+2c)}} - \frac{\mu b}{2\pi(1-\nu)}\left(\frac{(D+c)}{D(D+2c)} - \frac{1}{\sqrt{D(D+2c)}}\right)\right] b. \tag{9.60}$$

Thus, anti-shielding dislocations at $(D>0,0)$, with Burgers vectors $[0,-b,0]$, are attracted to the crack tip, whereas shielding dislocations, with Burgers vectors $[0,+b,0]$, move away from the crack tip under a sufficiently high applied stress σ to overcome the image force. It follows that dislocations emitted by the crack are shielding dislocations.[13] The square brackets contain the resultant normal stress acting on the edge dislocation at $(D,0)$. The dislocation will move only if this stress is greater in magnitude than the friction stress opposing dislocation motion.

Some comments are in order concerning the interpretation of the dislocation friction stress. Since we have focussed on mode I cracks the dislocations that formally make up the crack, and the real dislocation ahead of the crack, are edge dislocations with their Burgers vectors along x_2. Therefore if they are to move along x_1 they will do so by climb. But if the edge dislocation ahead of the crack had had a Burgers vector parallel to x_1 both the induced formal dislocations in the crack and the real dislocation would have been glide edge dislocations. Furthermore the integral equation defining the Burgers vector density of the induced glide dislocation density in the mode II crack would have been identical to the induced climb dislocation density in eqn. 9.50. The only difference would have beeen the involvement in the tractions on the crack faces of the stress component $\sigma_{12}(x_1 - x_1')$ for $\mathbf{b} = [b,0,0]$ rather than $\sigma_{22}(x_1 - x_1')$ for $\mathbf{b} = [0,b,0]$, but these stress components are mathematically identical. Therefore, from a mathematical point of view we can map the mode I loading of the slit crack involving climb dislocations onto a mode II loading of the slit crack involving glide dislocations. From a physical point of view we should acknowledge that the geometry of these one-dimensional models is a gross simplification of the reality of three-dimensional cracks interacting with dislocation loops. Therefore, it would not be appropriate to interpret the friction stress as anything other than the resistance to dislocation glide since that is how dislocations move in the

[13] This is analogous to the reduction in stress ahead of a pileup at a grain boundary by transmission of slip into the adjacent grain.

vicinity of a real crack, except possibly during creep rupture at elevated temperatures when climb may be involved.

Exercise 9.6

When $c \gg D$ in eqn. 9.60 show that the force $F_1^T(D,0)$ becomes

$$F_1^T(D,0) = \left(\frac{K_I}{\sqrt{D}} - \frac{\mu b}{4\pi(1-\nu)D}\right)b, \tag{9.61}$$

where $K_I = \sigma\sqrt{c/2}$ is the stress intensity factor of the bare, elastic crack. For a shielding dislocation b is positive, and for an anti-shielding dislocation b is negative.

9.7 Dugdale–Bilby–Cottrell–Swinden (DBCS) model

This model[14] extends the representation of an elastic slit crack as an array of dislocations to include regions of plasticity adjacent to the crack tips. These regions are the plastic zones and they are also represented by dislocation arrays. In contrast to the formal representation of a loaded crack as a continuous distribution of dislocations, the dislocations in the plastic zone are not formal but real objects, although they are treated mathematically in the same way as the dislocations representing the loaded crack. The idea is that the singularities in the elastic solution at the crack tips generate stresses that exceed the elastic limit and lead to plasticity in the plastic zones. In the absence of work hardening the stress in the plastic zones is assumed to be a constant, which is a friction stress that is equated to the yield stress, σ_1. As a result of the plastic zones the stress singularities at the crack tips are eliminated and the local stress intensity factors are zero. This section is based on the treatment by Lardner.[15]

Consider a slit crack loaded in mode I, as shown in Fig. 9.7. The crack is between $x_1 = -c$ and $x_1 = c$ on the $x_2 = 0$ plane and extends from $x_3 = -\infty$ to $x_3 = +\infty$. There are plastic zones between $x_1 = -a$ and $x_1 = -c$ and between $x_1 = c$ and $x_1 = a$. We shall find that the size of each plastic zone, $a - c$, is determined by the crack length $2c$, the yield stress σ_1 and the applied load normal stress σ at $x_2 = \pm\infty$.

The crack and the plastic zones are represented by a continuous distribution of edge dislocations along the x_1-axis with Burgers vectors along $\pm x_2$ and their positive line directions along the positive x_3-axis. As before, $f(x_1)\mathrm{d}x_1$ is the Burgers vector of dislocations between x_1 and $x_1 + \mathrm{d}x_1$. There are two boundary conditions to be satisfied:

[14] Dugdale, DS, *J. Mech. Phys. Solids* **8**, 100–4 (1960). https://doi.org/10.1016/0022-5096(60)90013-2. David S Dugdale, British engineer. Bilby, BA, Cottrell AH and Swinden KH, *Proc. R. Soc. A* **272**, 304–14 (1963). https://doi.org/10.1098/rspa.1963.0055. Sir Alan Howard Cottrell FRS 1919–2012, British metallurgist.

[15] Lardner, RW, *Mathematical theory of dislocations and fracture*, University of Toronto Press: Chapter 5 (1974), ISBN 0-8020-5277-0. Robin W Lardner, Canadian applied mathematician.

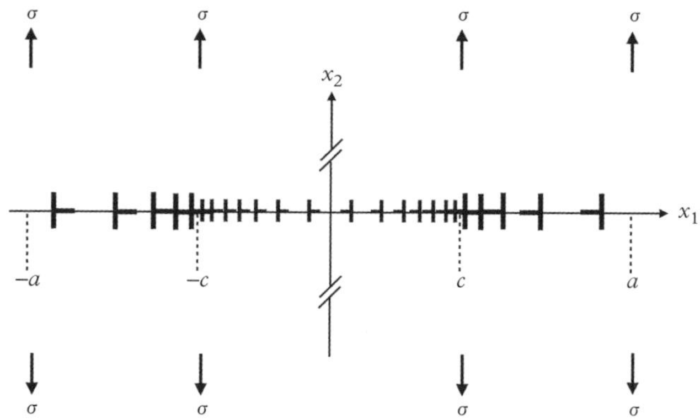

Figure 9.7 *To illustrate the geometry of the Dugdale–Bilby–Cottrell–Swinden (DBCS) model.*

$$\frac{\mu}{2\pi(1-\nu)} P\int_{-a}^{a} \frac{f(x_1')}{(x_1 - x_1')} dx_1' + \sigma = \begin{cases} 0 & \text{in } |x_1| < c, \\ \sigma_1 & \text{in } c < |x_1| < a. \end{cases} \tag{9.62}$$

At the ends of the plastic zones where $x_1 = \pm a$ the distribution $f(x_1) \to 0$ because there are no barriers for the dislocations to pile up against and eqns. 9.22 and 9.23 apply. The solution to the integral equation is then as follows:

$$f(x_1) = -\frac{1}{\pi^2}\frac{2\pi(1-\nu)}{\mu}\sqrt{a^2 - x_1^2}\left[\sigma P\int_{-c}^{c} \frac{1}{\sqrt{a^2 - x_1'^2}}\frac{dx_1'}{x_1' - x_1}\right.$$
$$\left. + (\sigma - \sigma_1)P\int_{-a}^{-c} \frac{1}{\sqrt{a^2 - x_1'^2}}\frac{dx_1'}{x_1' - x_1} + (\sigma - \sigma_1)P\int_{c}^{a} \frac{1}{\sqrt{a^2 - x_1'^2}}\frac{dx_1'}{x_1' - x_1}\right]. \tag{9.63}$$

Since

$$P\int_{-a}^{a} \frac{1}{\sqrt{a^2 - x_1'^2}}\frac{dx_1'}{x_1' - x_1} = 0, \tag{9.64}$$

eqn. 9.63 may be rewritten as follows:

$$f(x_1) = -\frac{\sigma_1}{\pi^2}\frac{2\pi(1-\nu)}{\mu}\sqrt{a^2 - x_1^2}\, P\int_{-c}^{c} \frac{1}{\sqrt{a^2 - x_1'^2}}\frac{dx_1'}{x_1' - x_1}. \tag{9.65}$$

The integral may be evaluated using the following indefinite integral, valid for $|x_1| < a$:

$$\int \frac{1}{\sqrt{a^2 - x_1'^2}} \frac{\mathrm{d}x_1'}{x_1' - x_1} = -\frac{1}{\sqrt{a^2 - x_1^2}} \ln\left\{\frac{\sqrt{(a-x_1)(a+x_1')} + \sqrt{(a+x_1)(a-x_1')}}{\sqrt{(a-x_1)(a+x_1')} - \sqrt{(a+x_1)(a-x_1')}}\right\}. \tag{9.66}$$

Thus, we obtain the solution:

$$f(x_1) = \frac{2(1-\nu)}{\pi} \frac{\sigma_1}{\mu} \ln\left|\frac{x_1\sqrt{a^2 - c^2} + c\sqrt{a^2 - x_1^2}}{x_1\sqrt{a^2 - c^2} - c\sqrt{a^2 - x_1^2}}\right|. \tag{9.67}$$

This solution is plotted in Fig. 9.8. The discontinuity in the stresses at the crack tips gives rise to logarithmic singularities at $|x_1| = c$.

The condition, eqn. 9.22, for this solution in eqn. 9.67 to exist is as follows:

$$\int_{-a}^{-c} \frac{(\sigma - \sigma_1)}{\sqrt{a^2 - x_1'^2}} \mathrm{d}x_1' + \int_{-c}^{c} \frac{\sigma}{\sqrt{a^2 - x_1'^2}} \mathrm{d}x_1' + \int_{c}^{a} \frac{(\sigma - \sigma_1)}{\sqrt{a^2 - x_1'^2}} \mathrm{d}x_1' = 0. \tag{9.68}$$

These integrals lead to the following expression for the size, $a - c$, of each plastic zone:

$$a - c = c\left(\sec\left(\frac{\pi\sigma}{2\sigma_1}\right) - 1\right), \text{ or } c/a = \cos(\pi\sigma/2\sigma_1). \tag{9.69}$$

Thus, as the yield stress σ_1 increases the size of the plastic zone decreases; when the plastic zone is small compared to the crack this condition is known as small scale yielding.

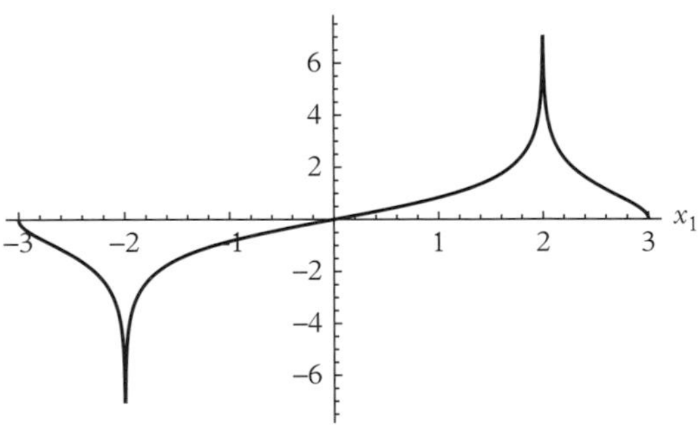

Figure 9.8 *Plot of the Burgers vector density $f(x_1)$ for the DBCS model, given by eqn. 9.67. In this example the half-length of the crack $c = 2$, and the plastic zones have length 1, so that $a = 3$. The vertical axis is in units of $2(1-\nu)\sigma_1/(\pi\mu)$. Note the logarithmic divergences at the crack tips at $x_1 = \pm 2$.*

As $\sigma \to \sigma_1$ the plastic zone becomes much larger than the crack, a condition known as general yielding.

As a result of the plastic zones the crack opening displacements are increased. This may be understood physically as a result of the dislocations in each plastic zone having been emitted from the adjoining crack tip, thereby opening the crack. From eqn. 9.26 the crack tip opening displacement at $x_1 = c$ is

$$s(c) = -\int_a^c f(x_1)\mathrm{d}x_1. \tag{9.70}$$

The Burgers vector density given by eqn. 9.67 may be integrated by making the change of variable $x_1 = a\cos\theta$. The integral in eqn. 9.70 then becomes

$$s(c) = -\frac{2(1-\nu)}{\pi}\frac{\sigma_1}{\mu}a\int_0^{\phi} \ln\left|\frac{\sin(\phi+\theta)}{\sin(\phi-\theta)}\right| \sin\theta\,\mathrm{d}\theta, \tag{9.71}$$

where $\cos\phi = c/a$, and $\phi = \pi\sigma/(2\sigma_1)$. After integrating by parts the following crack opening displacement is obtained:[16]

$$\begin{aligned} s(c) &= -\frac{2(1-\nu)}{\pi}\frac{\sigma_1}{\mu}a\left[\cos\theta\ln\left|\frac{\sin(\phi+\theta)}{\sin(\phi-\theta)}\right| + \cos\phi\ln\left|\frac{\sin\theta-\sin\phi}{\sin\theta+\sin\phi}\right|\right]_{\theta=0}^{\theta=\phi} \\ &= \frac{2(1-\nu)}{\pi}\frac{\sigma_1}{\mu}c\ln(\sec\phi) = \frac{2(1-\nu)}{\pi}\frac{\sigma_1}{\mu}c\ln(a/c). \end{aligned} \tag{9.72}$$

The physical significance of the crack tip opening displacement is that it is often observed that fracture in a material where plastic deformation can occur is preceded by a critical amount of plastic deformation at the crack tip. The crack tip opening displacement then has to reach a critical value for fracture to occur. This introduces a dependence of whether fracture will occur on the size of the sample: if it is less than the size of the plastic zone then fracture will occur only when yielding has occurred throughout the sample, which is the general yielding condition. But if the sample is much larger than the size of the plastic zone, fracture can occur at a smaller stress once the plastic zone has reached its critical size, which amounts to a critical value of the crack tip opening displacement. Understanding this size dependence of the fracture criterion in an elastic-plastic material was the initial motivation for the seminal paper by Bilby, Cottrell and Swinden (1963). They argued σ_1 depends on the size of the plastic zone. If the plastic zone is contained within one grain it may be identified with the friction stress opposing dislocation motion within that grain. That might consist of the Peierls stress and the friction stress created by impurities in solution or as precipitates. But when the plastic

[16] We note this is a factor of 2 less than the value given by Lardner (1974) in his equation 5.55, p.165.

zone is larger than a single grain plasticity has to propagate across grain boundaries, and that increases the value of σ_1 because it involves renucleating plasticity in each grain.

9.8 The dislocation free zone model

The assumption of the DBCS model that the plastic zone extends from the crack tip eliminates the elastic stress singularity at the crack tip. As a result the local stress intensity factor is zero. Furthermore, the stress acting at the crack tip in their model is σ_1, which is typically of the order of 10–100 MPa. A stress of 100 MPa creates forces between atoms of order 0.01 eVÅ^{-1}. As Thomson[17] points out on p.88 of his review article,[18] this is far too small to break bonds at the crack tip, which requires forces of order 0.1–1 eVÅ^{-1}. The shielding of the crack tip by the plastic zone in the DBCS model is too complete to enable the crack to propagate by breaking bonds at the crack tip. Instead the plastic zone increases in size as the applied stress σ increases, and the crack tip opening displacement increases indefinitely. The DBCS model always predicts ductile fracture.

The plastic zone is made up of shielding dislocations which reduce the local stress intensity factor. In section 9.6.1 we found that, at a sufficiently high applied stress to overcome the attraction of the image force, shielding dislocations are repelled from the crack tip. Anti-shielding dislocations are always attracted towards the crack where they can be absorbed, creating steps on the crack surfaces. If shielding dislocations were emitted from the crack tip at a sufficiently high applied stress to escape the attraction to the crack they will move from the crack tip until they are stopped by the friction stress σ_1. Alternatively, both shielding and anti-shielding dislocations may be generated from sources near the crack tip: anti-shielding dislocations run into the crack, while shielding dislocations move away from the crack and establish the plastic zone. These scenarios raise the possibility of a dislocation free zone between the crack tip and the plastic zone. Inside the dislocation free zone shielding dislocations are repelled into the plastic zone and anti-shielding dislocations are attracted into the crack.

If the plastic zone is located further from the crack tip the shielding of the crack tip decreases and the local stress intensity factor rises. Also, as the size of the dislocation free zone increases the size of the plastic zone decreases because the stress field of the crack decreases with distance from its tip. Eventually, shielding by the plastic zone diminishes to such an extent that we recover the elastic limit and the local stress intensity factor is that of the bare elastic crack, $\sigma\sqrt{c/2}$ (see eqn. 9.35). This describes the completely brittle limit. In the DBCS limit there is no dislocation free zone and the local stress intensity factor is zero. This describes the completely ductile limit. As the size of the dislocation free zone varies between these limiting cases we span the range of behaviour from purely brittle to purely ductile.

17 Robb Milton Thomson 1925–, US materials physicist.

18 Thomson, R, *Solid State Phys.* **39**, 2–129 (1986). https://doi.org/10.1016/S0081-1947(08)60368-9

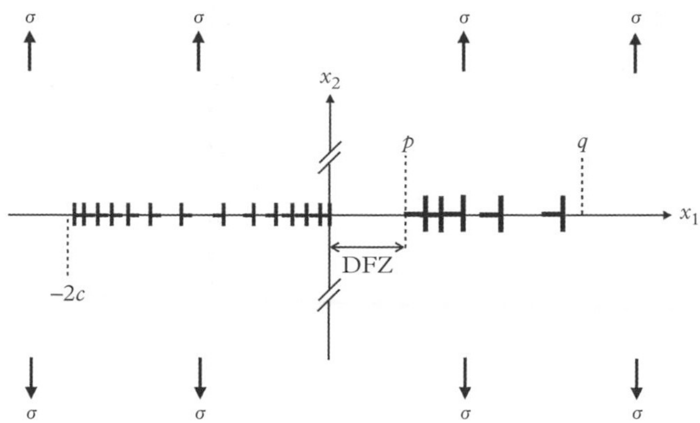

Figure 9.9 *To illustrate the geometry of the dislocation free zone model.*

Experimental evidence for the existence of dislocation free zones at cracks has been published.[19] Chang and Ohr[20] extended the DBCS model to include a dislocation free zone. Here we consider the force balance on dislocations in a plastic zone separated from the crack by a dislocation free zone. This section is based on the paper[21] by Majumdar[22] and Burns[23] (1983), but for a mode I crack rather than a mode III crack.

As before we consider a slit crack in the $x_2 = 0$ plane between $x_1 = -2c$ and $x_1 = 0$. It is loaded by a tensile stress $\sigma_{22} = \sigma$ at $x_2 = \pm\infty$. There is a plastic zone between $x_1 = p$ and $x_1 = q$, where $q > p > 0$. There is a dislocation free zone between $x_1 = 0$ and $x_1 = p$. The crack length $2c$ is assumed to be much larger than p and q, which amounts to the assumption of small scale yielding. This set-up is illustrated in Fig. 9.9. The friction stress, which we again equate to the yield stress, is σ_1.

Let $f(x_1)$ be the continuous distribution of Burgers vector density in the plastic zone. The infinitesimal Burgers vector between x_1 and $x_1 + \mathrm{d}x_1$, where $p \leq x_1 < q$, is $\mathrm{d}b(x_1) = f(x_1)\mathrm{d}x_1$. Equation 9.61 gives the force on the dislocation at x_1 arising from the loaded crack and from the image interaction. This force is

$$\mathrm{d}F(x_1) = \frac{K_I \mathrm{d}b(x_1)}{\sqrt{x_1}} - \frac{\mu\, \mathrm{d}b(x_1)\, \mathrm{d}b(x_1)}{4\pi(1-\nu)x_1}. \tag{9.73}$$

If this dislocation were the only dislocation in the plastic zone this would be the total force acting on it in the presence of the crack, apart from the friction force. But it also experiences a force arising from the stresses σ_{22} generated in the presence of the crack

19 Horton JA and Ohr SM, *J. Mater. Sci.* **17**, 3140–8 (1982). https://doi.org/10.1007/BF01203476; Chia, KY and Burns, SJ, *Scripta Metall.* **18**, 467–72 (1984). https://doi.org/10.1016/0036-9748(84)90423-X

20 Chang, S-J and Ohr, SM, *J. Appl. Phys.* **52**, 7174–81 (1981). http://dx.doi.org/10.1063/1.328692; Chang, S-J and Ohr, SM, *Int. J. Fract.* **23**, R3–R6 (1983). https://doi.org/10.1007/BF00020160

21 Majumdar, BS and Burns, SJ, *Int. J. Fract.* **21**, 229–40 (1983). https://doi.org/10.1007/BF00963390

22 Bhaskar S Majumdar, Indian and US engineer.

23 Stephen J Burns, US engineer.

by other dislocations in the plastic zone. These stresses are given by eqn. 9.56. Thus the force balance on the dislocation at x_1 becomes the following:

$$\frac{K_I \mathrm{d}b(x_1)}{\sqrt{x_1}} - \frac{\mu \mathrm{d}b(x_1)\mathrm{d}b(x_1)}{4\pi(1-\nu)x_1} + \mathrm{d}b(x_1)P\int_{x_1'=p}^{q} \frac{\mu}{2\pi(1-\nu)} \frac{f(x_1')}{x_1 - x_1'} \sqrt{\frac{x_1'}{x_1}}\mathrm{d}x_1' = \sigma_1 \mathrm{d}b(x_1). \tag{9.74}$$

Since the image force involves the product of two infinitesimal quantities it may be neglected. Thus we arrive at the following integral equation:

$$\int_p^q \frac{g(x_1')}{x_1 - x_1'}\mathrm{d}x_1' = \frac{2\pi(1-\nu)}{\mu}\left(\sigma_1\sqrt{x_1} - K_I\right), \tag{9.75}$$

where $g(x_1) = f(x_1)\sqrt{x_1}$. Since the distribution must be zero at $x_1 = p, q$, the solution of this integral equation is given by eqn. 9.23, provided the condition embodied in eqn. 9.22 is satisfied. Thus,

$$f(x_1) = \frac{2(1-\nu)}{\pi\mu}\sqrt{\frac{(x_1-p)(q-x_1)}{x_1}}\,P\int_p^q \frac{(\sigma_1\sqrt{x_1'} - K_I)}{\sqrt{(x_1'-p)(q-x_1')}}\frac{1}{(x_1'-x_1)}\mathrm{d}x_1', \tag{9.76}$$

provided the following condition is satisfied:

$$\int_p^q \frac{(K_I - \sigma_1\sqrt{x_1})}{\sqrt{(x_1-p)(q-x_1)}}\mathrm{d}x_1 = 0. \tag{9.77}$$

This condition leads to the following relationship:

$$\frac{\pi K_I}{2\sigma_1\sqrt{q}} = E\left(\frac{q-p}{q}\right), \tag{9.78}$$

where $E(z)$ is the complete elliptic integral of the second kind:

$$E(z) = \int_0^{\pi/2} \sqrt{\left(1 - z\sin^2\theta\right)}\,\mathrm{d}\theta. \tag{9.79}$$

For given values of K_I and σ_1 we can use eqn. 9.78 to show that as the plastic zone is located further from the crack, the width of the plastic zone decreases. The argument is as follows. When $p = 0$ the plastic zone begins at the crack tip, which is the DBCS limit. Then $(q - p)/q = 1$ and $E(1) = 1$. When $p = q$ the plastic zone vanishes and we have the elastic limit. In that case $(q - p)/q = 0$ and $E(0) = \pi/2$. Let $q = q_{BCS}$ in the DBCS limit, and $q = q_{el}$ in the elastic limit. Using eqn. 9.78 we find $q_{el} = (4/\pi^2)q_{BCS}$. Thus, as p increases from zero to q_{el}, q decreases monotonically from q_{BCS} to $(4/\pi^2)q_{BCS}$, and the width of the plastic zone decreases monotonically from q_{BCS} to zero. From eqn. 9.78 the width, q_{BCS}, of the plastic zone in the DBCS limit is $\pi^2\sigma^2 c/(8\sigma_1^2)$. This agrees with eqn. 9.69 in the limit $\sigma \ll \sigma_1$, which is the appropriate limit when the crack is much longer than the plastic zone.

The integral in eqn. 9.76 may be expressed in terms of elliptic integrals:

$$f(x_1) = \frac{4(1-\nu)}{\pi}\frac{\sigma_1}{\mu}\sqrt{\frac{(x_1-p)}{(q-x_1)}}\sqrt{\frac{q}{x_1}}\,P\int_0^{\pi/2}\frac{1-\frac{q-p}{q}\sin^2\theta}{\left(1-\frac{q-p}{q-x_1}\sin^2\theta\right)\sqrt{1-\frac{q-p}{q}\sin^2\theta}}\,d\theta$$

$$= \frac{4(1-\nu)}{\pi}\frac{\sigma_1}{\mu}\sqrt{\frac{(x_1-p)}{(q-x_1)}}\sqrt{\frac{q}{x_1}}\left\{\frac{x_1}{q}\Pi\left(\frac{q-p}{q-x_1},\frac{q-p}{q}\right)+\left(1-\frac{x_1}{q}\right)K\left(\frac{q-p}{q}\right)\right\}, \tag{9.80}$$

where

$$\Pi(n,z) = \int_0^{\pi/2}\frac{d\theta}{\left(1-n\sin^2\theta\right)\sqrt{1-z\sin^2\theta}} \tag{9.81}$$

is the complete elliptic integral of the third kind, and

$$K(z) = \int_0^{\pi/2}\frac{d\theta}{\sqrt{1-z\sin^2\theta}} \tag{9.82}$$

is the complete elliptic integral of the first kind, not to be confused with K_I the applied stress intensity factor.

Exercise 9.7

Derive eqn. 9.80 from eqn. 9.76.

In the previous section we found that in the DBCS model the local stress intensity factor at the crack tip is zero—the crack tip is completely screened by the plastic zone. We may use eqn. 9.57 and eqn. 9.80 to calculate the local stress intensity factor in the dislocation free zone model:

$$
\begin{aligned}
k_I^{DFZ} &= K_I - \frac{\mu}{2\pi(1-\nu)} \int_p^q \frac{f(x_1)}{\sqrt{x_1}}\, dx_1 \\
&= K_I - \frac{2\sigma_1}{\pi^2} \int_0^{\pi/2} d\theta \sqrt{q-(q-p)\sin^2\theta}\; P\!\int_p^q dx_1 \frac{\sqrt{(x_1-p)(q-x_1)}}{x_1(q-(q-p)\sin^2\theta - x_1)} \\
&= K_I - \frac{2\sigma_1}{\pi} \int_0^{\pi/2} d\theta \sqrt{q-(q-p)\sin^2\theta} \left(1 - \frac{\sqrt{pq}}{q-(q-p)\sin^2\theta}\right) \\
&= K_I - \frac{2\sigma_1\sqrt{q}}{\pi}\left(E\left(\frac{q-p}{q}\right) - \sqrt{p/q}\, K\left(\frac{q-p}{q}\right)\right).
\end{aligned}
\tag{9.83}
$$

If we now use eqn. 9.78 the local stress intensity factor may be rewritten more succinctly as follows:

$$
k_I^{DFZ} = \frac{\sqrt{p/q}\, K(1-(p/q))}{E(1-(p/q))} K_I. \tag{9.84}
$$

When $p = 0$ we recover the DBCS limit and eqn. 9.84 confirms the local stress intensity factor on the crack tip is zero. When $p = q$ the plastic zone vanishes and we recover the elastic limit where eqn. 9.84 confirms the local stress intensity factor is the applied stress intensity factor, K_I.

Figure 9.10 shows a plot of k_I^{DFZ}/K_I against p/q, obtained using eqn. 9.84. It is seen that k_I^{DFZ}/K_I increases very rapidly for small values of p/q and approaches unity asymptotically. The asymptotic expansion for the elliptic function of the first kind near $z = 1$ is as follows:[24]

$$
K(1-\varepsilon) \to \frac{1}{2}\ln\left(\frac{16}{\varepsilon}\right) + O(\varepsilon \ln \varepsilon), \tag{9.85}
$$

where $0 < \varepsilon \ll 1$. Using this expansion, eqns. 9.78 and 9.84 and $E(z) \to 1$ as $z \to 1$ we obtain the following relations for small values of p/q:

[24] http://functions.wolfram.com/EllipticIntegrals/EllipticK/introductions/CompleteEllipticIntegrals/05/

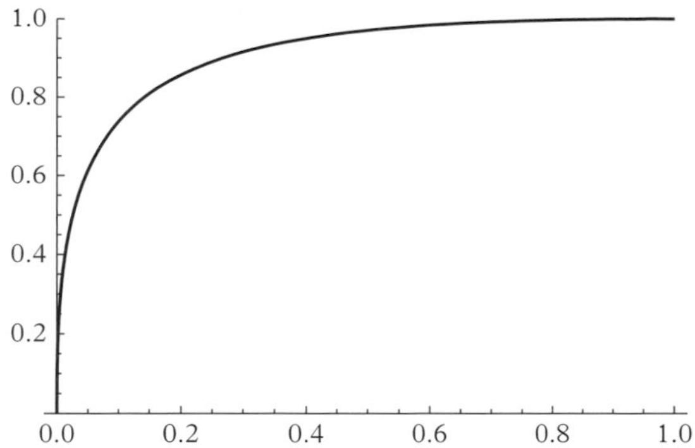

Figure 9.10 *A plot of eqn. 9.84. The vertical axis is* k_I^{DFZ}/K_I. *The horizontal axis is* p/q.

$$K_I = \frac{2\sigma_1\sqrt{q}}{\pi} \tag{9.86}$$

$$k_I^{DFZ} = \frac{2\sigma_1}{\pi}\sqrt{p}\left(\ln 4 - \frac{1}{2}\ln\frac{p}{q}\right) \tag{9.87}$$

$$\frac{k_I^{DFZ}}{K_I} = \sqrt{\frac{p}{q}}\left(\ln 4 - \frac{1}{2}\ln\frac{p}{q}\right). \tag{9.88}$$

Equation 9.86 is an equation for q (which is approximately the width of the plastic zone when $p/q \ll 1$) in terms of the applied stress intensity factor K_I and the yield stress σ_1. Equation 9.87 expresses the local stress intensity factor of the crack tip in terms of the width p of the dislocation free zone, the approximate width of the plastic zone q, and the yield stress σ_1. It is seen in this equation, and graphically in Fig. 9.10 at small values of p/q, that small changes in the width p of the dislocation free zone have a more marked influence on the local stress intensity factor than changes in the width of the plastic zone. Equations similar to eqns. 9.86 to 9.88 were published by Weertman et al. in 1983.[25] The accuracy of the approximation in eqn. 9.88 may be gauged by plotting it and the exact result, eqn. 9.84, on the same graph at small values of p/q. This is done in Fig. 9.11 where it is seen that the approximation underestimates the screening by the plastic zone. Nevertheless it is reasonably accurate up to $p/q \approx 0.1$, which spans the range $0 \le k_I^{DFZ}/K_I \lesssim 0.7$.

Both the DBCS and dislocation free zone models assume dislocations are available and mobile to populate the plastic zone. One possibility is that shielding dislocations are emitted from the crack tip, under the influence of the high stresses present there. Another is that sources operate near the crack, also driven by the stress field near the

[25] Weertman, J, Lin, I-H and Thomson, R, *Acta Metall.* **31**, 473–82 (1983). https://doi.org/10.1016/0001-6160(83)90035-4. Johannes Weertman 1925–2018, US materials scientist and geophysicist.

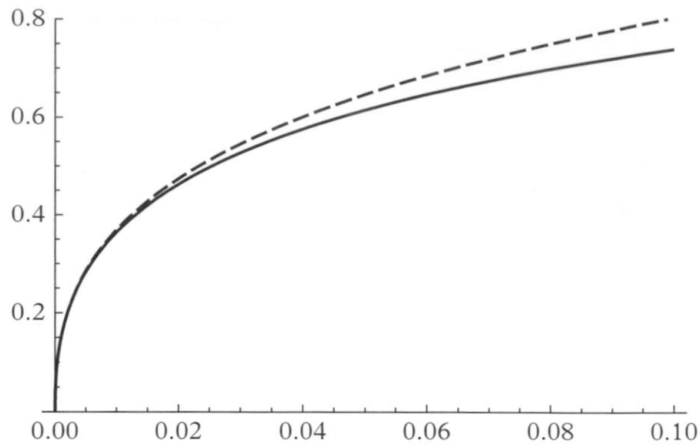

Figure 9.11 *Comparison of k_I^{DFZ}/K_I (vertical axis) as a function of p/q for $0 \leq p/q \leq 0.1$ as given exactly (solid line) by eqn. 9.84 and approximately (dashed line) by eqn. 9.88.*

tip, sending anti-shielding dislocations into the crack and shielding dislocations away from it to form the plastic zone. However, if the operation of dislocation sources, or dislocation glide, is thermally activated,[26] this may give rise to a transition from brittle behaviour at low temperatures, where a plastic zone cannot form, to ductile behaviour at higher temperatures. Such a transition may occur over a narrow range of temperatures if there are very few available sources and the dependence of the dislocation mobility on temperature is strong. Dislocation mobility is influenced by other microstructural features too, such as solute atoms raising the friction stress, and obstacles such as precipitates and grain boundaries. The brittle to ductile transition temperature is not a material parameter.

9.9 The influence of interatomic forces on slit cracks

The stress singularities at the crack tips of section 9.4 are a consequence of the use of linear elasticity to describe the elastic field. These singularities lead to infinite stresses at the crack tips, which are unrealistic. Even the elliptical shape of the crack is doubtful, particularly at the crack tips where it is rounded. In reality interatomic forces at the tip will pull the faces of the crack towards each other, so that far from being rounded the profile is more like a cusp. In other words, the Burgers vector density $f(x_1)$ should

[26] See Samuels, J and Roberts, SG, *Proc. R. Soc. A* **421**, 1–23 (1989). https://doi.org/10.1098/rspa.1989.0001; Hirsch, PB, Roberts, SG and Samuels, J, *Proc. R. Soc. A* **421**, 25–53 (1989). https://doi.org/10.1098/rspa.1989.0002. Sir Peter Bernhard Hirsch FRS 1925–, British physicist and materials scientist born in Germany.

tend to zero at the crack tips, not the infinite value predicted by linear elasticity in eqn. 9.31.

The Barenblatt[27] model[28] of slit cracks addresses these shortcomings of the theory of the elastic crack presented in section 9.4 by including cohesive forces acting between the crack faces. In a metal these forces are significant only when the crack faces are separated by no more than a few angstroms. The regions where these cohesive forces are significant are limited to 'cohesive zones' of the order of 10 Å in length from the crack tips. Where the cohesive zones begin the attraction between the crack faces is considered just about negligible. It quickly rises towards the cracks tips, becoming much larger than the tractions on the crack faces arising from the applied normal loads. It is reasonable to assume that the cohesive forces in the cohesive zones are independent of the surface tractions arising from the applied normal loads.

As shown[29] by Willis[30] and more generally by Rice[31] the Barenblatt model does not alter the Griffith criterion for fracture. This is because, regardless of the existence of the cohesive zones, if the crack length increases by δc the surface area of the two crack faces increases by $2\delta c$ per unit length, and thus the energy cost remains $2\gamma\delta c$.

The Griffith and Barenblatt models are continuum models. For a given applied load they predict a unique crack length where the crack is in equilibrium: if it is smaller than the equilibrium length the crack will tend to close completely, if it is larger it will tend to continue to grow. This picture changes when we take into account the discrete atomic structure of a crack. Recall that in the treatment of a single dislocation in a continuum the energy of the dislocation is independent of its position. But the energy of a single dislocation in a crystal is a periodic function of its position in the slip plane. The peaks in the energy oscillations are the Peierls barriers discussed in section 7.5. Similar barriers may exist for cracks in a crystal lattice, and if they become sufficiently large cracks may become trapped in the energy minima. This phenomenon is known as 'lattice trapping'.[32] The Peierls barriers for dislocations decrease as the core width increases. We may expect a similar trend in the case of cracks: the longer the cohesive zone the smaller the barrier to crack growth. The Frenkel–Kontorova model would suggest that the length of the cohesive zone increases with the stiffness of the bonds between the crack faces and decreases with the amplitude of the periodic variations of the energy of the crack as a function of its position.

[27] Grigory Isaakovich Barenblatt ForMemRS 1927–2018, Russian mathematician and physicist.

[28] Barenblatt, GI, *J. Appl. Math. Mech.* **23**, 622–36 (1959). https://doi.org/10.1016/0021-8928(59)90157-1; Barenblatt, GI, *J. Appl. Math. Mech.* **23**, 1009–29 (1959). https://doi.org/10.1016/0021-8928(59)90036-X; Barenblatt, GI, *Adv. Appl. Mech.* **7**, 55–129 (1962). https://doi.org/10.1016/S0065-2156(08)70121-2

[29] Willis, JR, *J. Mech. Phys. Solids* **15**, 151–62 (1967). https://doi.org/10.1016/0022-5096(67)90029-4; Rice, JR, *J. Appl. Mech.* **35**, 379–86 (1968). https://doi.org/10.1115/1.3601206

[30] John Raymond Willis FRS, British mathematician.

[31] James Robert Rice ForMemRS 1940–, US engineer and geophysicist.

[32] Thomson, R, Hsieh, C and Rana, V, *J. Appl. Phys.* **42**, 3154–60 (1971). https://doi.org/10.1063/1.1660699

9.10 Problem set 9

1. Consider a slit crack between $x_1 = -2c$ and $x_1 = 0$ loaded in mode I by a tensile stress $\sigma_{22} = \sigma$ far from the crack. Using eqns. 9.27 and 9.31 show that at $(x_1, x_2) = (\rho\cos\alpha, \rho\sin\alpha)$, where $\rho \ll c$, the stress field of the crack is as follows:

$$\sigma_{11}(\rho\cos\alpha,\rho\sin\alpha) = \frac{K_I}{\sqrt{\rho}}\left(\cos(\alpha/2) - \frac{1}{2}\sin(3\alpha/2)\sin\alpha\right)$$

$$\sigma_{22}(\rho\cos\alpha,\rho\sin\alpha) = \frac{K_I}{\sqrt{\rho}}\left(\cos(\alpha/2) + \frac{1}{2}\sin(3\alpha/2)\sin\alpha\right)$$

$$\sigma_{12}(\rho\cos\alpha,\rho\sin\alpha) = \frac{K_I}{2\sqrt{\rho}}\cos(3\alpha/2)\sin\alpha, \quad (9.89)$$

where $K_I = \sigma\sqrt{c/2}$.

Hint: Write down an integral for each stress component like eqn. 9.36. Evaluate these integrals using contour integration, noting the second order poles at $\rho e^{\pm i\alpha}$. Then identify the dominant terms when $c \gg \rho$.

Hence show that the elastic energy density close to the crack tip is given by

$$W = \frac{K_I^2}{4\mu\rho}\left(2(1-2\nu)\cos^2(\alpha/2) + \frac{1}{2}\sin^2\alpha\right). \quad (9.90)$$

2. Show that the stress field components in eqn. 9.89 are related to the following displacement field components through Hooke's law:

$$u_1(\rho\cos\alpha,\rho\sin\alpha) = \frac{K_I\sqrt{\rho}}{\mu}\cos(\alpha/2)\left(1 - 2\nu + \sin^2(\alpha/2)\right)$$

$$u_2(\rho\cos\alpha,\rho\sin\alpha) = \frac{K_I\sqrt{\rho}}{\mu}\sin(\alpha/2)\left(2(1-\nu) - \cos^2(\alpha/2)\right). \quad (9.91)$$

Hint: Show that the displacement gradients are as follows, and then apply Hooke's law:

$$u_{1,1} = \frac{K_I}{2\mu\sqrt{\rho}}\left[(1-2\nu)\cos\left(\frac{\alpha}{2}\right) - \frac{1}{2}\sin\alpha\sin\left(\frac{3\alpha}{2}\right)\right]$$

$$u_{2,2} = \frac{K_I}{2\mu\sqrt{\rho}}\left[(1-2\nu)\cos\left(\frac{\alpha}{2}\right) + \frac{1}{2}\sin\alpha\sin\left(\frac{3\alpha}{2}\right)\right]$$

$$u_{1,2} = \frac{K_I}{2\mu\sqrt{\rho}}\left[2(1-\nu)\sin\left(\frac{\alpha}{2}\right) + \frac{1}{2}\sin\alpha\cos\left(\frac{3\alpha}{2}\right)\right]$$

$$u_{2,1} = \frac{K_I}{2\mu\sqrt{\rho}}\left[-2(1-\nu)\sin\left(\frac{\alpha}{2}\right) + \frac{1}{2}\sin\alpha\cos\left(\frac{3\alpha}{2}\right)\right].$$

Comment: The force per unit length on the mode I crack tip at $x_1 = 0$, tending to make it advance along the x_1-axis, may be obtained by evaluating the integral in eqn. 8.17, where S is any surface enclosing the tip. In the context of fracture this integral is called the J-integral, where it was discovered independently by Rice (1968). To use the stress components of eqn. 9.89 we should choose S to be close to the crack tip. Since the crack geometry is invariant along x_3 the surface integral becomes a line integral around the crack tip. The terms required to evaluate the line integral are contained in eqns. 9.89, 9.90 and the four displacement gradients above. The following result is obtained:

$$F_1 = \frac{\pi(1-\nu)K_I^2}{\mu} \equiv J_I, \tag{9.92}$$

where $K_I = \sigma\sqrt{c/2}$ as before. When K_I reaches the critical value for the crack to grow according to Griffith's criterion, eqn. 9.48, the force on the crack tip, J_I, reaches 2γ. At this critical condition the force J_I, which arises from the change in the sum of the potential energy of the external loading mechanism and the elastic energy of the crack, exactly balances the force 2γ required to increase the area of the two crack faces. Thus, for an elastic crack, criteria for crack growth based on critical values of K_I and J_I are equivalent. But this equivalence breaks down when there is plasticity because J_I becomes dependent on whether the contour S includes the plastic zone. However, when there is a dislocation free zone the elastic field at the crack tip again becomes singular. The singularity is characterised by a local stress intensity factor k_I (see eqn. 9.83) which reflects the screening of the applied stress by the plastic zone. S may then be chosen to enclose the crack tip only in which case eqn. 9.92 is recovered with K_I replaced by the local stress intensity factor k_I.

3. Consider a slit crack between $x_1 = -2c$ and $x_1 = 0$ on $x_2 = 0$ loaded in mode III by a shear stress $\sigma_{23} = \sigma^A$ at $x_2 = \pm\infty$. The tractions created on the crack faces by the applied stress must be eliminated. This is achieved by introducing a distribution of screw dislocations in $-2c \leq x_1 \leq 0$, with lines parallel to x_3, and a Burgers vector density $f(x_1)$. Using the stress field of a screw dislocation in eqn. 6.42 show that the integral equation governing the distribution $f(x_1)$ is as follows:

$$P\int_{-2c}^{0} \frac{f(x_1')}{x_1 - x_1'}\mathrm{d}x_1' = -\frac{2\pi\sigma^A}{\mu}. \tag{9.93}$$

Show that solution of this equation is

$$f(x_1) = \frac{2\sigma^A}{\mu} \frac{(x_1 + c)}{\sqrt{(x_1 + 2c)(-x_1)}}. \tag{9.94}$$

Show that non-zero stress field components of the crack at (x_1, x_2) are as follows:

$$\sigma_{13}(x_1, x_2) = -\frac{\sigma^A x_2}{\pi} \int_{-2c}^{0} \frac{(x_1' + c)}{\sqrt{(x_1' + 2c)(-x_1')}} \frac{1}{(x_1 - x_1')^2 + x_2^2} \mathrm{d}x_1'$$

$$\sigma_{23}(x_1, x_2) = \frac{\sigma^A}{\pi} \int_{-2c}^{0} \frac{(x_1' + c)}{\sqrt{(x_1' + 2c)(-x_1')}} \frac{(x_1 - x_1')}{(x_1 - x_1')^2 + x_2^2} \mathrm{d}x_1'. \tag{9.95}$$

Evaluate these integrals by contour integration to obtain the following expressions:

$$\sigma_{13}(x_1, x_2) = \sigma^A \operatorname{Im}\left\{ \frac{x_1 + ix_2 + c}{\sqrt{(x_1 + ix_2 + 2c)(x_1 + ix_2)}} \right\}$$

$$\sigma_{23}(x_1, x_2) = \sigma^A \operatorname{Re}\left\{ \frac{x_1 + ix_2 + c}{\sqrt{(x_1 + ix_2 + 2c)(x_1 + ix_2)}} \right\}. \tag{9.96}$$

For $(x_1, x_2) = (\rho\cos\alpha, \rho\sin\alpha)$, where $\rho/c \ll 1$, show that these stress components become the following close to the crack tip at $(0, 0)$:

$$\sigma_{13} = -\sigma^A \sqrt{\frac{c}{2\rho}} \sin(\alpha/2)$$

$$\sigma_{23} = \sigma^A \sqrt{\frac{c}{2\rho}} \cos(\alpha/2). \tag{9.97}$$

It follows that the stress intensity factor for the mode III slit crack is also $\sigma^A\sqrt{c/2}$. Show that the crack opening displacement $s(x_1)$ is as follows:

$$s(x_1) = \frac{2\sigma^A}{\mu} \sqrt{c^2 - (c + x_1)^2}. \tag{9.98}$$

Show that the change in the total potential energy following the introduction of the crack is

$$\Delta U = -\frac{\pi \left(\sigma^A\right)^2 c^2}{2\mu}. \tag{9.99}$$

Hence show that the Griffith criterion for the growth of a mode III slit crack is as follows:

$$\sigma^A \sqrt{c/2} = \sqrt{\frac{2\gamma\mu}{\pi}}. \tag{9.100}$$

4. Show the force along the x_1-axis on unit length of the tip of the mode III slit crack of the previous question is $J_{III} = \left(\sigma^A\right)^2 \pi c/(2\mu)$. When Griffith's criterion is satisfied show that $J_{III} = 2\gamma$.
5. A Frank–Read source in the centre of a rectangular grain emits dislocation loops under the action of a shear stress resolved on the slip plane in the direction of the Burgers vector. If the boundaries surrounding the grain are impenetrable obstacles the loops pile up at them, as illustrated in Fig. 9.12. If one side of the grain is

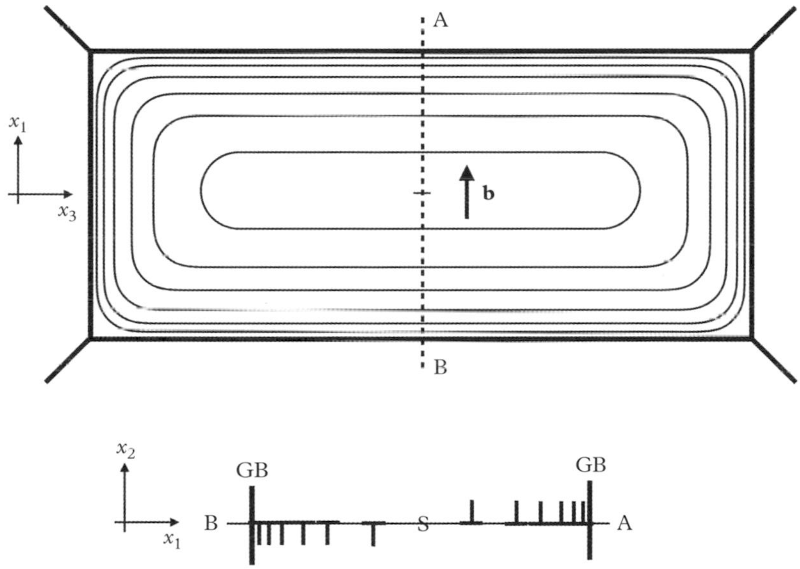

Figure 9.12 *Upper: Plan view of a rectangular grain in a polycrystal containing a Frank–Read source at its centre which is emitting dislocation loops that pile up at the grain boundaries, shown as thicker lines. The Burgers vector* **b** *is along* x_1*, and the grain sides are parallel to* x_1–x_2 *and* x_3–x_2 *planes of a Cartesian coordinate system. The normal to the page is along* x_2*. The broken line AB passes through edge dislocations piled up at the upper and lower grain boundaries. Lower: Side view of the edge dislocation pileups along AB looking along the* x_3*-axis. S signifies the Frank–Read source, GB signifies a grain boundary. The same coordinate system is used in both figures.*

parallel to the Burgers vector the dislocations parallel to that side have pure screw character, and on the perpendicular sides they have pure edge character. If there are many dislocations in each pileup we may make the continuum approximation to derive the stress field created by the pileups.

Consider the two pileups of edge dislocations and approximate the edge dislocations lines as having infinite length. Let the normal to the slip plane be along the positive x_2-axis, and the Burgers vector of each dislocation be parallel to x_1 (see Fig. 9.12). The edge dislocation line directions are then along the $\pm x_3$-axis. The edge dislocations on either side of the source have opposite signs because their line directions are reversed (since they form loops). We may treat them equivalently as having the same line sense but opposite sign Burgers vectors. Let σ^A be the effective resolved component of the applied stress acting on the slip plane in the direction of the Burgers vector.[33] Let $f(x_1)$ be the Burgers vector density of dislocations in the two pileups. The source continues to emit dislocations until σ^A is counteracted by the shear stress created by the pileups. The system is then in mechanical equilibrium.

Using the stress field of an edge dislocation given in eqn. 6.34 show that the condition for mechanical equilibrium of the two pileups is as follows:

$$\frac{\mu}{2\pi(1-\nu)} P \int_{-d}^{d} \frac{f(x_1')}{x_1 - x_1'} \mathrm{d}x_1' = -\sigma^A, \tag{9.101}$$

where d is the length of each pileup. By now this should look very familiar. The solution follows immediately from eqn. 9.31:

$$f(x_1) = \frac{2(1-\nu)\sigma^A}{\mu} \frac{x_1}{\sqrt{d^2 - x_1^2}}. \tag{9.102}$$

If b is the magnitude of the Burgers vector of the discrete dislocations in the pileups show that the number n of dislocations in each pileup is

$$n = \frac{2(1-\nu)\sigma^A}{\mu} \frac{d}{b}. \tag{9.103}$$

Following the same procedure that led to the stress components in eqn. 9.89, it may be shown that at $(x_1, x_2) = (d + \rho\cos\alpha, \rho\sin\alpha)$, where $\rho/d \ll 1$ the stress field close to the tip of the pileup has the following components:

[33] By 'effective' we mean after the shear stress required to operate the Frank–Read source has been subtracted from the resolved component of the applied stress.

$$\sigma_{11}(d+\rho\cos\alpha,\rho\sin\alpha) = -\sigma^A\sqrt{\frac{d}{2\rho}}\left(2\sin\left(\frac{\alpha}{2}\right)+\frac{1}{2}\sin\alpha\cos\left(\frac{3\alpha}{2}\right)\right)$$

$$\sigma_{22}(d+\rho\cos\alpha,\rho\sin\alpha) = \sigma^A\sqrt{\frac{d}{2\rho}}\left(\frac{1}{2}\sin\alpha\cos\left(\frac{3\alpha}{2}\right)\right)$$

$$\sigma_{12}(d+\rho\cos\alpha,\rho\sin\alpha) = \sigma^A\sqrt{\frac{d}{2\rho}}\left(\cos\left(\frac{\alpha}{2}\right)-\frac{1}{2}\sin\alpha\sin\left(\frac{3\alpha}{2}\right)\right). \quad (9.104)$$

These stress components $\sigma_{ij}(\rho,\alpha)$ are expressed in the coordinate system of the grain containing the pileup.[34] In the adjacent grain let the normal to a slip plane and the corresponding slip direction be $\hat{\mathbf{n}}(\beta) = [-\sin\beta, \cos\beta, 0]$ and $\hat{\mathbf{s}}(\beta) = [\cos\beta, \sin\beta, 0]$ respectively, expressed in the coordinate system of the grain containing the pileup. Let there be a source in the adjacent grain at a distance $\rho = \rho_s$ from the end of the pileup, where $\rho_s/d \ll 1$. If the resolved shear on the adjacent slip system required to activate the source is τ_s show that the following condition must be satisfied if the source is to be activated:

$$\frac{\sigma^A}{4}\sqrt{\frac{d}{2\rho_s}}\left[\cos\left(\frac{\beta}{2}\right)+3\cos\left(\frac{3\beta}{2}\right)\right] = \tau_s \quad (9.105)$$

The average of value of $[\cos(\beta/2)+3\cos(3\beta/2)]$ between $\beta = -\pi/2$ and $\beta = +\pi/2$ is $4\sqrt{2}/\pi$. Writing $\sigma^A = \sigma_y - \sigma_f$, where σ_y is the yield stress at which plastic deformation is propagated from one grain to another, and σ_f is the friction stress opposing dislocation motion, eqn. 9.105 may be written in the following form for a polycrystal:

$$\sigma_y = \sigma_f + k_y d^{-1/2}, \quad (9.106)$$

where $k_y = \pi\tau_s\sqrt{\rho_s}$ is a factor that determines how easily slip is transmitted from one grain to the next. Equation 9.106 is called the Hall–Petch relation after Hall.[35] and Petch[36] who found it experimentally.[37] From a technological point of view it is useful because it shows the yield stress may be raised by decreasing the grain size. A smaller grain size may be achieved by a combination of plastic deformation and carefully controlled annealing to induce and arrest recrystallisation to freeze in a small grain size.

6. Griffith's criterion for crack growth assumes a pre-existing crack, which raises the question of how cracks form in a metal. Following an earlier suggestion[38] by

[34] They also describe the stress field close to the tip of a mode II slit crack.

[35] Eric O Hall, New Zealand physicist.

[36] Norman James Petch FRS 1917–92, British metallurgist.

[37] Hall, EO, *Proc. Phys. Soc.* **B64**, 747–53 (1951). https://doi.org/10.1088/0370-1301/64/9/303; Petch, NJ, *J. Iron Steel Inst.* **174**, 25 (1953).

[38] Mott, NF, *Proc. R. Soc. A* **220**, 1 (1953). https://doi.org/10.1098/rspa.1953.0167

Mott,[39] Stroh[40] showed[41] during his doctorate how the stress concentration ahead of a dislocation pileup may nucleate a crack. Consider a mode I crack nucleated ahead of a pileup. We assume this happens when the tensile stress across some plane just ahead of the pileup exceeds a critical value. For this to happen the total energy of the system must decrease as a result of the nucleation of the crack. Using eqn. 9.46 for the potential energy released when a mode I slit crack of length $2c$ forms, and taking into account the energy $4\gamma c$ of creating the two crack surfaces, show that the total energy decreases provided:

$$\sigma > \sqrt{\frac{8\gamma\mu}{\pi(1-\nu)c}}. \tag{9.107}$$

The stress field ahead of the pileup is given by eqn. 9.104. The next step is to identify the angle α in eqn. 9.104 associated with the largest tensile stress along the normal $\hat{\mathbf{n}} = [-\sin\alpha, \cos\alpha]$. This is given by $\sigma(\alpha) = n_i\sigma_{ij}n_j$. Show that

$$\sigma(\alpha) = -3\sigma^A\sqrt{\frac{d}{2\rho}}\sin(\alpha/2)\cos^2(\alpha/2). \tag{9.108}$$

$\sigma(\alpha)$ is a maximum when $\alpha = -\cos^{-1}(1/3)$, which is $-70.5°$. Show that the corresponding maximum tensile stress is

$$\sigma_{max} = \frac{2}{\sqrt{3}}\sqrt{\frac{d}{2\rho}}\sigma^A. \tag{9.109}$$

If the crack is nucleated its length $2c$ will equal ρ. Hence show that the condition for the pileup to nucleate a mode I crack is as follows:

$$\sigma^A\sqrt{d} > 2\sqrt{\frac{6\gamma\mu}{\pi(1-\nu)}}. \tag{9.110}$$

In brittle metals, where fracture occurs before slip is transmitted across grain boundaries, the fracture stress has an inverse square root dependence on the grain size. This has also been confirmed experimentally.

[39] Sir Nevill Francis Mott FRS 1905–96, British Nobel Prize-winning theoretical physicist, who established solid state physics in the UK.

[40] Alan N Stroh 1926–62, South African theoretical physicist, studied with Eshelby and Mott for his PhD in Bristol, creator of the elegant and widely used sextic formalism of anisotropic elasticity, his career cut short at age 36 by a fatal car accident in Colorado. A more extensive biography is available on pp.159–61 of *Anisotropic elasticity* by Ting, TCT, Oxford University Press: Oxford and New York (1996). ISBN: 0195074475

[41] Stroh, AN, *Proc. R. Soc. A* **223**, 404–14 (1954). https://doi.org/10.1098/rspa.1954.0124

10 Open questions

10.1 Introduction

In this chapter four areas of current research in the physics of crystal defects are introduced. The presentation differs from earlier chapters in being more like research seminars than detailed expositions. References to the literature are given where further information may be found. There are other large areas of current research in the physics of defects such as very high strain rate deformation and irradiation damage. The four topics selected here have been chosen because they can be introduced relatively briefly, and questions to frame further research can be formulated.

10.2 Work hardening

10.2.1 The nature of the problem

Work hardening, or strain hardening as it is sometimes called, is the increase in the flow stress of a metal as a result of plastic deformation. As Cottrell noted[1] in 1953 it is a spectacular effect enabling the yield strengths of pure copper and aluminium to be raised a hundredfold. The goal of a theory of work hardening is to explain and predict the stress–strain relation of a material as a function of temperature, strain rate, grain size, alloy composition and microstructural features such as second phase particles and the distribution of grain orientations known as texture. Since dislocations are the agents of plastic deformation it is their collective behaviour that leads to the observed stress–strain relation. Their collective behaviour involves short- and long-range interactions with each other, and with other microstructural features. In 1953 Cottrell pointed out that work hardening was the first problem to be tackled by dislocation theory, and may well prove to be the last to be solved. Almost fifty years later he expressed the magnitude of the enduring challenge of developing a successful theory of work hardening as follows:

[1] Cottrell, AH, *Dislocations and plastic flow in crystals*, Clarendon Press: Oxford (1953), Chapter 10.

Physics of elasticity and crystal defects. Adrian P. Sutton, Oxford University Press (2020). © Adrian P. Sutton.
DOI: 10.1093/oso/9780198860785.001.0001

> It is sometimes said that the turbulent flow of liquids is the most difficult remaining problem in classical physics. Not so. Work hardening is worse. . . . Whereas fluid dynamics can be treated by continuum methods, so that everything can be reduced to the purely mathematical problem of solving standard differential equations, there is no similar escape in work hardening, for the discrete structures of dislocations render the theory intrinsically atomistic, even though in their lengthwise dimensions dislocations are macroscopic objects, governed mainly by a classical physics which is unusual in not being reducible to continuum theory. . . . Another unusual and extremely complicating feature is of course that dislocations are lines, not the familiar point particles of mainstream physics—and flexible lines at that—so that the standard methods of particle theory are inapplicable. . . . Furthermore, neither of the two main strategies of theoretical many-body physics—the statistical mechanical approach; and the reduction of the many-body problem to that of the behaviour of a single element of the assembly—is available to work hardening. The first fails because the behaviour of the whole system is governed by that of weakest links, not the average, and is thermodynamically irreversible. The second fails because dislocations are flexible lines, interlinked and entangled, so that the entire system behaves more like a single object of extreme structural complexity and deformability. . . . Of course, the properties of single dislocations have long been well-established: glide, climb, cross-slip and sessile kinematic modes; reactions, combinations and dissociations; Frank–Read sources, etc. These properties provide the alphabet in which the story of work hardening must be written. But only the alphabet, no more than that.[2]

10.2.2 Work hardening of fcc single crystals

Early experiments on tensile stress–strain relations of single crystals of ductile face-centred cubic (fcc) metals such as copper and aluminium revealed at least three stages of work hardening. If the crystal is oriented so that only one slip system is activated initially the crystal enters stage I where the slope $d\sigma/de$ is very small at about $10^{-4}\mu$. This small hardening, often called 'easy glide', arises from having to overcome attractive multipolar interactions between trains of dislocations on parallel slip planes. The trains are moving in opposite directions because the Burgers vectors in one train have the opposite sign to those in the other train. As plastic strain increases the crystal axes rotate and the resolved shear stress due to the applied load on a second slip system increases. When the resolved shear stress due to the applied load on the second slip system becomes comparable to that on the first, the two slip systems are equally active, and the stress–strain curve has then fully entered stage II. However, secondary slip is activated by local internal stresses before the resolved stress due to the applied load is sufficient to activate it. Consequently the transition from stage I to stage II occurs gradually over a range of plastic strains. Stage II is the dominant part of the stress–strain curve, where the slope is remarkably constant at about $\mu/300$ and independent of temperature. Slip takes place on inclined slip planes and dislocations on one slip plane have to cut through those on inclined planes forming

[2] Reprinted from *Dislocations in solids* **11**, ed. FRN Nabarro and MS Duesbery, *Commentary: a brief view of work hardening*, by AH Cottrell, pages vii–xvii, Copyright 2002, with permission from Elsevier. Elsevier: Amsterdam, ISBN 0-444-50966-6.

steps or 'jogs' on the dislocation lines. If the jogs are sessile additional stress leads to the formation of dislocation dipoles and point defects. Dislocations in inclined slip planes may also combine and form sessile dislocations which may block the passage of further dislocations on either slip plane. These interactions between dislocations on inclined slip planes are collectively called 'forest' interactions and they are responsible for the large increase in the hardening rate in stage II compared to stage I. The Orowan flow stress predicts that in stage II the increase in the flow stress due to these forest interactions is $\alpha\mu b\sqrt{\rho_f}$ where ρ_f is the density of forest interactions[3] and $\alpha \approx 0.3$. Stage II ends and stage III begins when the slope starts to decrease and ductile fracture eventually occurs at a stress which depends on temperature. If the crystal is oriented so that two or more slip systems are activated by the external load as soon as plasticity begins there is no stage I and stage II is entered straight away. For the same reason in a fcc polycrystal stage II is entered as soon as plasticity begins. Conversely, in some hexagonal crystals where slip occurs only on one slip plane, stage II is never entered before fracture occurs.

It is not difficult to show that during stage II most of the work of plastic deformation is not stored in the deformed body as potential energy but is dissipated as heat.[4] During stage II the stress–strain relation is $\sigma = (\mu/300)e$. The work done per unit volume during an increment de of strain is $\mathrm{d}W = \sigma \mathrm{d}e = 300(\sigma/\mu)\mathrm{d}\sigma$. The increment in the density of stored potential energy is $\mathrm{d}E \approx \frac{1}{2}\mu b^2 \mathrm{d}\rho$, where ρ is the average dislocation density. Assuming $\sigma = 0.3\mu b\sqrt{\rho_f}$, where ρ_f is the density of forest interactions, and $\rho_f = \beta\rho$ where[5] $\beta \approx \frac{1}{2}$, we obtain $\mathrm{d}E/\mathrm{d}W \approx 1/14$. Work hardening is a dissipative process.

10.2.3 The Cottrell–Stokes law

In Fig. 10.1 we see two schematic stress–strain curves for a polycrystalline sample under uniaxial tension measured at the same strain rate and at temperatures T_1 and T_2, where $T_2 > T_1$. Consider the sample at the higher temperature T_2 strained to the point A. The flow stress is then at the point B. If the sample is unloaded and immediately reloaded at the lower temperature T_1 plastic flow recommences at D, above B, and rises to join the curve for T_1. If the change of temperature from T_2 to T_1 is made quickly BD is the reversible thermal change of flow stress, since cycling between T_1 and T_2 the flow stress alternates between D and B. DC is the athermal difference in the levels of work hardening reached at T_1 and T_2 at the same strain and reflects the differences in the densities and distributions of dislocations in the two samples at these temperatures. The Cottrell–Stokes law[6] is that AB/AD is found experimentally to be independent of strain and dependent only on T_1 and T_2. It follows that BD/AB, the ratio of the reversible

[3] It is a sobering fact that the dependence of the flow stress on the square root of dislocation density first appeared in GI Taylor's paper of 1934.

[4] Nabarro, FRN, Basinski, ZS and Holt, DB, *Adv. Phys.* **13**, 193–323 (1964). https://doi.org/10.1080/00018736400101031. Zbigniew Stanislaw Basinski FRS 1928–99.

[5] Basinski, ZS and Basinski, SJ, *Phil. Mag.* **9**, 51–80 (1964). https://doi.org/10.1080/14786436408217474

[6] Cottrell, AH and Stokes, RJ, *Proc. R. Soc. A* **233**, 17–34 (1955). https://doi.org/10.1098/rspa.1955.0243

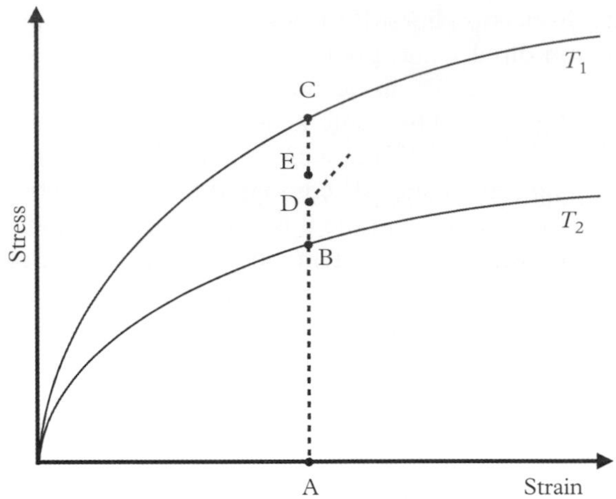

Figure 10.1 *Schematic of two stress–strain curves obtained at the same strain rate and temperatures T_1 and T_2, where $T_2 > T_1$. After Cottrell and Stokes (1955).*

thermal component of the flow stress to the total flow stress, must also be independent of strain and dependent only on T_1 and T_2

Similarly, when the sample at the lower temperature T_1 is strained to the point A the flow stress rises to the point C. If the sample is unloaded and reloaded quickly at the higher temperature T_2 plastic flow recommences at E, beneath the stress at C, and on further straining it eventually reaches the T_2 curve. This is called work-softening, since the flow stress decreases with increasing strain. Cottrell and Stokes found the ratio AE/AC, at a given strain rate, was also independent of strain and dependent only on T_1 and T_2 over a wide range of temperatures and plastic strains.

Basinski[7] made equivalent experimental observations where the strain rate was changed abruptly at constant temperature. The explanation put forward by Basinski (1959) for both the Cottrell–Stokes law and his own observations was that both the thermal and athermal contributions to the flow stress arise from the same source, namely the forest interactions between dislocations on inclined slip planes. This suggestion was developed further by Nabarro[8] and Brown,[9] who showed that the Cottrell–Stokes law could be explained provided just one type of obstacle is present at all plastic strains, which on average has the same finite strength and range, and provided dislocations advance

[7] Basinski, ZS, *Phil. Mag.* **4**, 393–432 (1959). https://doi.org/10.1080/14786435908233412

[8] Nabarro, FRN, *Acta Metall.* **38**, 161–4 (1990). https://doi.org/10.1016/0956-7151(90)90044-H

[9] Brown, LM, in *Dislocations in solids*, ed. FRN Nabarro and MS Duesbery, Elsevier: Amsterdam (2002), pp.193–210. ISBN 0-444-50966-6. Lawrence Michael Brown FRS, 1936–, Canadian and British materials physicist.

rapidly over these obstacles so that they do not have time to relax to their equilibrium shapes.

10.2.4 Work hardening and self-organised criticality

The density of dislocations in a heavily deformed ductile metal, such as copper, may reach 10^{12} cm^{-2}. It is remarkable that this is equivalent to ten million kilometres of dislocation line in each cubic centimetre, or roughly one in every thousand atoms is in a dislocation core. However, one of the most significant discoveries by transmission electron microscopy of deformed samples is that dislocations are not distributed uniformly during plastic deformation. A cellular structure is formed, with the cell walls made of dense bundles or 'braids' of dislocations separated by regions where the dislocation density is much smaller. This is an example of self-organisation of dislocations. During work-hardening dislocations tend to organise themselves into low energy configurations in which their long-range elastic fields are mutually screened.[10] For example, dislocations of opposite sign tend to form dipoles, while edge dislocations of the same sign tend to form small angle grain boundaries, often called subgrain boundaries. Screw dislocations of opposite sign will also tend to form dipoles and they may also annihilate each other by cross-slip. If a group of dislocations within a region has a net Burgers vector its elastic field will stimulate the migration of neighbouring dislocations into the region to reduce the net Burgers vector to zero, possibly through the activation of sources. Observations have also shown that most dislocations are at rest most of the time in metals subjected to heavy deformation at moderate strain rates, indicating that dislocations are in temporary local states of mechanical equilibrium. When dislocation motion occurs it is in bursts or avalanches involving groups of dislocations forming slip bands. This has led Brown to describe plasticity as 'constantly intermittent',[11] and it contrasts with the assumption made in continuum plasticity of continuous deformation. These characteristic features of plasticity are also common to systems displaying self-organised criticality.

Self-organised criticality (SOC) combines the two concepts of self-organisation and criticality.[12] Self-organisation is the ability of a non-equilibrium system to develop structures and patterns without any external interference. Criticality is precisely defined in statistical mechanics of second order phase transitions where at the transition temperature a localised disturbance can propagate across the entire system even though interactions between constituent particles have much shorter range. In the words of Bak,[13] who developed the concept of SOC,

[10] Kuhlmann-Wilsdorf, D, *Mater. Sci. Eng.* **86**, 53–66 (1987). https://doi.org/10.1016/0025-5416(87)90442-3. Doris Kuhlmann-Wilsdorf 1922–2010, US metallurgist born in Germany.

[11] Brown, LM, *Mater. Sci. Technol.* **28**, 1209–32 (2012). https://doi.org/10.1179/174328412X13409726212768

[12] Jensen, HJ, *Self-organized criticality*, Cambridge University Press: Cambridge (1998). ISBN: 0521483719.

[13] Per Bak 1948–2002, Danish theoretical physicist who worked in Denmark, the US and the UK.

> ... complex behaviour in nature reflects the tendency of large systems with many components to evolve into a poised, 'critical' state, way out of balance, where minor disturbances may lead to events, called avalanches, of all sizes. Most of the changes take place through catastrophic events rather than by following a smooth gradual path. The evolution to this very delicate state occurs without design from any outside agent. The state is established solely because of the dynamical interactions among individual elements of the system: the critical state is self-organised.[14]

In the context of work hardening SOC describes self-organisation of the dislocation structure resulting in every part of the structure having an equal probability over time of being a site where further plastic deformation is initiated. Brown describes[15] this self-organised, critical state as one of maximum 'suppleness'. Before the material reaches this state plastic deformation is initiated at particular susceptible sites. As deformation proceeds all regions are 'invaded' by slip bands. Eventually the material enters a critical state where all regions are on the verge of an avalanche of plastic deformation. In the language of dynamical systems the self-organised critical state is an 'attractor'[16] towards which the dislocation structure of the metal evolves. When it is slightly disturbed by the creation of a new slip band it returns to the self-organised critical state. Brown (2016) argued that ductile crystals provide model systems to study SOC free from uncontrollable variables. One of the features of SOC is that it leads to power laws between various observable quantities. Brown (2016) discusses a number of such power laws in work hardening.

10.2.5 Slip lines and slip bands

Slip lines are traces of a slip plane where dislocation glide has introduced a relative displacement on either side of the slip plane. The slip *line* is created where the slip plane exits at a free surface and it is made visible by the formation of a step on the surface. Slip *bands* are long blade-like regions with an aspect ratio of about 50:1 that have been sheared by the collective motion of dislocation loops on parallel slip planes. Thus slip bands comprise groups of slip lines. Brown (2012) has compiled the following list of experimental observations, with references to the literature in his paper, about slip lines and slip bands during work hardening of pure metals not subjected to cyclic loading:

- The average plastic displacement associated with a slip line, as determined by the height of steps where they emerge at a surface, varies by at most a factor of three, typically between 3 and 10 nm, as the stress level changes by a factor of fifty. This behaviour is found in Cu, Zn and NaCl. Although there is much greater variability in the plastic displacement associated with individual slip lines the *average* plastic

[14] Reproduced with permission by Oxford University Press from Bak, P, *How nature works*, Oxford University Press: Oxford (1997), pp.1–2. ISBN 0198501641. Copyright 1997.

[15] Brown, LM, *Phil. Mag.* **96**, 2696–713 (2016). https://doi.org/10.1080/14786435.2016.1211330

[16] Prigogine, I, *From being to becoming*, W.H. Freeman and Co.: San Francisco (1980), pp.7–8. ISBN 0-7167-1108-7. Ilya Romanovich Prigogine 1917–2003, Russian born, Nobel Prize-winning physical chemist, who worked in Belgium and the US.

displacement is observed to be roughly constant and independent of the material, throughout the deformation.

- Over the same range of stress and strain the average length of slip lines decreases inversely with stress by a factor of sixty, from about 20 μm at the beginning of stage II.
- Acoustic emissions from plastically deformed copper single crystals indicate that the bursts of slip last a few microseconds at the onset of stage II. The intensity of an acoustic emission is proportional to the slipped area multiplied by the plastic displacement. Since the average plastic displacement is approximately constant, and the slipped area is inversely proportional to the stress squared, the intensity of the acoustic emission is expected to vary with the inverse square of the stress, which is observed in acoustic emission experiments.
- To produce the observed plastic displacements and the observed intensity of acoustic emissions some tens of dislocations move during a slip burst, which implies they move in a cooperative manner. The speed of the dislocations during a burst is of order metres per second.
- Once a slip line has emerged at a surface the height of the surface step is stable and does not increase with time. The steps formed when slip bands intersect inside the crystal also do not change after they have been formed. These observations indicate that once a slip band has been formed it becomes inactive and does not continue to grow. Further slip occurs through the formation of new slip bands. The crystal becomes a palimpsest,[17] where each region of the crystal is sheared repeatedly through participation in slip bands. Brown (2012) estimated that after 10% plastic strain each atom has participated in a slip band about 100 times.
- There are observations of slip bands inclined by 1–5° to the slip direction. A slip band is nearly planar. By analogy with aerodynamics the 'pitch' is the angle between the slip band and the slip direction. The angle of 'roll' is between the perpendicular to the slip direction and the slip band. Thus, slip bands have two more degrees of freedom than dislocation pileups on single slip planes.

10.2.6 Slip bands as the agents of plastic deformation

The observations listed in the previous section suggest a coarse-grained model in which slip bands are the agents of plastic deformation, an idea apparently first suggested[18] by Jackson[19] in 1985 who treated slip bands as ellipsoidal regions undergoing simple shear. As Jackson noted the great advantage of treating slip bands as ellipsoids is that the stresses and strains inside and outside the shear band are readily calculated using Eshelby's theory of ellipsoidal inclusions (Eshelby (1957)). It follows that dislocation loops making up the slip band experience the same stress provided they are located

[17] A palimpsest is a document that has been cleaned and written on again, such as a slate written on with chalk and wiped clean to be written on again.

[18] Jackson, PJ, *Acta Metall.* **33**, 449–54 (1985). https://doi.org/10.1016/0001-6160(85)90087-2

[19] Paul J Jackson, South African materials physicist.

on the surface of an ellipsoid, a condition that imposes collective motion on the loops. In particular a slip band comes to rest when all dislocation loops on its surface simultaneously come to a stop, because they experience the same Peach–Koehler force. Brown models slip bands as long blade-like ellipsoids. During the few microseconds it takes for a slip band to form it lengthens and it can change the orientation of its principal axes so long as it remains ellipsoidal. *An as yet unanswered question is whether the intermittency of slip and self-organised criticality observed experimentally can be captured by a model in which slip bands, treated as ellipsoids, interact with each other through long- and short-range forces.*

Consider first a shear band represented by an ellipsoid with no pitch or roll. Let the principal axes of the ellipsoid define a right-handed Cartesian coordinate system x_1, x_2, x_3, with the origin at the centre of the ellipsoid, and let the shortest axis of the ellipsoid be along x_3 and the longest along x_1. The slip planes of the dislocation loops are parallel to the x_1-x_2 plane, and their Burgers vectors have magnitude b and they are parallel to x_1. This is the primary slip system. The equation of the ellipsoid is $x_1^2/a_1^2 + x_2^2/a_2^2 + x_3^2/a_3^2 = 1$, where $a_1 \geq a_2 \gg a_3$. The material inside the ellipsoid undergoes a simple shear on the x_1-x_2 plane of magnitude ε. This simple shear produces a pure shear $e_{13}^T = e_{31}^T = \varepsilon/2$, and all other strain components are zero. The T superscript indicates this is a transformation, or stress-free, strain. It is the strain the material inside the ellipsoid would undergo if it were not constrained by the surrounding matrix. Eshelby (1957) showed how to calculate the uniform constrained strain inside the ellipsoid assuming isotropic elasticity, and hence the uniform stress inside the ellipsoid. Using the detailed solutions to the Eshelby ellipsoidal inclusion problem provided by Mura,[20] the only non-zero components of the stress tensor inside the ellipsoid are

$$\sigma_{13}(\text{in}) = \sigma_{31}(\text{in}) = -2\mu e_{13}^T \frac{a_3}{a_2}\left(\frac{\nu}{1-\nu} a_2^2 \frac{K(k^2) - E(k^2)}{a_1^2 - a_2^2} + E(k^2)\right), \tag{10.1}$$

where $k^2 = (a_1^2 - a_2^2)/a_1^2$ and $K(k^2)$ and $E(k^2)$ are complete elliptic integrals of the first and second kind as defined in eqns. 9.82 and 9.79 respectively. If the ellipsoid is in mechanical equilibrium the total shear stress acting on each dislocation loop has to be zero. This requires $\sigma_{13}(\text{in}) + \sigma_{13}^A + \sigma_f = 0$ where σ_{13}^A is the applied shear stress and σ_f is the friction stress opposing glide of dislocations. In the absence of a friction stress $\sigma_{13}(\text{in}) = -\sigma_{13}^A$.

When the ellipsoid is a long blade-like structure $a_1 \gg a_2 \gg a_3$ and k^2 is slightly less than unity. The following asymptotic expansions[21] of the complete elliptic integrals apply:

[20] Mura, T, *Micromechanics of defects in solids*, 2nd edn., Chapter 2, Kluwer Academic Publishers: Dordrecht (1991). ISBN 90-247-3256-5.

[21] http://functions.wolfram.com/EllipticIntegrals/EllipticK/introductions/CompleteEllipticIntegrals/05/

$$K(1-\delta) = \frac{1}{2}\ln\left(\frac{16}{\delta}\right) - \frac{\delta}{8}\ln\delta + O(\delta^2\ln\delta)$$

$$E(1-\delta) = 1 + \frac{\delta}{4}\left[\ln\left(\frac{16}{\delta}\right) - 1\right] + O(\delta^2\ln\delta), \tag{10.2}$$

where $\delta \ll 1$. With $\delta = a_2^2/a_1^2 \ll 1$ we find $\sigma_{13}(\text{in})$ is as follows:

$$\sigma_{13}(\text{in}) = -2\mu e_{13}^T \frac{a_3}{a_2} + \mu O(a_2 a_3/a_1^2). \tag{10.3}$$

When the ellipsoid is penny-shaped we have $a_1 = a_2 \gg a_3$. Taking the limit $a_1 \to a_2$ in eqn. 10.1 we obtain

$$\sigma_{13}(\text{in}) = -2\mu e_{13}^T \left(\frac{2-\nu}{1-\nu}\right)\frac{\pi}{4}\frac{a_3}{a_1} = -\mu\varepsilon\left(\frac{2-\nu}{1-\nu}\right)\frac{\pi}{4}\frac{a_3}{a_1}. \tag{10.4}$$

Returning to the more general case of eqn. 10.1, the stress immediately outside the slip band may also be calculated using the theory described by Eshelby (1957). If n_i is the local unit normal to the surface of the ellipsoid the stress $\sigma_{il}(\text{out})(\hat{\mathbf{n}})$ immediately outside the slip band is as follows:

$$\sigma_{il}(\text{out}) = \left(\sigma_{il}(\text{in}) + 2\mu e_{il}^T\right)(\delta_{i1}\delta_{l3} + \delta_{i3}\delta_{l1}) + \frac{4\mu e_{13}^T n_1 n_3}{1-\nu}(n_i n_l - \nu\delta_{il})$$
$$- \; 2\mu e_{13}^T(\delta_{i1}n_3 n_l + \delta_{l1}n_i n_3 + \delta_{i3}n_1 n_l + \delta_{l3}n_1 n_i). \tag{10.5}$$

It is not difficult to show that $\sigma_{il}(\text{out})n_l = \sigma_{il}(\text{in})n_l$, as required for continuity of tractions across the interface. It follows that

$$\sigma_{13}(\text{out}) = \sigma_{13}(\text{in}) + 2\mu e_{13}^T\left[n_2^2 + \frac{2n_1^2 n_3^2}{1-\nu}\right]. \tag{10.6}$$

The maximum values of $\sigma_{13}(\text{out})$ are found at $\hat{\mathbf{n}} = (0, \pm 1, 0)$ where $\sigma_{13}(\text{out}) = \sigma_{13}(\text{in}) + 2\mu e_{13}^T$. If we ignore the friction stress σ_f the stress concentration factor at these points is

$$\frac{\sigma_{13}(\text{out})}{\sigma_{13}^A} = \frac{(a_2/a_3)(1-\nu)}{\nu a_2^2(K(k^2) - E(k^2))/(a_1^2 - a_2^2) + (1-\nu)E(k^2)} - 1. \tag{10.7}$$

In contrast to the infinite stress concentration factor of a planar dislocation pileup this stress concentration factor is finite and therefore more realistic. For example, for a penny-shaped ellipsoid it is as follows:

$$\frac{\sigma_{13}(\text{out})}{\sigma_{13}^A} \approx \frac{4}{\pi}\frac{(1-\nu)}{(2-\nu)}\frac{a_1}{a_3}. \tag{10.8}$$

Consider the case where the ellipsoid has a small angle of pitch equal to θ. The pitch is a small rotation of the slip band about the x_2-axis. In the coordinate system of the rotated ellipsoid the transformation strain tensor becomes

$$e_{ij}^T = \frac{\varepsilon}{2}\begin{bmatrix} -\sin 2\theta & 0 & \cos 2\theta \\ 0 & 0 & 0 \\ \cos 2\theta & 0 & \sin 2\theta \end{bmatrix} \approx \frac{\varepsilon}{2}\begin{bmatrix} -2\theta & 0 & 1 \\ 0 & 0 & 0 \\ 1 & 0 & 2\theta \end{bmatrix} + O(\theta^2). \tag{10.9}$$

Note the diagonal components. To first order in θ the shear strain $e_{13}^C(\text{in})$, and hence the shear stress $\sigma_{13}(\text{in})$, is unaffected by the rotation. However, normal strains appear as follows:

$$e_{11}^C(\text{in}) = -\left\{\left(\frac{1-2\nu}{1-\nu}\right) a_2 a_3 \frac{(K(k^2)-E(k^2))}{a_1^2 - a_2^2} + O(a_3^2/a_1^2)\right\}\varepsilon\theta$$

$$e_{22}^C(\text{in}) = -\left\{\frac{1}{2(1-\nu)}\frac{a_1^2}{a_1^2-a_2^2}\left(\frac{a_3 E(k^2)}{a_2} - \frac{2a_2 a_3(K(k^2)-E(k^2))}{a_1^2-a_2^2}\right) + O(a_3^2/a_2^2)\right\}\varepsilon\theta$$

$$e_{33}^C(\text{in}) = \frac{1}{2(1-\nu)}\left\{2(1-2\nu) + \frac{a_2 a_3(K(k^2)-E(k^2))}{a_1^2-a_2^2} + (4\nu-1)\frac{a_3 E(k^2)}{a_2}\right\}\varepsilon\theta. \tag{10.10}$$

These normal strains give rise to normal stresses inside the ellipsoid as follows:

$$\sigma_{11}(\text{in}) = 2\mu\left(e_{11}^C(\text{in}) + \varepsilon\theta\right) + \frac{2\mu\nu}{1-2\nu}e^C(\text{in})$$

$$\sigma_{22}(\text{in}) = 2\mu e_{22}^C(\text{in}) + \frac{2\mu\nu}{1-2\nu}e^C(\text{in})$$

$$\sigma_{33}(\text{in}) = 2\mu\left(e_{33}^C(\text{in}) - \varepsilon\theta\right) + \frac{2\mu\nu}{1-2\nu}e^C(\text{in}), \tag{10.11}$$

where $e^C(\text{in}) = e_{11}^C(\text{in}) + e_{22}^C(\text{in}) + e_{33}^C(\text{in})$ is the dilation in the ellipsoid. $\sigma_{11}^C(\text{in})$ is called the fibre stress by analogy with the normal stress carried by a fibre in a fibre-reinforced composite material. It changes sign with θ, and it increases linearly with θ and with the magnitude of the simple shear ε in the slip band. If the ellipsoid has a penny shape ($a_1 = a_2 \gg a_3$) the non-zero components of the stress tensor expressed in the coordinate system of the tilted ellipsoid are as follows:

$$\sigma_{11}(\text{in}) = \frac{2\mu\varepsilon\theta}{1-\nu} + \mu O\left(\frac{a_3}{a_1}\varepsilon\theta\right) = \frac{Y}{(1-\nu^2)}\varepsilon\theta + O\left(\frac{a_3}{a_1}\varepsilon\theta\right) \approx Y\varepsilon\theta + O\left(\frac{a_3}{a_1}\varepsilon\theta\right)$$

$$\sigma_{22}(\text{in}) = \frac{2\mu\nu\varepsilon\theta}{1-\nu} + \mu O\left(\frac{a_3}{a_1}\varepsilon\theta\right)$$

$$\sigma_{33}(\text{in}) = 0 + \mu O\left(\frac{a_3}{a_1}\varepsilon\theta\right)$$

$$\sigma_{13}(\text{in}) = -\mu\varepsilon\left(\frac{2-\nu}{1-\nu}\right)\frac{\pi}{4}\frac{a_3}{a_1}, \tag{10.12}$$

where we have treated a_3/a_1, ε and θ as comparably small quantities, and retained only the terms containing products of two of them at a time. Brown (2012) has derived similar expressions for when a penny-shaped ellipsoid undergoes a small angle of roll and mixed pitch and roll.

The fibre stress σ_{11} inside the ellipsoid has no influence on slip on the primary slip system. But as first noted by Jackson (1985) it can activate secondary slip systems within the ellipsoid, that is, slip on planes and in directions inclined to those of the primary slip system. Activation of secondary slip systems reduces the fibre stress inside the ellipsoid. When loops of the primary and secondary slip systems meet on the periphery of the ellipsoid they interact and form obstacles—these are the 'forest interactions'. Thus the loops of the primary slip system block further slip, or contraction, of loops on the secondary slip systems and vice versa. The ellipsoid is thus stabilised by these forest interactions on its periphery. The walls of the slip band consist of arrays of primary and secondary dislocations. They are almost parallel to the primary slip plane because of the long thin shape of the slip band. The interior of the slip band is virtually free of dislocations. In cross section this produces the cellular structure observed in transmission electron microscopy. Brown[22] has estimated the linear hardening rate on the primary slip system in this ellipsoidal slip band model. On the primary slip plane it involves the spacing of obstacles arising from forest interactions with secondary dislocations. These obstacles act as pinning points between which the primary dislocations have to bow out to escape. The stress required for them to bow out is inversely proportional to the spacing of the secondary dislocations, which is directly proportional to the simple shear ε on the primary slip system, and hence linear hardening results.

Because the thickness (a_3) of the ellipsoid is so much less than its width (a_2) and its length (a_1) the extent of secondary slip is much less than primary slip. This is why the rotation of the crystal is due almost entirely to slip on the primary system. As the crystal rotates the resolved shear stress on one or more secondary slip systems increases, while the resolved shear stress on the primary slip system decreases. Eventually the sum of the applied stress and the fibre stress resolved onto one or more secondary slip systems enables secondary dislocations to bow out between the obstacles provided by the primary dislocations and create shear bands on secondary systems.

[22] Brown, LM, *Metall. Trans. A*, **22**, 1693–708 (1991). https://doi.org/10.1007/BF02646493

We have here the rudiments of a coarse-grained model for work hardening. It needs considerably more work to formulate a protocol for a simulation in two or three dimensions in which slip bands are created self-consistently and overwrite each other. It would be very interesting, not only as a model of work hardening but as a possible paradigm for SOC, to see whether the model leads to a self-organised critical state and whether it captures the 'constantly intermittent' nature of slip.

10.3 Electroplasticity

10.3.1 Introduction and experimental observations

In 1963 in the former Soviet Union it was discovered that a direct electric current reduces the flow stress of a metal and improves its overall ductility. During irradiation of zinc single crystals with 1 MeV electrons, with the crystals being deformed in uniaxial tension at liquid nitrogen temperatures, Troitskii and Likhtman[23] observed a significant decrease in the flow stress and higher ductility when the electron beam was in the basal slip plane, compared to when it was normal to the basal plane. The elongation at fracture also increased when the electron beam was in the basal plane compared to when it was normal to the plane, and compared to when there was no irradiation. The predominant slip systems in zinc have Burgers vectors in the basal plane. These observations suggested the possibility that the passage of a directed electron current through a metal increases the mobility of dislocations.

Troitskii and other Soviet scientists carried out further experiments revealing the influence of pulses ($\lesssim 100\ \mu$s) of direct currents (10^3-10^5 A cm^{-2}) on the flow stress, stress relaxation, creep, dislocation generation and mobility, brittle fracture, fatigue and metal-working. For references see the reviews[24] by Conrad[25] and co-workers. The pulses were short to minimise Joule heating and the current densities were high to maximise their effect. The influence of the current pulses on plasticity was called the electroplastic effect. Figure 10.2 shows the load against displacement for a wire single crystal of zinc,[26] diameter 1 mm and length 15 mm, stretched at a constant speed of 0.01 cm min^{-1} equivalent to a strain rate of 1.1×10^{-5} s^{-1}. Pulses of direct current of duration $\sim 10^{-4}$ s and between 600 A and 1800 A were passed through the wire. The current between the pulses did not exceed 0.3 A. As soon as each pulse was applied the load dropped, and the size of the drop increased with increasing current. Load drops were not observed during the elastic deformation, or during stress relaxation when the load was turned off, and they began only when the sample had started to deform plastically. Troitskii and co-workers found the load drop ΔP varied between 10% and 40% of the applied load, increased approximately linearly with the current density **J**, varied with the direction of

[23] Troitskii, OA and Likhtman, VI, *Dokl. Akad. Nauk SSSR* **148**, 332–4 (1963). In Russian.

[24] Sprecher, AF, Mannan, SL and Conrad, H *Acta Metall.* **34**, 1145–62 (1986). https://doi.org/10.1016/0001-6160(86)90001-5; Conrad, H and Sprecher, AF in *Dislocations in solids*, **8**, North-Holland: Amsterdam (1989), pp.497–541. ISBN 978-0444705150; Conrad, H, *Mater. Sci. Eng. A* **287**, 276–87 (2000). https://doi.org/10.1016/S0921-5093(00)00786-3

[25] Hans Conrad 1922–. US materials scientist.

[26] Troitskii, OA, *JETP Lett.* **10**, 11–14 (1969). http://www.jetpletters.ac.ru/ps/1686/article_25672.shtml

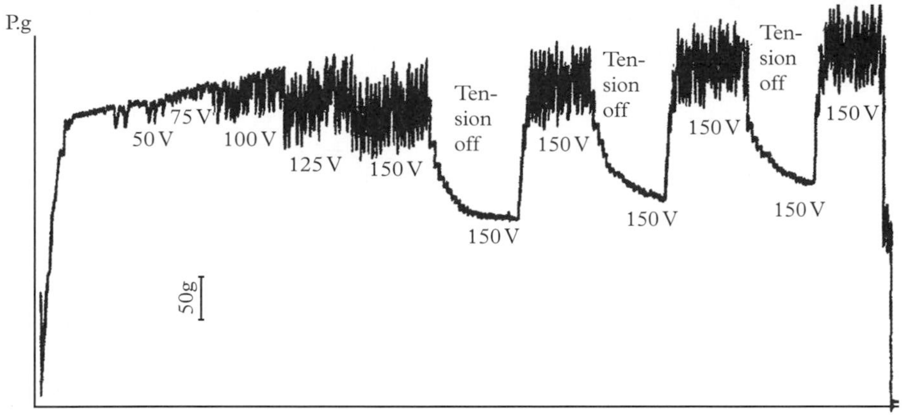

Figure 10.2 *Load–displacement curve for a single crystal of zinc, subjected to DC current pulses while it is deformed in uniaxial tension at 78 K at a strain rate of* $1.1 \times 10^{-5}\ s^{-1}$. *The horizontal axis is time. The voltage is increased in steps of 25 V from 50 V to 150 V, where 100 V corresponds to a current of approximately* $1.5 \times 10^{5}\ Acm^{-2}$. *The load falls with each current pulse, and rises again when the pulse ends. The size of the load drops increases with the voltage (and current) of the current pulses. When the test machine is shut off there is stress relaxation and the current pulses then have a much smaller influence. Reproduced with permission from Troitskii, OA,* JETP Lett. **10**, *11–14 (1969).*

J relative to crystal axes, increased with decreasing temperature, tended to decrease with increasing strain rate and occurs in compression as well as tension.

Stimulated by these results Conrad and others in the US confirmed and extended these studies to polycrystalline samples of Al, Cu, Pb, Ni, Fe, Nb, W, Sn and Ti. Figure 10.3 from Okazaki et al.[27] shows the relation between true stress and true strain for a polycrystalline wire of Ti (99.97% purity), ~50 mm gauge length and 0.51 mm diameter, subjected to a uniaxial tensile loading and pulses of current of duration $\lesssim$100 μs at a nominal temperature of 300 K with forced air cooling. The current density has to exceed a threshold value of about 1000 A mm^{-2} before a load drop is seen. Note the immediate stress relaxation when the pulse is applied and the return to a continuation of the upper envelope of the stress–strain curve after the pulse. They also found the upper envelope of the stress-strain curve with current pulses was the same as that for a separate specimen which had been deformed without current pulses. These observations indicate that whatever changes occur in the wire, as a result of the current pulses, they must be reversible.

10.3.2 Possible mechanisms

The reversibility of the load drops seen clearly in Fig. 10.3 suggests that they are associated with the thermal component of the flow stress, as discussed in section 10.2.3 in

[27] Okazaki, K, Kagawa, M and Conrad, H, *Scripta Metall.* **12**, 1063–8 (1978). https://doi.org/10.1016/0036-9748(78)90026-1

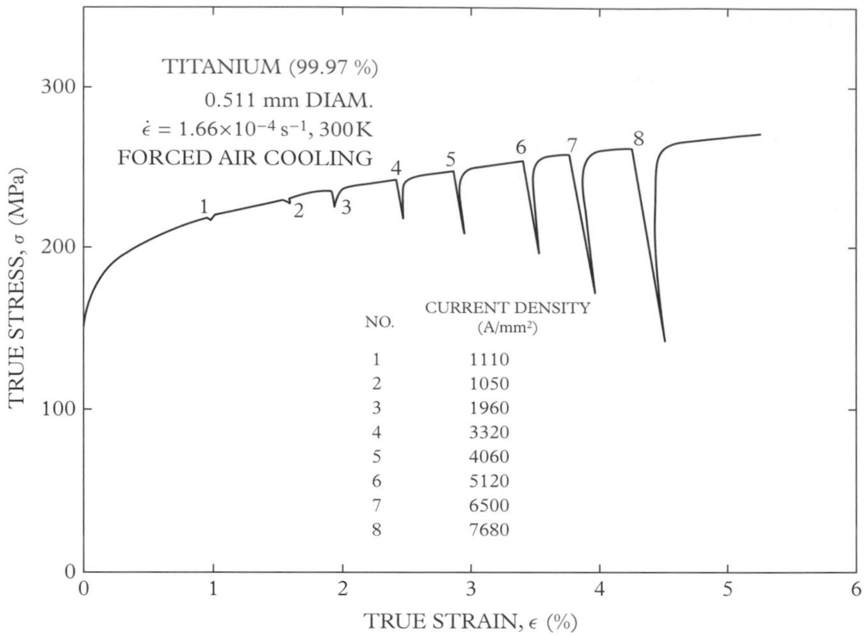

Figure 10.3 *True stress–true strain relation for a Ti wire deformed in tension and subjected to current pulses of increasing current density. Reprinted from Okazaki, K, Kagawa, M and Conrad, H,* Scripta Metall. **12**, *1063–8 Copyright 1978, with permission from Elsevier.*

connection with the Cottrell–Stokes law. The picture is as follows. Dislocation segments are pinned at obstacles, such as forest interactions. To overcome the free energy barrier these obstacles present dislocations require the assistance of thermal activation. Conrad (2000) argued that the primary influence of the current is to alter the pre-exponential factor in the rate at which these obstacles are overcome. This factor includes the frequency of vibration made by pinned dislocation segments and the entropy of the activated state. It also includes geometrical terms unlikely to be influenced by the current. To increase the frequency of vibration of the dislocation the local shear modulus would have to be increased by the current. This seems unlikely because electrons in states below the Fermi level are excited into states above the Fermi energy, which will tend to make the material softer. It is more likely that the current increases the entropy of the activated state. Current-carrying electrons may reduce the stiffnesses of bonds in the dislocation core, increasing the entropy associated with the atomic rearrangements in the activated state. Secondly the activated state may excite current-carrying electrons to higher energy states, increasing their entropy. Both contributions would reduce the free energy barrier to dislocation motion. Troitskii[28] also suggested, in more general terms, that the principal mechanism of the electroplastic effect is to facilitate dislocations overcoming barriers to slip.

[28] Troitskii, OA, *Strength Mater.* 7, 804–9 (1975). https://doi.org/10.1007/BF01522653

Since the principal influence of the current pulses is on the thermal component of the flow stress the question arises whether they inject pulses of heat. This has been carefully considered by Troitskii (1975) and by Conrad's reviews. If the load drops were due to Joule heating the magnitude of the drops would vary with the square of the current density. The linear dependence of the load drops on the current density rules out Joule heating as a mechanism.

The orientational dependence of the electroplastic effect, observed by Troitskii and Likhtman (1963), and in other experiments reviewed by Antolovich[29] and Conrad,[30] suggests the effect of the current is greatest when it is parallel to the Burgers vector. Indeed, when the current is perpendicular to the Burgers vector it is difficult to see how it can have any influence on the dislocation mobility. In the following we consider how the current may contribute to the glide force on a dislocation.

The Peach–Koehler glide force on a dislocation is a configurational force. In contradistinction, a current-induced glide force is a body force. Electromigration is a related phenomenon where point defects experience current-induced body forces making them drift in particular directions. For example, consider the current-induced force on a vacancy. It arises from current-induced changes in the forces between its neighbours, which alter the probabilities of them jumping into the vacancy slightly from when there is no current. A point defect presents a scattering potential V to the electron current, measured from the bottom of the conduction band. If $V > 0$ the point defect obstructs the current and electrons impinging on the point defect are reflected back creating an excess of electrons on the electron source side of the defect and a deficit on the electron drain side. The resulting dipole of charge establishes an electric field in the opposite direction to the current flow. If $V < 0$ the defect locally facilitates the current flow and the current-induced electric field is in the same direction as the electron current flow. The current-induced electric field exerts a body force on the defect. This picture of the current-induced force is equivalent to the 'electron-wind force', the origin of which is the transfer of momentum from current-carrying electrons to the point defect.[31] For a vacancy the dipole introduces an asymmetry in the forces between neighbouring atoms. Bonds on one side of the vacancy become slightly stronger while bonds on the opposite side become slightly weaker. This is the origin of the slight bias in the probability of where the vacancy jumps next. Conversely, if the local conductivity is enhanced, for example through the formation of a shorter bond enabling faster electron hopping between atoms, the sense of the dipole is reversed.[32]

[29] Stephen D Antolovich, US materials engineer.

[30] Antolovich, SD and Conrad, H, *Mater. Manuf. Process.* **19**, 587–610 (2004). http://dx.doi.org/10.1081/AMP-200028070

[31] Bosvieux, C and Friedel, J, *J. Phys. Chem. Solids* **23**, 123–36 (1962). https://doi.org/10.1016/0022-3697(62)90066-5

[32] The Bosvieux–Friedel dipole is not to be confused with the Landauer residual resistivity dipole. The Landauer dipole accounts for the contribution to the macroscopic resistivity of the sample caused by the defect. The Bosvieux–Friedel dipole does not contribute to the macroscopic resistivity. The Bosvieux–Friedel dipole depends on V to first order, and hence it changes sign when V changes sign. The Landauer dipole depends on V^2 and is thus independent of the sign of V. For further discussion of the Landauer and Bosvieux–Friedel dipoles see Sorbello, RS, *Phys. Rev. B* **23**, 5119–27 (1981). https://doi.org/10.1103/PhysRevB.23.5119; Sorbello, RS, and Chu, CS, *IBM J. Res. Dev.* **32**, 58–62 (1988). http://dx.doi.org/10.1147/rd.321.0058; Sorbello, RS, *Solid State Phys.* **51**, 159–231 (1998). https://doi.org/10.1016/S0081-1947(08)60191-5. Rolf William Landauer 1927–99, German born US physicist. Richard S Sorbello, US physicist.

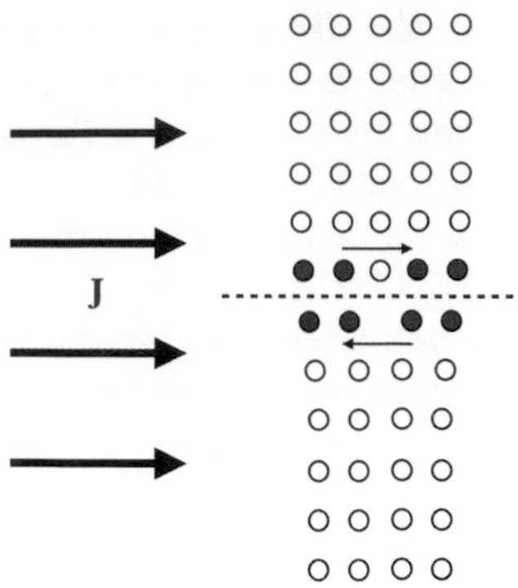

Figure 10.4 *Sketch to illustrate a current-induced charge quadrupole at an edge dislocation. The large arrows show an incident electron current of density* **J** *from the left. The red atoms signify an excess of electronic charge and the blue a deficit. The quadrupole generates forces, shown by small arrows, on atoms either side of the slip plane (broken line) in the dislocation core. These body forces create a shear stress on the dislocation core.*

For an electron current to exert a glide force on a dislocation it has to generate a resolved shear stress on atoms in the dislocation core. This requires atoms in the core on either side of the slip plane to experience resolved forces parallel and anti-parallel to the Burgers vector **b**. Since a dipole of electronic charge creates a net force on a point defect the current has to induce a *quadrupole* of charge along the dislocation line to create a shear stress. The quadrupole of charge generates a dipole of forces, with the directions of the forces parallel and anti-parallel to **b**, as sketched in Fig. 10.4 for an edge dislocation. At the termination of the extra half plane atoms are forced closer together which facilitates current flow leading to a dipole of charge as shown. Just below the termination of the extra half-plane atoms are forced apart, obstructing current flow which creates a dipole of the opposite sense. Thus, the current-induced charge distribution is a quadrupole.[33]

To establish whether there is a current-induced charge quadrupole, leading to a current-induced force dipole, in a dislocation core requires a self-consistent treatment of the electronic structure of the dislocation in the presence of an electron current flow. At the time of writing (2019) this does not appear to have been done. In Fig. 10.3 the flow stress in Ti reduces by ~100 MPa when the current density is of order 10^6 Amm^{-2}. For atoms ~2 Å apart a shear stress of order ~100 MPa requires current-induced forces on atoms in the core of the order of 2×10^{-3} eVÅ^{-1}. This force is about two orders of magnitude larger[34] than might be expected to be produced by a current density of order 10^6 Amm^{-2}. We conclude it is unlikely that the electroplastic effect is caused by

[33] It may be thought that since a kink on a dislocation is effectively a point defect only a current-induced dipole of charge is needed to make it drift. But the movement of a kink also requires a local shear stress and hence a quadrupole of current-induced charge.

[34] Hoekstra, J, Sutton, AP, Todorov, TN and Horsfield A P, *Phys. Rev. B* **62**, 8568–71 (2000). https://doi.org/10.1103/PhysRevB.62.8568

current-induced shear stresses on the dislocation core augmenting the Peach–Koehler glide force. This conclusion is consistent with Conrad (2000) who maintained that the current does not affect significantly the activation enthalpy for dislocations to overcome barriers.

Finally in this sub-section we consider tensile stresses in the wire created by the electrodynamics of current pulses. There is experimental evidence that a pulse of sufficient current density may create *tensile* stresses along the wire. These stresses may be sufficient to fracture the wire in the solid state, for example, see Graneau[35] (1983).[36] This raises the intriguing possibility that the load drops seen in Figs. 10.2 and 10.3 are a result of additional plastic strain caused by transient tensile stresses due to the current pulses. It is a complex magnetothermoelastodynamic problem.[37] It involves the coupling of elastodynamic equations with thermal expansion caused by Joule heating and a radial body force due to the Lorentz force, while allowing the radial distribution of the current to diffuse through the skin effect. However, it is not obvious that this effect can explain the reversibility of the load drops seen most clearly in Fig. 10.3, or the very small effect of the current pulses seen in Fig. 10.2 when the externally applied load is removed.

10.3.3 Recommendations for further research

Further *systematic* experimental research on fully characterised single crystals of the highest purity, with the current flow along defined crystal orientations, is needed to elucidate the mechanism of the electroplastic effect in metals. Self-consistent electronic structure calculations of dislocations, with and without kinks, in the presence of electron currents along defined directions will provide realistic values of the current-induced charge quadrupoles and associated shear stresses in the core. It would be useful to model scattering of current-carrying electrons by stationary dislocations with DFT methods, to elucidate its effect on softening interatomic forces in the dislocation core and its effect on the electronic entropy. Finally the influence of electromagnetic induction associated with current pulsing, including the skin effect and the inverse skin effect,[38] on *plasticity* in the wires is an area rich with possibilities.

10.4 The mobility of dislocations in pure single crystals

In this section we consider the factors limiting the mobility of a single dislocation in a pure single crystal where there are no other defects or microstructure to limit its motion. We may think of this mobility as an intrinsic dislocation mobility, limited only by properties of the dislocation itself and the surrounding pure crystal.

[35] Peter Graneau 1921–2014, electrical engineer and physicist who worked in the UK and the US, born in Silesia.

[36] Graneau, P, *Phys. Lett.* **97A**, 253–55 (1983). https://doi.org/10.1016/0375-9601(83)90760-0

[37] Wall, DP, Allen, JE and Molokov, S, *J. Appl. Phys.* **98**, 023304 (2005). https://doi.org/10.1063/1.1924871. John Edward Allen 1928–, British plasma physicist.

[38] Haines, MG, *Proc. Phys. Soc.* **74**, 576–84 (1959). https://doi.org/10.1088/0370-1328/74/5/310. Malcolm Golby Haines 1936–2013. British plasma physicist, born in Northern Ireland.

10.4.1 Static vs. dynamic friction

In section 7.5 the Peierls–Nabarro model was used to estimate the stress required to enable a dislocation to glide in a crystal lattice. According to this model, once the Peierls stress has been overcome a dislocation continues to move indefinitely until it meets an obstacle, or it exits the crystal. As the dislocation glides over the landscape of the periodic Peierls potential the rises and falls of its potential energy are exactly compensated by the falls and rises of the kinetic energy. The Peierls–Nabarro model treats the motion of a dislocation in the absence of any energy dissipation.

Experimental reality could not be more different. Observations indicate that even in the purest single crystals dislocation glide is overdamped. As a dislocation glides in a metal the breaking and making of bonds between atoms in the core excites crystal lattice vibrations and electronic transitions to states above the Fermi energy. The Peierls stress may be thought of as the *static* frictional stress that has to be overcome to set the dislocation in motion. But once it is moving the physics of the *dynamic* frictional stress is different, its origin being the coupling to the elementary excitations of the crystal. The dissipative nature of dislocation glide is consistent with the observation in section 10.2.2 that more than 90% of the work done on a metal during plastic deformation is converted into heat. It also gives rise to energy absorption in internal friction experiments.

In the elastodynamic theory of moving dislocations the self-energy of a dislocation comprises an elastic potential energy and a kinetic energy associated with the motion of volume elements throughout the continuum. The self-energy diverges at the shear wave speed c_t suggesting that the shear wave speed is an upper limit on the speed of a dislocation. However, the dependence of the self-energy on the speed of the dislocation is a separate matter from dissipation of its energy due to friction. In elastodynamics an isolated dislocation in an infinite continuum moving at constant velocity experiences no energy dissipation: it continues to move at the same velocity indefinitely. That is because in the inertial frame of the dislocation core the system is time-invariant. No elastic waves are radiated from a dislocation moving at a constant velocity in elastodynamics, and therefore the self-energy of the dislocation is constant. Elastic waves are radiated from a dislocation in elastodynamics only when it is accelerated or created or annihilated in the continuum.

In an attempt to include the physics that is missing in elasticity theory of a dislocation moving at a constant speed v it is usual to introduce an ad hoc viscous drag force f_{drag} per unit length through an empirical relation of the form

$$f_{drag} = B \cdot v, \tag{10.13}$$

where B is the drag coefficient, with units $\mathrm{Nm^{-2}s} = \mathrm{Pa\ s}$. The drag coefficient is thus an inverse mobility. When $v \ll c_t$ a dislocation moves at a constant speed of $\tau b/B$, where τb is the usual Peach–Koehler glide force on the dislocation. At such low dislocation speeds it is reasonable to assume the drag coefficient is independent of v and dependent only on whether the dislocation is edge or screw or mixed, and dependent on the vibrational and electronic properties of the crystal. In the following we estimate contributions to the drag coefficient from fundamental physical principles.

10.4.2 Drag forces due to the phonon wind and fluttering mechanisms

Much of the early work focused on the drag coefficient at a finite temperature where the dislocation scatters phonons impinging on it. Two mechanisms were considered. A rigid, stationary dislocation in a temperature gradient experiences a 'phonon wind'. The distortion of the crystal lattice created by a rigid dislocation scatters incident phonons, which diminishes the thermal conductivity.[39] The resulting force imparted to the dislocation is called a phonon wind force. Conversely, a rigid dislocation moving in a crystal at a uniform finite temperature, with the same structure as when it was stationary, experiences the same phonon wind force. This is the first mechanism. Of course, dislocations are not rigid objects and an incident phonon induces atoms in a moving dislocation to vibrate emitting secondary waves, and the dislocation loses energy. This is called the 'fluttering mechanism'.[40] It is expected to dominate over the phonon wind force only at low temperatures. At temperatures above the Debye temperature the two contributions add to give[41]

$$B_{wind} \approx \frac{0.02\mu b}{c_t}. \tag{10.14}$$

For lattice dislocations in aluminium $B_{wind} \approx 5 \times 10^{-5}$ Pa s.

Calculations of phonon scattering in the continuum elastic field of a dislocation, such as those described in the previous paragraph, predict the drag force increases linearly with temperature. This disagrees with molecular dynamics simulations that find a temperature-independent contribution to the drag force for highly mobile dislocations in body-centred cubic (bcc) metals and for nanoscale defects, such as crowdions and nanoscale loops. A new theory for the drag coefficient on defects in crystals arising from their interaction with crystal vibrations has been developed by Swinburne and Dudarev.[42] It treats the discrete atomic structure of the defect and the full nonlinear nature of atomic interactions in the defect core. Contrary to the assumption of phonon-scattering theory they find the vibrational modes of the crystal change as the defect moves, and this is the origin of the temperature-independent contribution to the drag coefficient.

10.4.3 Drag forces due to electronic excitations

As lattice vibrations propagate in a crystal they excite electrons at the Fermi surface of a metal. The vibrations decay partly as a result of this energy transfer to the electrons.

[39] For a recent theoretical treatment of the contribution of dislocations to the thermal resistance and a review of the experimental and theoretical literature see Lund, F and Scheihing, B, *Phys. Rev. B* **99** 214102 (2019). https://doi.org/10.1103/PhysRevB.99.214102. This paper recognises that the interaction between phonons and a dislocation involves the resolved shear stress, which leads to the Peach–Koehler force. Fernando Lund, Chilean materials physicist.

[40] Alshits, VI, in *Elastic strain fields and dislocation mobility*, ed. VL Indenbom and J Lothe, North-Holland: Amsterdam (1992), pp.625–97. ISBN 0444887733. Vladimir Iosifovich Alshits 1941–. Russian materials physicist.

[41] Hirth, JP and Lothe, J, *Theory of dislocations*, 2nd edn., Krieger: Malabar, FL (1982), p.209. ISBN 0-89464-617-6.

[42] Swinburne, TD and Dudarev, SL, *Phys. Rev. B* **92**, 134302 (2015). https://doi.org/10.1103/PhysRevB.92.134302

The strain field of a gliding dislocation may be viewed as a collection of lattice vibrations moving at the speed of the dislocation. They are also capable of exciting electrons at the Fermi surface and this electronic excitation is another source of drag on a dislocation. The modern approach to electron–phonon interactions is based on first principles DFT methods and has predictive capabilities.[43] As far as I know these methods have not been applied to a gliding dislocation.

A more direct approach would be to use time-dependent DFT to simulate the excitation of electrons by a gliding dislocation with Ehrenfest dynamics. In Ehrenfest dynamics the atomic nuclei satisfy Newtonian equations of motion with forces coupled explicitly to the evolving electronic structure, while the electrons are treated fully quantum mechanically and self-consistently by solving the time-dependent Schrödinger equation.[44] I am not aware this has been done (in May 2019) either.

Existing approaches in the literature have been superseded by the capabilities of modern methods, although they have not been applied to dislocations in metals. For example, consider the analysis by Holstein,[45] which appeared in the appendix to a paper[46] by Tittmann and Bömmel. His analysis is based on the concept of a deformation potential to provide the link between the moving strain field of a dislocation and a potential that excites electrons in first order time-dependent perturbation theory. For an approach based on deformation potentials to be valid the strain field has to be slowly varying relative to the atomic spacing, which breaks down in the dislocation core. This is particularly serious in metals because the rapidly changing potentials in the dislocation core are screened in reality, which cannot be treated adequately by deformation potentials. Modern methods do not have these deficiencies. But perhaps the most significant concern about the use of deformation potentials is that in a cubic crystal they have no shear components to first order in the strain. Since shear is the key feature of the movement of any dislocation this is indeed a serious weakness. For example, the application of Holstein's analysis to a screw dislocation in isotropic elasticity would predict no electronic excitations.

To calculate the electronic excitations in simulations of moving dislocations using modern DFT methods would be a major undertaking. It would provide, for the first time, reliable information about the contributions of electronic excitations to the drag on dislocations as a function of their speed. It would also be useful to investigate electronic excitations in a metal at moving kinks on a dislocation, which might be more tractable.

10.4.4 Radiation of atomic vibrations from a moving dislocation

In section 6.9 we introduced the concept of a force dipole to introduce the displacement by the Burgers vector of a dislocation. As a dislocation glides, atoms on either of its slip plane experience impulses to raise them over the potential barrier provided by the

[43] See Giustino, F, *Rev. Mod. Phys.* **89** 015003 (2017). https://doi.org/10.1103/RevModPhys.89.015003; Erratum *Rev. Mod. Phys.* **91**, 019901 (2019). https://doi.org/10.1103/RevModPhys.91.019901

[44] See, for example, Mason, DR, Le Page, J, Race, CP, Foulkes, WMC, Finnis, M W and Sutton, AP, *J. Phys. Condens. Matter* **19**, 436209 (2007). https://doi.org/10.1088/0953-8984/19/43/436209

[45] Theodore David Holstein 1915–85, US theoretical physicist.

[46] Tittmann, BR and Bömmel, HE, *Phys. Rev.* **151**, 178–89 (1966). https://doi.org/10.1103/PhysRev.151.178

γ-surface and bring about the relative displacement across the slip plane by the Burgers vector. Once they have surmounted the potential barrier they fall into the next potential well and their potential energy is converted into kinetic energy which is radiated into the surrounding crystal. This is the origin of acoustic emission of dislocations when they glide. If the potential barrier has a height V_B, the dislocation speed is v, its core width is w and the atomic mass is m_a, the minimum average force imparted to an atom is $\sqrt{2m_a V_B}(v/w)$. Each atom on either side of the slip plane moves by $\pm b/2$ but the maximum in the potential barrier is reached in approximately half this distance $b/4$. Therefore the work done by the force on each atom is $\sim \sqrt{2m_a V_B}(bv/(4w))$. If the spacing of atoms along the dislocation line is s the power injected into the crystal lattice per unit length of dislocation is $\sim \sqrt{2m_a V_B}(v^2/(2sw))$, which yields a drag coefficient of

$$B_{rad} \approx \frac{\sqrt{2m_a V_B}}{2sw}. \tag{10.15}$$

Putting in numbers for aluminium, if we assume V_B is the energy of a metallic bond ~ 0.3 eV, $s \sim 3$ Å, $w \sim 5$ Å, we obtain $B_{rad} \sim 2.2 \times 10^{-4}$ Pa s. It arises at all temperatures because the vibrations are created by the dislocation itself, not by thermal excitation of crystal lattice vibrations. It is clearly a consequence of the atomic structure of the dislocation.

Recent research[47] has suggested that the crystal lattice vibrations created by a dislocation gliding at a sufficiently high speed can spawn the creation of further dislocations. This is a result of a resonance between the vibrations created by the dislocation and the natural vibrations of the crystal lattice. Although these resonances were identified in earlier papers,[48] it was not recognised that they may lead to mechanical instabilities resulting in the formation of new dislocations. A resonance arises when the group and phase velocities of the lattice waves travelling in the same direction as the dislocation equal the dislocation velocity. When this condition is satisfied the energy dissipated in the dislocation core cannot escape and its accumulation eventually leads to the creation of two further dislocations with equal and opposite Burgers vectors, thus conserving the total Burgers vector. This mechanism of generating dislocations kinematically was confirmed by molecular dynamics simulations in the NVE-ensemble of screw dislocations in tungsten by Verscheuren et al. (2018). A resonance was found at $0.24c_t$, where c_t is the transverse wave speed, accompanied by an avalanche of dislocation loops. The kinematic generation mechanism takes a small but finite time to act. A dislocation accelerated from rest will behave in one of two ways. If there is sufficient time for the generation mechanism to act when resonances occur, it will act repeatedly and avalanches of dislocations will be created, limiting the speeds of the dislocations to those at which

[47] Verschueren, J, Gurrutxaga-Lerma, B, Balint, DS, Sutton, AP and Dini D, *Phys. Rev. Lett.* **121**, 145502 (2018). https://doi.org/10.1103/PhysRevLett.121.145502

[48] Atkinson, W and Cabrera, N, *Phys. Rev.* **138**, A763–6 (1965). https://doi.org/10.1103/PhysRev.138.A763; Celli, V and Flytzanis, N, *J. Appl. Phys.* **41**, 4443–7 (1970). https://doi.org/10.1063/1.1658479; Ishioka, S, *J. Phys. Soc. Jpn.* **30**, 323–7 (1971). https://doi.org/10.1143/JPSJ.30.323; Caro, JA and Glass N, *J. Phys. Lett.* **45**, 1337–45 (1984). https://doi.org/10.1051/jphys:019840045080133700

the resonances occur.[49] Alternatively, if the dislocations are accelerated so rapidly there is insufficient time for the generation mechanism to act their speeds will not be limited by the resonance speeds. Of course if dislocations are injected at speeds above those of the resonances they are avoided altogether.

The kinematic generation of dislocations at resonant speeds may be relevant to adiabatic shear. Failure of metallic components subjected to very high rate deformation often occurs through the formation of narrow shear bands, 5–500 μm wide, in which the shear strain is extremely high. These shear bands are 'adiabatic' because they are formed so rapidly there is insufficient time for most of the heat generated within them to escape. It would be interesting and useful to test this idea by simulating an adiabatic shear failure allowing for the possibility of kinematic generation of dislocations at resonant dislocation speeds.

10.5 Hydrogen-assisted cracking in metals

10.5.1 Introduction

Hydrogen-assisted cracking is one of the most common forms of embrittlement of metals and alloys. In this section we review briefly the mechanisms that have been proposed in the light of experimental observations. It is fair to say that after nearly a century and a half of research there is no single widely accepted mechanism for hydrogen-assisted cracking. But hydrogen-assisted cracking remains a major concern for many industries including energy production, automobiles, aerospace, shipping, construction, defence, oil extraction and the burgeoning hydrogen economy.

In 1875 Johnson[50] presented a remarkable paper[51] to the Royal Society in which he described a series of elegant experiments that established a good deal of what is known today about the embrittlement of iron and steel by hydrogen. He observed that when iron is immersed for a few minutes in acids that result in the generation of hydrogen the toughness and ductility of the metal are reduced significantly. When a fractured surface of the embrittled iron was moistened it frothed as bubbles of hydrogen gas were released from the metal. The embrittlement was reversible because when the iron was left for sufficiently long for the hydrogen to escape, its ductility returned. The time taken for the ductility to return decreased if the iron was warmed. These observations showed it is hydrogen that can diffuse in the metal that is responsible for the embrittlement. If there were any remaining hydrogen trapped in the metal it was harmless, although as discussed below this is not true if the hydrogen is trapped in brittle hydride phases. He observed that if the iron was exposed to hydrogen gas its toughness and ductility were not reduced, proving that hydrogen in its molecular form was not responsible for the embrittlement. Johnson also proved that the embrittlement was due to the ingress of nascent hydrogen

[49] These limiting speeds may be much less than the shear wave speed and they have nothing to do with the divergence of the self-energy of the dislocation at the shear wave speed in elastodynamics.

[50] William H Johnson, British metallurgist.

[51] Johnson, WH, *Proc. R. Soc.* **23**, 168–79 (1875). https://doi.org/10.1098/rspl.1874.0024

and not to the acid by connecting iron wires to the electrodes of an electrochemical cell in which the electrolyte was Manchester town water so that there was no acid present. The iron attached to the electrode where hydrogen was released became embrittled while the iron attached to the electrode where oxygen was released retained its ductility.

10.5.2 Possible mechanisms

Hydrogen-enhanced decohesion (HEDE) was proposed by Pfeil[52] in his 1926 paper[53] regarding tensile tests during 'pickling' of ferritic steels. He observed intergranular and transgranular failures, and he suggested they were caused by a reduction in strength of metal–metal bonds. He did not offer an explanation for the reduction, and even today there does not seem to be any direct experimental evidence for weakening of metallic bonds caused by hydrogen.[54] Nevertheless hydrogen-induced decohesion remains a plausible mechanism, particularly for some intergranular fractures.

In 1969 Westlake[55] noticed that whenever hydrides form they are almost always brittle.[56] A mechanism arises in which hydrogen is attracted elastically to a loaded crack tip, leading to supersaturation and precipitation of a hydride particle. The hydride particle grows and it becomes brittle. It provides an easy path for crack growth. The crack pauses at the interface between the hydride and matrix until more hydrogen accumulates ahead of the crack and the process repeats. There are many observations of such a mechanism, and there does not appear to be any controversy about it. In the remainder of this section we consider the influence of hydrogen when hydrides are not formed.

Three years later Beachem[57] published a paper[58] that has proved seminal. Before the widespread use of high resolution scanning electron microscopes to image fracture surfaces he had the brilliant idea of looking at them with carbon replica techniques in a transmission electron microscope. At higher stress intensity factors he observed dimples on the fracture surfaces, indicating that the crack grew as a result of microvoid coalescence. As the stress intensity factor was reduced the extent of the plastic deformation decreased gradually and the fracture mode changed to quasi-cleavage, and eventually, at the lowest stress intensity, to intergranular fracture. In a sense, Beachem turned the thinking up to that point about hydrogen embrittlement on its head. He wrote

> The flat, brittle fractures produced at surprisingly low stresses in the laboratory or in service are therefore thought to be caused by severe, localized crack-tip deformation even when the cracks are propagating along prior austenite grain boundaries, and are

[52] Leonard Bessemer Pfeil FRS 1898–1969, British metallurgist.

[53] Pfeil, LB, *Proc. R. Soc. A* **112**, 182–95 (1926). https://doi.org/10.1098/rspa.1926.0103

[54] Several theories of HEDE are based on the assumption that hydrogen attracts electrons from neighbouring metallic bonds and that those bonds are always weakened as a result. But if the valence band of the metal is more than half full then taking electrons from metallic bonds will *strengthen* them because electrons are removed from *anti*-bonding states. The assumption on which these theories are based is valid only when the valence band is less than half full, so that electrons occupy bonding states in the metal only.

[55] Donald G Westlake, US metallurgist.

[56] Westlake, DG, *ASM Trans. Q.* **62**, 1000–6 (1969).

[57] Cedric D Beachem, US metallurgist.

[58] Beachem, CD, *Metall. Trans.* **3**, 441–55 (1972). https://doi.org/10.1007/BF02642048

not believed to be the result of a cessation, restriction, or exhaustion of ductility. Therefore, the term hydrogen-assisted cracking is probably more descriptive than hydrogen 'embrittlement' cracking.

(Reprinted by permission from Springer Nature Customer Service Centre GmbH. Published by Springer Nature in *Metallurgical Transactions* **3**, 441–55, A new model for hydrogen-assisted cracking (hydrogen "embrittlement"), by CD Beachem. Copyright 1972. Journal homepage: https://www.tms.org/pubs/journals/mt/.)

In the late 1970s, Lynch[59] proposed a mechanism of adsorption-induced dislocation emission (AIDE).[60] He suggested that *hydrogen at crack tips facilitates the emission of dislocations*, enabling the crack to advance into microvoids thereby limiting the degree of ductile tearing. This was based in part on his remarkable observations of similarities between the fracture surfaces seen in hydrogen-embrittled materials, and those seen in liquid metal embrittlement. In liquid metal embrittlement there is limited mutual solubility of the liquid and host metals, for example mercury and aluminium alloys, and cracks can grow at speeds up to several hundred $mm\,s^{-1}$. Therefore, the liquid metal must be confined to the crack surface. The liquid metal enhances the emission of dislocations at the crack, enabling the crack to advance into microvoids, resulting in fracture surfaces displaying small dimples. The same features are seen on the crack faces in Beachem's experiments. Lynch argued it is adsorption of hydrogen at the crack tip, not hydrogen in solution ahead of the crack tip, that leads to plasticity being more localised than during fracture in inert environments. This occurs because more dislocations are emitted from crack tips promoting the advance of cracks towards voids.[61] In inert environments little or no dislocation emission occurs from crack tips, and crack growth occurs by plastic deformation around crack tips. It is the extent of dislocation emission from crack tips compared with the extent of plasticity around crack tips that determines the degree of embrittlement. His papers contain many examples in fcc, bcc and hexagonal close-packed (hcp) metals. The generality of his observations is compelling.

In the late 1980s, Birnbaum[62] and co-workers proposed[63] hydrogen-enhanced localised plasticity, or HELP. In this case, hydrogen in solution ahead of the crack tip, not hydrogen adsorbed at the crack tip, is responsible for the localisation of plasticity reported by Beachem. Beachem (1972) had argued on the basis of his torsion tests that hydrogen in solid solution lowered the macroscopic flow stress, and he presumed it would also reduce the microscopic flow stress at crack tips. Birnbaum et al. (1994) presented experimental evidence that the activation energies for dislocation motion in Ni and Ni–C alloys decrease in the presence of hydrogen. Birnbaum et al. observed highly

[59] Stanley Peter Lynch 1945–, British and Australian metallurgist.

[60] Lynch, SP, *Scripta Metall.* **13**, 1051–6 (1979). http://dx.doi.org/10.1016/0036-9748(79)90202-3. The mechanism was first proposed in Lynch, SP and Ryan, NE, in *Proceedings of the 2nd International Congress on Hydrogen in Metals*, Paper 3D12, Pergamon Press: Oxford (1977). ISBN 0080221084.

[61] Lynch, S, *Corros. Rev.* **37**, 377–95 (2019). https://doi.org/10.1515/corrrev-2019-0017

[62] Howard Kent Birnbaum 1932–2005, US metallurgist.

[63] Shih, DS, Robertson, IM and Birnbaum, HK, *Acta Metall.* **36**, 111–24 (1988). http://dx.doi.org/10.1016/0001-6160(88)90032-6; Birnbaum, HK and Sofronis, P, *Mater. Sci. Eng. A* **176**, 191–202 (1994). http://dx.doi.org/10.1016/0921-5093(94)90975-X

localised ductile failure by microvoid coalescence, and they presumed it was facilitated by enhanced mobility of dislocations (edge, screw and mixed types) in fcc, bcc and hcp metals where hydrogen is in solution.

Following in-situ observations by transmission electron microscopy, Birnbaum and co-workers put forward an explanation of the enhanced dislocation mobility based on the screening of elastic interactions resulting from segregation of hydrogen to dislocations. The degree of segregation is sensitive to strain rate and temperature. There seems little reason to doubt that hydrogen is attracted elastically to dislocations forming Cottrell atmospheres. Hydrogen may also be segregated to dislocation cores. Once hydrogen has segregated to dislocations it will certainly reduce the strength and range of elastic interactions between dislocations.

However, the connection between reduced elastic interactions and enhanced dislocation mobility is less obvious. Obstacles to dislocation motion such as second phase particles and high angle grain boundaries are short range and the ability of these obstacles to impede dislocation motion is largely unaffected by elastic interactions. Intrinsic dislocation mobility is determined by properties of the core, such as the formation and migration energies of kinks, the mobilities of which are also largely unaffected by elastic interactions. One scenario where reduced elastic interactions certainly would enhance dislocation mobility is in stage I work hardening, where elastic multipole interactions between dislocations on parallel slip planes inhibit their motion. But these elastic interactions are relatively weak compared to the short-range interactions operating in stage II. Once dislocations are saturated with hydrogen they will be less able to attract other impurities through their elastic fields, but the hydrogen may pin a dislocation rather than enhance its mobility.

There are some experimental observations that cannot be explained by the HELP mechanism. When hydrogen enters some metals at the crack tip the crack may grow at speeds much faster than hydrogen can diffuse in the metal.[64] But the fracture surfaces may still display dimples indicating localised plasticity. It has also been shown experimentally by a number of researchers[65] that abrupt changes in the environment of a fracture test, such as changing the partial pressure of hydrogen gas or adding oxygen to the hydrogen gas, have an immediate effect on the crack growth rate and the appearance of the fracture surfaces. These observations are consistent with the view that hydrogen enhances dislocation emission at the crack tip, but not with the view it enhances dislocation mobility ahead of the crack tip.

Segregation of hydrogen to dislocations brings about a reduction in the free energy of the system, as described by the Gibbs adsorption isotherm. As Kirchheim[66] has noted[67] this may be viewed as a reduction in the free energy of *formation* of dislocation kinks. In that case hydrogen may increase the mobility of dislocations moving by a kink mechanism. But it may also raise the free energy of *migration* of kinks, since hydrogen atoms would either have to be dragged along with the kinks or kinks would have to break

[64] Lynch, SP, *Acta Metall.* **36**, 2639–61 (1988). http://dx.doi.org/10.1016/0001-6160(88)90113-7

[65] See references in Lynch (2019).

[66] Reiner Kirchheim 1943–, German materials scientist.

[67] Kirchheim, R, *Scripta Mater.* **67**, 767–70 (2012). https://doi.org/10.1016/j.scriptamat.2012.07.022

free from them. In bcc metals it is usually the case that the Peierls valleys are so deep that dislocations do move by a kink mechanism. But in some fcc metals, such as copper, which can deform by dislocation motion at liquid helium temperatures,[68] it seems unlikely that kinks are involved in dislocation motion, or they are so wide they are unlikely to attract more hydrogen atoms than straight dislocations. It is therefore unclear whether hydrogen segregation to dislocation kinks is necessary to increase dislocation mobility.

In 2001 Nagumo proposed that hydrogen is attracted to vacancies, based on interpretations of thermal desorption experiments.[69] This work suggests that hydrogen lowers the free energy of formation of vacancies, similar to Kirchheim's idea that hydrogen lowers the free energy of formation of dislocation kinks, so that hydrogen enhances the equilibrium concentration of vacancies. DFT calculations by Hickel et al.[70] also showed that the energy of hydrogen in solution in iron is reduced when the hydrogen sits in a vacancy. If the vacancies cluster this may lead to greater ease of formation of microvoids.

10.5.3 Theory of hydrogen in metals

If a hydrogen molecule is adsorbed chemically at a metal surface the anti-bonding orbital of the molecule becomes occupied and the hydrogen atoms separate and enter the metal. In 1980 Stott[71] and Zaremba[72] calculated the energy of inserting a hydrogen atom into a homogeneous, paramagnetic electron gas as a function of its density.[73] They found the energy of the hydrogen atom was minimised when the number density of the electron gas was approximately 0.017 Å^{-3} (i.e. $r_s = 4.6\, a_0$ where r_s is the radius of a sphere containing one electron and a_0 is the Bohr radius, which is 0.529 Å). This is a relatively small electron density, comparable to that of potassium metal, and it is consistent with the observation that hydrogen atoms tend to segregate to surfaces and vacancies in metals where they can find such relatively low electron densities. In these calculations they found the s-state of the hydrogen atom is doubly occupied so that it is present as H^-. The negative charge on this ion is screened by a depletion in the electron density surrounding it. The diameter of the screened hydrogen ion must be at least the Fermi wavelength, since this determines the smallest size of any local variation of the electronic charge density. But as with any point defect in a free electron gas there are also long-range Friedel oscillations in the charge density around it.

It has been found in DFT calculations of single hydrogen interstitials that the energy of the 1s-state of the hydrogen atom is less than the bottom of the s-p-d conduction band in iron.[74] When this occurs the 1s-orbital of the hydrogen atom is a localised state forming a delta-function in the density of electronic states, occupied by two electrons. In this

[68] For example, Basinski, ZS and Basinski, SJ, *Prog. Mater. Sci.* **36**, 89–148 (1992). http://dx.doi.org/10.1016/0079-6425(92)90006-S

[69] Nagumo, M, *ISIJ Int.* **41**, 590–8 (2001). https://doi.org/10.2355/isijinternational.41.590

[70] Hickel, T, Nazarov, R, McEniry, EJ, Leyson, G, Grabowski, B and Neugebauer, J, *JOM* **66**, 1399–405 (2014). http://dx.doi.org/10.1007/s11837-014-1055-3

[71] Malcolm J Stott, Canadian physicist.

[72] Eugene Zaremba, Canadian physicist.

[73] Stott, MJ and Zaremba, E, *Phys. Rev. B* **22**, 1564–83 (1980). https://doi.org/10.1103/PhysRevB.22.1564

[74] Paxton, AT (2018), private communication. Anthony Thomas Paxton 1953–, British materials physicist.

configuration the screened H^- ion cannot form covalent bonds to the metal because it is isoelectronic to helium. The local bonding is then an electrostatic interaction between the imperfectly screened H^- ion and the neighbouring imperfectly screened positive metal ions that have transferred some of their electronic charge to the proton. The screening is imperfect because the screening clouds overlap as their size is comparable to the spacing between the host metal atoms. This picture is consistent with the repulsion that is often observed between hydrogen atoms in transition metals at normal pressures. It is also consistent with hydrogen atoms acting as centres of dilation of the host metal lattice because the screened H^- ion is significantly larger than a neutral hydrogen atom. It is quite different from bonding in a transition metal hydride, where metal atoms are further apart than in the pure metal, enabling s-states on the hydrogen atoms to hybridise with d-states of the metal atoms forming a band of hybridised states. It is conceivable that the increase in the separation of transition metal atoms surrounding a hydrogen interstitial weakens those metal–metal bonds. The bonds may also be weakened electrostatically by the metal atoms surrounding the interstitial becoming slightly positively charged as a result of the transfer of one electron from them to the proton. It would be useful to test all these ideas with accurate electronic structure calculations.

Nazarov et al.[75] carried out DFT calculations for hydrogen atoms in vacancies in twelve fcc metals. They found hydrogen was bound to vacancies in all twelve metals, with binding energies from 0.17 to 0.86 eV. In all cases the hydrogen atom was off-centre in the vacancy. As the chemical potential of hydrogen is raised, for example by exposing the metal to higher partial pressures of hydrogen gas, they found H_2 molecules are more stable in a vacancy than two H-atoms in silver, gold and lead. At extremely high hydrogen partial pressures they found up to eight stable H_2 molecules inside a vacancy in silver. In contrast two H-atoms in nickel and rhodium never bonded to each other in a vacancy.

10.5.4 Suggestions for further research

This brief review suggests the following points for further research:

1. Is there any *direct* experimental evidence for the weakening of metal–metal bonding by hydrogen with respect to either tensile or shear deformation? Dislocation glide involves bond switching events where bonds crossing the slip plane are broken and reformed. Haydock[76] coined the term 'bond mobility'[77] to describe the ease with which bonds are broken and reformed in such a process. Strong directional bonds like those in diamond have very limited mobility at room temperature, whereas those in pure copper are much more mobile. Therefore, a sharper question is to ask whether hydrogen increases the mobility of bonds at the surface and within the interior of a metal.

[75] Nazarov, R, Hickel, T and Neugebauer, J, *Phys. Rev. B* **89**, 144108 (2014). https://doi.org/10.1103/PhysRevB.89.144108

[76] Roger Haydock, US physicist.

[77] Haydock, R, *J. Phys. C: Solid State Phys.* **14** 3807 (1981). https://doi.org/10.1088/0022-3719/14/26/016

2. Liquid lithium is a potent embrittling element of some transition metals and like hydrogen it has a half-filled valence s-shell. But lithium is a larger atom owing to its full 1s-core. A detailed comparison of the local electronic structures of relaxed configurations of hydrogen, helium and lithium interstitials in solid solution would highlight the unique chemistry of hydrogen in these metals.
3. Replacing hydrogen with deuterium will highlight the role of the mass of the atom. For example, if dislocation mobility is enhanced by hydrogen then replacing hydrogen with deuterium should reduce dislocation mobility at a given temperature and stress state, assuming the hydrogen/deuterium is transported with the dislocation.
4. How does hydrogen enhance the emission of dislocation loops from cracks? This question is germane to the AIDE mechanism. Answering this question requires a multi-million atom 3D simulation with a realistic crack geometry including steps on the crack front where loops can nucleate heterogeneously. It also requires a credible model of interatomic forces for hydrogen in the metal.

Recommended books

Elasticity

A treatise on the mathematical theory of elasticity Love, AEH, Dover Publications: New York (1944), reprint of the 4th edition of this book published in 1927. ISBN 0-486-60174-9. A classic, not so easy to read because of the dated notation, but well worth dipping into including the appendices.

Theory of elasticity Landau, LD and Lifshitz, EM, 3rd edn. Pergamon Press: Oxford (1986). Volume 7 of the famous *Course on Theoretical Physics*. ISBN 0-08-033916-6. An advanced treatment with not one superfluous word. Well worth careful study. The chapter on dislocations was written jointly with AM Kosevich and covers a lot of ground in just 24 pages.

Physical properties of crystals Nye, JF, Oxford University Press: Oxford (1985). ISBN 0-19-851165-5. An exceptionally clear account of tensor properties of crystals. Chapter 8 treats the elastic constant tensor in different crystal symmetries.

Dynamical theory of crystal lattices Born, M and Huang, K, Oxford Classic Texts in the Physical Sciences, Oxford University Press: Oxford and New York (1998). ISBN 978-0198503699. This was one of the first books to attempt to derive properties of (non-metallic) crystals from fundamental principles at the atomic scale (it was first published by Oxford University Press in 1954). Chapter 3 deals with elasticity.

Statistical mechanics of elasticity Weiner, JH, Dover Publications: New York (2002). ISBN 0-486-42260-7. First published in 1983, this is another book that treats elasticity at the atomic scale.

Thermodynamics of crystals Wallace, DC, John Wiley & Sons Inc.: New York (1972). ISBN 978-0471918554. This book provides an exceptionally clear treatment of the thermodynamics of crystals including elasticity. It was republished by Dover Publications in 2003 with ISBN 978-0486402123.

Theory of elasticity and crystal defects

Introduction to elasticity theory for crystal defects Balluffi, RW, World Scientific: Singapore (2017). ISBN 9814749729. A comprehensive and systematic treatment of defects and their interactions in anisotropic elasticity, with many worked examples. An outstanding pedagogical text.

Elastic models of crystal defects Teodosiu, C, Springer-Verlag: Berlin (1982). ISBN 0-387-11226-X. A wonderful book providing a rigorous treatment of elasticity of defects.

Micromechanics of defects in solids Mura, T, Kluwer Academic Publishers: Dordrecht (1991). ISBN 90-247-3256-5. An excellent, wide-ranging mathematical treatment of elasticity and defects.

Collected works of J. D. Eshelby Markenscoff, X and Gupta, A (eds.), Springer: Dordrecht (2006). ISBN 9781402044168 hard back, ISBN 9789401776448 soft cover. Eshelby wrote only 57 papers, but every one of them is a gem. They are all here.

Anisotropic continuum theory of lattice defects Bacon, DJ, Barnett, DM and Scattergood, RO, *Prog. Mater. Sci.* **23**, 51–262 (1978). ISBN 0080242472. An excellent introduction to the modern theory of anisotropic elasticity and its application to defects in crystals.

Theory of crystal defects Gruber, B (ed.), Proceedings of the Summer School held in Hrazany in September 1964, Academia Publishing House of Czechoslovak Academy of Sciences: Prague (1966). Kroupa's chapter on dislocation loops is particularly illuminating.

The physics of metals. 2. Defects Hirsch, PB (ed.), Cambridge University Press: Cambridge (1975). ISBN 0 521 200776. Contains six chapters on different aspects of the physics of defects in metals written by leaders in the field. The absence of heavy mathematical analysis and the focus on physics renders it very readable.

Mechanics of solid materials Asaro, RJ and Lubarda, VA, Cambridge University Press: Cambridge (2006). ISBN 978-0-521-85979-0. A comprehensive treatment from an engineering perspective of continuum mechanics, linear elasticity, micromechanics, thin films and interfaces, plasticity and viscoplasticity and biomechanics, with solved problems.

Dislocations

Introduction to dislocations Hull, D and Bacon, DJ, Pergamon Press: Oxford (1984). ISBN 0-08-028720-4. This book should be read in conjunction with Chapter 6. Very readable, covering the basic phenomenology of dislocations.

Elementary dislocation theory Weertman, J and Weertman, JR, Oxford University Press: Oxford (1992). ISBN 0-19-506900-5. Another very readable introduction to dislocations.

Dislocations and plastic flow in crystals Cottrell, AH, The Clarendon Press: Oxford (1953). The first monograph on dislocations, and still one of the best, written by a pioneer of the field. Few authors can match Cottrell's clarity and insight.

Dislocations in crystals Read, WT, McGraw-Hill Book Company: New York (1953). Another early monograph on dislocations, written by a pioneer of the field.

Theory of dislocations Hirth, JP and Lothe, J, Krieger Publishing Company: Malabar, FL (1992). ISBN 0-89464-617-6. For many this is the last word on dislocations.

Dislocations Friedel, J, Addison Wesley: Reading, MA (1964). A classic written by a physicist for metallurgists and engineers in industry as well as researchers in universities. It is an expanded translation of his book *Les dislocations*, which was written in French and published by Gauthier-Villars: Paris in 1956.

Theory of crystal dislocations Nabarro, FRN, Dover Publications: New York (1987). ISBN 0-486-65488-5. A major treatise on dislocations written by one of the central figures in the development of the subject.

Anisotropic elasticity theory of dislocations Steeds, JW, Oxford University Press: Oxford (1973). Systematic treatment of dislocations in different crystal symmetries using anisotropic elasticity.

Dislocations in solids Nabarro, FRN, Duesbery, MS, Hirth, JP and Kubin, L (eds.), Vols. 1–16, North-Holland Publishing Company: Amsterdam (1979–2009). Each volume contains reviews written by leading researchers on various aspects of dislocations. Eye-wateringly expensive books.

Dislocations, mesoscale simulations and plastic flow Kubin, LP, Oxford University Press: Oxford (2013). ISBN 978-0-19-852501-1. Modern survey of how dislocations control plasticity in elemental metals, from the atomic scale to the microstructural level and the connection to continuum plasticity. Strong on experimental observations with an incisive review of different techniques for modelling dislocation dynamics in three dimensions (3D).

Computer simulations of dislocations Bulatov, VV and Cai, W, Oxford University Press: Oxford (2006). ISBN 0-19-852614-8. A practical 'how to do it' introduction to computer simulations of dislocation statics and dynamics in 3D.

Strengthening mechanisms of crystal plasticity Argon, AS, Oxford University Press: Oxford (2008). ISBN 978-0-19-851600-2. Systematic treatment of the physics of strengthening mechanisms including the lattice resistance, solid-solution strengthening, precipitation strengthening and strain hardening, all supported by experimental observations.

Cracks

Mathematical theory of dislocations and fracture Lardner, RW, University of Toronto Press: Toronto (1974). ISBN 0-8020-5277-0. Particularly good on continuous distributions of dislocations and cracks.

Physics of fracture Thomson, R, *Solid State Phys.* **39**, 2–129 (1986). A physicist's view on cracks and dislocations. Particularly good on shielding of cracks by dislocations, and limitations of continuum theories of cracks and fracture.

Dislocation based fracture mechanics Weertman, J, World Scientific: Singapore (1996). ISBN 9810226209. A nice introduction to the treatment of the modelling of cracks by distributions of dislocations. Also very good on shielding of cracks by dislocations.

Solution of crack problems Hills, DA, Kelly, PA, Dai, DN and Korsunsky, AM, Kluwer Academic Publishers: Dordrecht (1996). ISBN 978-90-481-4651-2. An advanced treatment of a wide range of crack problems using distributions of dislocations.

History of elasticity

History of strength of materials Timoshenko, SP, Dover Publications: New York (1983), reprint of the book published originally in 1953. ISBN 0-486-61187-6. A fascinating account of the historical development of elasticity and the people who worked on it from the seventeenth century to 1950.

References

http://blogs.nature.com/news/2014/05/global-scientific-output-doubles-every-nine-years.html

http://functions.wolfram.com/EllipticIntegrals/EllipticK/introductions/CompleteEllipticIntegrals/05/

http://www.math.cmu.edu/~wn0g/

Alshits, VI, in *Elastic strain fields and dislocation mobility*, ed. VL Indenbom and J Lothe, North-Holland: Amsterdam (1992), pp.625–97. ISBN: 0444887733.

Antolovich, SD and Conrad, H, *Mater. Manuf. Process.* **19**, 587–610 (2004). http://dx.doi.org/10.1081/AMP-200028070

Ashcroft, NW and Mermin, ND, *Solid state physics*, Brooks/Cole Publishing (1976), pp.437–40. ISBN 978-0030839931.

Atkinson, W and Cabrera, N, *Phys. Rev.* **138**, A763–6 (1965). https://doi.org/10.1103/PhysRev.138.A763

Bak, P, *How nature works*, Oxford University Press: Oxford (1997), pp.1–2. ISBN 0198501641.

Bacon, DJ, Barnett, DM and Scattergood RO, *Anisotropic continuum theory of lattice defects*, *Prog. Mater. Sci.* **23**, 51–262 (1979). ISBN 0080242472. https://doi.org/10.1016/0079-6425(80)90007-9

Barenblatt, GI, *J. Appl. Math. Mech.* **23**, 622–36 (1959). https://doi.org/10.1016/0021-8928(59)90157-1

Barenblatt, GI, *J. Appl. Math. Mech.* **23**, 1009–29 (1959). https://doi.org/10.1016/0021-8928(59)90036-X

Barenblatt, GI, *Adv. Appl. Mech.* **7**, 55–129 (1962). https://doi.org/10.1016/S0065-2156(08)70121-2

Basinski, ZS, *Phil. Mag.* **4**, 393–432 (1959). https://doi.org/10.1080/14786435908233412

Basinski, ZS and Basinski, SJ, *Phil. Mag.* **9**, 51–80 (1964). https://doi.org/10.1080/14786436408217474

Basinski, ZS and Basinski, SJ, *Prog. Mater. Sci.* **36**, 89–148 (1992). http://dx.doi.org/10.1016/0079-6425(92)90006-S

Beachem, CD, *Metall. Trans.* **3**, 441–55 (1972). https://doi.org/10.1007/BF02642048

Bilby, BA, *Proc. Phys. Soc. A* **63**, 191 (1950). https://doi.org/10.1088/0370-1298/63/3/302

Bilby, BA, Cottrell AH and Swinden KH, *Proc. R. Soc. A* **272**, 304–14 (1963). https://doi.org/10.1098/rspa.1963.0055

Bilby, BA, and Eshelby, JD, Dislocations and the theory of fracture, in *Fracture*, ed. H Liebowitz, Vol. 1, Academic Press: New York (1968), Chapter 2.

Birnbaum, HK and Sofronis, P, *Mater. Sci. Eng. A* **176**, 191–202 (1994). http://dx.doi.org/10.1016/0921-5093(94)90975-X

Bosvieux, C and Friedel, J, *J. Phys. Chem. Solids* **23**, 123–36 (1962). https://doi.org/10.1016/0022-3697(62)90066-5

Brown, JM, Abramson, EH and Angel, RJ, *Phys. Chem. Minerals* **33**, 256–65 (2006). https://doi.org/10.1007/s00269-006-0074-1

Brown, LM, *Metall. Trans. A* **22**, 1693–708 (1991). https://doi.org/10.1007/BF02646493

Brown, LM, in *Dislocations in solids*, ed. FRN Nabarro and MS Duesbery, Elsevier: Amsterdam (2002), pp.193–210. ISBN 0-444-50966-6.
Brown, LM, *Mater. Sci. Technol.* **28**, 1209–32 (2012). https://doi.org/10.1179/174328412X13409726212768
Brown, LM, *Phil. Mag.* **96**, 2696–713 (2016). https://doi.org/10.1080/14786435.2016.1211330
Burgers, JM, *Koninklijke Nederlandsche Akademie van Wetenschappen* **42**, 293 (1939). http://www.dwc.knaw.nl/toegangen/digital-library-knaw/?pagetype=publDetail&pId=PU00014649
Burridge, R and Knopoff, L, *Bull. Seismol. Soc. Am.* **54**, 1875–88 (1964). http://bssa.geoscienceworld.org/content/ssabull/54/6A/1875.full.pdf
Caro, JA and Glass N, *J. Phys. Lett.* **45**, 1337–45 (1984). https://doi.org/10.1051/jphys:019840045080133700
Celli, V and Flytzanis, N, *J. Appl. Phys.* **41**, 4443–7 (1970). https://doi.org/10.1063/1.1658479
Chang, S-J and Ohr, SM, *J. Appl. Phys.* **52**, 7174–81 (1981). http://dx.doi.org/10.1063/1.328692
Chang, S-J and Ohr, SM, *Int. J. Fract.* **23**, R3–R6 (1983). https://doi.org/10.1007/BF00020160
Chia, KY and Burns, SJ, *Scripta Metall.* **18**, 467–72 (1984). https://doi.org/10.1016/0036-9748(84)90423-X
Christian, JW, and Vitek, V, *Rep. Prog. Phys.* **33**, 307 (1970). https://doi.org/10.1088/0034-4885/33/1/307
Conrad, H and Sprecher, AF, in *Dislocations in solids*, Vol. 8, North-Holland: Amsterdam (1989), pp.497–541. ISBN 978-0444705150
Conrad, H, *Mater. Sci. Eng. A* **287**, 276–287 (2000). https://doi.org/10.1016/S0921-5093(00)00786-3
Cottrell, AH, *Dislocations and plastic flow in crystals*, Clarendon Press: Oxford (1953), Chapter 10.
Cottrell, AH and Stokes, RJ, *Proc. R. Soc. A* **233**, 17–34 (1955). https://doi.org/10.1098/rspa.1955.0243
Cottrell, AH, in *Dislocations in solids*, Vol. 11, ed. FRN Nabarro and MS Duesbery, Elsevier: Amsterdam (2002), pp.vii–xvii. ISBN 0-444-50966-6.
Dehlinger, U, *Ann. Phys.* **394** 749–93 (1929). https://doi.org/10.1002/andp.19293940702
Dudarev, SL and Sutton, AP, *Acta Mater.* **125**, 425–30 (2017). https://doi.org/10.1016/j.actamat.2016.11.060.
Dugdale, DS, *J. Mech. Phys. Solids* **8**, 100–4 (1960). https://doi.org/10.1016/0022-5096(60)90013-2
Economou, EN, *Green's functions in quantum physics*, Springer-Verlag: Berlin (1983). ISBN 978-3642066917.
Eshelby, JD, *Phil. Mag.* **40**, 903–12 (1949). https://doi.org/10.1080/14786444908561420
Eshelby, JD, *Phil. Trans. R. Soc. A* **244**, 87–111 (1951). https://doi.org/10.1098/rsta.1951.0016
Eshelby, JD, *Solid State Phys.* **3**, 79–144 (1956). https://doi.org/10.1016/S0081-1947(08)60132-0
Eshelby, JD, *Proc. R. Soc. A* **241**, 376–96 (1957). https://doi.org/10.1098/rspa.1957.0133
Eshelby, JD, *Proc. R. Soc. A* **252**, 561–9 (1959). https://doi.org/10.1098/rspa.1959.0173.
Eshelby, JD, in *Inelastic behaviour of solids*, ed. MF Kanninen, WF Adler, AR Rosenfield and RI Jaffee, McGraw-Hill: New York (1970), pp.77–115.
Eshelby, JD, *J. Elast.* **5**, 321–35 (1975). https://doi.org/10.1007/BF00126994
Eshelby, JD, in *Point defect behaviour and diffusional processes*, ed. RE Smallman and E Harris, The Metals Society, London (1977), pp.3–10.
Eshelby, JD, in *Continuum models of discrete systems*, ed. E Kröner and K-H Anthony, University of Waterloo Press: Waterloo, ON (1980), pp.651–65.

Eshelby, JD, Frank, FC and Nabarro, FRN, *Phil. Mag.* **42**, 351–64 (1951). https://doi.org/10.1080/14786445108561060

Fehlner, WR and Vosko, SH, *Can. J. Phys.* **54**, 2159–69 (1976). https://doi.org/10.1139/p76-256

Finnis, MW, *Interatomic forces in condensed matter*, Oxford University Press: Oxford (2003). ISBN 978-0198509776.

Frank, FC, *Phil. Mag.* **42**, 809 (1951). http://dx.doi.org/10.1080/14786445108561310

Frank, FC, *Proc. R. Soc. A* **371**, 136 (1980). http://dx.doi.org/10.1098/rspa.1980.0069

Frank, FC, and Read, WT, *Phys. Rev.* **79**, 722 (1950). https://doi.org/10.1103/PhysRev.79.722

Frenkel, J, and Kontorova, T, *J. Phys. Acad. Sci. USSR* **1**, 137–49 (1939).

Friedel, J, *Les dislocations*, Gauthier-Villars: Paris (1956), p.215.

Giustino, F, *Rev. Mod. Phys.* **89** 015003 (2017). https://doi.org/10.1103/RevModPhys.89.015003; Erratum *Rev. Mod. Phys.* **91**, 019901 (2019). https://doi.org/10.1103/RevModPhys.91.019901.

Goldstein, H, *Classical mechanics*, Addison-Wesley: Reading, MA (1980). ISBN 0-201-02969-3.

Gradshteyn, IS and Ryzhik, IM, *Table of integrals, series and products*, 5th edition corrected and enlarged by Alan Jeffrey, Academic Press Inc.: Orlando FL (1980). ISBN 0-12-294760-6. Expressions for I_1, I_2 and I_3 appear in 3.133, numbers 1, 7 and 13 respectively on pp.220–2.

Green, G, *Trans. Camb. Phil. Soc.*, **7**, 1 (1839). https://archive.org/details/transactionsofca07camb

Griffith, AA, *Phil. Trans. R. Soc. A* **221**, 163–98 (1921). https://doi.org/10.1098/rsta.1921.0006

Haines, MG, *Proc. Phys. Soc.* **74**, 576–84 (1959). https://doi.org/10.1088/0370-1328/74/5/310

Hall, EO, *Proc. Phys. Soc.* **B64**, 747–53 (1951). https://doi.org/10.1088/0370-1301/64/9/303

Hammad, A, Swinburne, TD, Hasan, H, Del Rosso, S, Iannucci, L and Sutton, AP, *Proc. R. Soc. A* **471** 20150171 (2015). http://dx.doi.org/10.1098/rspa.2015.0171.

Haydock, R, *J. Phys. C: Solid State Phys.* **14** 3807 (1981). https://doi.org/10.1088/0022-3719/14/26/016

Head, AK, and Louat, N, *Aust. J. Phys.* **8**, 1 (1955). https://doi.org/10.1071/PH550001

Heine, V, *Solid State Physics* eds. H Ehrenreich, F Seitz and D Turnbull **35**, p.114-120 (1980). https://doi.org/10.1016/S0081-1947(08)60503-2

Hickel, T, Nazarov, R, McEniry, EJ, Leyson, G, Grabowski, B and Neugebauer, J, *JOM* **66**, 1399–405 (2014). http://dx.doi.org/10.1007/s11837-014-1055-3

Hirsch, PB, Roberts, SG and Samuels, J, *Proc. R. Soc. A* **421**, 25–53 (1989). https://doi.org/10.1098/rspa.1989.0002

Hirth, JP, and Lothe, J, *Theory of dislocations*, 2nd edn., Krieger Publishing Company: Malabar, FL (1992), pp.117 and 209. ISBN 0-89464-617-6.

Hirth, JP, and Lothe, J, *Theory of dislocations*, 2nd edn., Krieger Publishing Company: Malabar, FL (1992), section 15-2. ISBN 0-89464-617-6.

Hoekstra, J, Sutton, AP, Todorov, TN and Horsfield, AP, *Phys. Rev. B* **62**, 8568–71 (2000). https://doi.org/10.1103/PhysRevB.62.8568

Horton, JA and Ohr, SM, *J. Mater. Sci.* **17**, 3140–8 (1982). https://doi.org/10.1007/BF01203476

Ishioka, S, *J. Phys. Soc. Jpn.* **30**, 323–7 (1971). https://doi.org/10.1143/JPSJ.30.323

Jackson, PJ, *Acta Metall.* **33**, 449–54 (1985). https://doi.org/10.1016/0001-6160(85)90087-2

Jardine, L, *The curious life of Robert Hooke*, Harper Collins: London (2003). ISBN 978-0007151752.

Jensen, HJ, *Self-organized criticality*, Cambridge University Press: Cambridge (1998). ISBN: 0521483719

Jinha, AE, *Learned Publishing* **23**, 258–63 (2010). https://doi.org/10.1087/20100308.

Johnson, WH, *Proc. R. Soc.* **23**, 168–79 (1875). https://doi.org/10.1098/rspl.1874.0024

Jokl, ML, Vitek, V and McMahon Jr, CJ, *Acta Metall.* **28**, 1479–88 (1980). https://doi.org/10.1016/0001-6160(80)90048-6

Kellogg, OD, *Foundations of potential theory*, Dover Publications: New York (1954), pp.192–6. ISBN 0486601447.

Kirchheim, R, *Scripta Mater.* **67**, 767–70 (2012). https://doi.org/10.1016/j.scriptamat.2012.07.022

Kröner, E, *Kontinuumstheorie der Versetzungen und Eigenspannungen*, Springer-Verlag: Berlin (1958). ISBN 978-3540022619.

Kroupa, F, in *Theory of crystal defects*, Proceedings of the Summer School held in Hrazany in September 1964, Academia Publishing House of the Czechoslovak Academy of Sciences (1966), pp.275–316.

Kubin, LP, *Dislocations, mesoscale simulations and plastic flow*, Oxford University Press: Oxford (2013), section 2.4. ISBN 978-0-19-852501-1.

Kuhlmann-Wilsdorf, D, *Mater. Sci. Eng.* **86**, 53–66 (1987). https://doi.org/10.1016/0025-5416(87)90442-3.

Landau, LD and Lifshitz, EM, *Theory of elasticity*, 3rd. edn., Pergamon Press: Oxford (1986), p.111. ISBN 0-08-033916-6.

Lardner, RW, *Mathematical theory of dislocations and fracture*, University of Toronto Press: Toronto (1974), Chapter 5. ISBN 0-8020-5277-0.

Leibfried, G, *Z. Phys.* **130**, 244 (1951). https://doi.org/10.1007/BF01337695

Leibfried, G and Breuer, N, *Point defect in metals I*, Springer-Verlag: Berlin (1978), p.146. ISBN 978-3662154489.

Lund, F and Scheihing B, *Phys. Rev. B* **99** 214102 (2019). https://doi.org/10.1103/PhysRevB.99.214102.

Lynch, S, *Corros. Rev.* **37**, 377–95 (2019). https://doi.org/10.1515/corrrev-2019-0017.

Lynch, SP, *Scripta Metall.* **13**, 1051–6 (1979). http://dx.doi.org/10.1016/0036-9748(79)90202-3

Lynch, SP, *Acta Metall.* **36**, 2639–61 (1988). http://dx.doi.org/10.1016/0001-6160(88)90113-7

Lynch, SP and Ryan, NE, in Proceedings of the 2nd International Congress on Hydrogen in Metals, Paper 3D12, Pergamon Press: Oxford (1977). ISBN 0080221084.

Majumdar, BS and Burns, SJ, *Int. J. Fract.* **21**, 229–40 (1983). https://doi.org/10.1007/BF00963390

Markenscoff, X, *J. Elast.* **49**, 163–6 (1998). https://doi.org/10.1023/A:1007474108433

Mason, DR, Le Page, J, Race, CP, Foulkes, WMC, Finnis, MW and Sutton, AP, *J. Phys. Condens. Matter* **19**, 436209 (2007). https://doi.org/10.1088/0953-8984/19/43/436209

Maxwell, JC, *Phil. Mag.* **27**, 294–99 (1864). http://www.tandfonline.com/doi/abs/10.1080/14786446408643668

Mott, NF, *Proc. R. Soc. A* **220**, 1 (1953). https://doi.org/10.1098/rspa.1953.0167

Mura, T, *Micromechanics of defects in solids*, 2nd edn., Kluwer Academic Publishers: Dordrecht (1991). ISBN 90-247-3256-5.

Muskhelishvili, NI, *Singular integral equations*, Dover Publications: New York (2008). ISBN 9780486462424.

Nabarro, FRN, *Proc. Phys. Soc.* **59**, 256 (1947). https://doi.org/10.1088/0959-5309/59/2/309

Nabarro, FRN, *Phil. Mag.* **42**, 1224 (1951). http://dx.doi.org/10.1080/14786444108561379

Nabarro, FRN, *Adv. Phys.* **1**, 269–394 (1952). https://doi.org/10.1080/00018735200101211

Nabarro, FRN, *Acta Metall.* **38**, 161–4 (1990). https://doi.org/10.1016/0956-7151(90)90044-H

Nabarro, FRN, Basinski, ZS and Holt, DB, *Adv. Phys.* **13**, 193–323 (1964). https://doi.org/10.1080/00018736400101031

Nagumo, M, *ISIJ Int.* **41**, 590–8 (2001). https://doi.org/10.2355/isijinternational.41.590

Nazarov, R, Hickel, T and Neugebauer, J, *Phys. Rev. B* **89**, 144108 (2014). https://doi.org/10.1103/PhysRevB.89.144108

Neumann, FE, *Vorlesungen über die Theorie der Elastizität der festen Körper und des Lichtäthers*, ed. OE Meyer, BG Teubner-Verlag: Leipzig (1885).

Nye, JF, *Physical properties of crystals*, Oxford University Press: Oxford (1957), p.21. ISBN 0-19-851165-5.

Okazaki, K, Kagawa, M and Conrad, H, *Scripta Metall.* **12**, 1063–8 (1978). https://doi.org/10.1016/0036-9748(78)90026-1

Orowan, E, *Z. Phys.* **89**, 634 (1934). https://doi.org/10.1007/BF01341480

Orowan, E, *Trans. Inst. Eng. Shipbuilders, Scotland* **89**, 165 (1945).

Orowan, E, *Rep. Prog. Phys.* **12**, 185–232 (1949). https://doi.org/10.1088/0034-4885/12/1/309

Peach, MO and Koehler, JS, *Phys. Rev.* **80**, 436 (1950). https://doi.org/10.1103/PhysRev.80.436

Peierls, R, *Proc. Phys. Soc.* **52**, 34 (1940). https://doi.org/10.1088/0959-5309/52/1/305

Peierls, R, in *Dislocation dynamics*, ed. AR Rosenfield, GT Hahn, AL Bement Jr and RI Jaffee, McGraw-Hill: New York (1968). A more complete account of the genesis of the Peierls–Nabarro model is given by Peierls on pages xiii–xiv. Peierls' original derivation of the integral equation and his solution is transcribed on pages xvii–xx.

Peierls, RE, *Proc. R. Soc. A* **371**, 28–38 (1980). https://doi.org/10.1098/rspa.1980.0053

Petch, NJ, *J. Iron Steel Inst.* **174**, 25 (1953).

Pettifor, DG, *Commun. Phys.* **1**, 141–6 (1976).

Pfeil, LB, *Proc. R. Soc. A* **112**, 182–95 (1926). https://doi.org/10.1098/rspa.1926.0103.

Polanyi, M, *Z. Phys.* **89**, 660 (1934). https://doi.org/10.1007/BF01341481

Prigogine, I, *From being to becoming*, WH Freeman and Co.: San Francisco (1980), pp.7–8. ISBN 0-7167-1108-7.

Read, WT, and Shockley, W, *Phys. Rev.* **78**, 275 (1950). https://doi.org/10.1103/PhysRev.78.275

Ready, AJ, Haynes, PD, Rugg, D, and Sutton, AP, *Phil. Mag.* **97**, 1129–43 (2017). http://dx.doi.org/10.1080/14786435.2017.1292059

Rice, JR, *J. Appl. Mech.* **35**, 379–86 (1968). https://doi.org/10.1115/1.3601206

Robinson, A, *The last man who knew everything*, One World: Oxford (2006). ISBN 978-0452288058.

Routh, EJ, *A treatise on analytical statics*, Vol. 2, Cambridge University Press: London (1892), pp.106–8. https://archive.org/details/treatiseonanalyt02routiala/page/n8

Samuels, J, and Roberts, SG, *Proc. R. Soc. A* **421**, 1–23 (1989). https://doi.org/10.1098/rspa.1989.0001

Shih, DS, Robertson, IM and Birnbaum, HK, *Acta Metall.* **36**, 111–24 (1988). http://dx.doi.org/10.1016/0001-6160(88)90032-6

Siems, R, *Phys. Stat. Sol.* **30**, 645–58 (1968). https://doi.org/10.1002/pssb.19680300226

Sokolnikoff, IS, *Mathematical theory of elasticity*, 2nd edn., McGraw-Hill Book Company Inc.: New York (1956), section 48. ISBN 978-0070596290.

Sorbello, RS, *Phys. Rev. B* **23**, 5119–27 (1981), https://doi.org/10.1103/PhysRevB.23.5119

Sorbello, RS, *Solid State Phys.* **51**, 159–231 (1998). https://doi.org/10.1016/S0081-1947(08)60191-5

Sorbello, RS, and Chu, CS, *IBM J. Res. Dev.* **32**, 58–62 (1988). http://dx.doi.org/10.1147/rd.321.0058

Sprecher, AF, Mannan, SL and Conrad, H, *Acta Metall.* **34**, 1145–62 (1986). https://doi.org/10.1016/0001-6160(86)90001-5

Stokes, GG, *Trans. Camb. Phil. Soc.* **9**, 1–62 (1849). https://archive.org/stream/transactionsofca09camb

Stott, MJ, and Zaremba, E, *Phys. Rev. B* **22**, 1564–83 (1980). https://doi.org/10.1103/PhysRevB.22.1564

Stroh, AN, *Proc. R. Soc. A* **223**, 404–14 (1954). https://doi.org/10.1098/rspa.1954.0124

Sutton, AP and Balluffi, RW, *Interfaces in crystalline materials*, Oxford classic texts in the physical sciences, Clarendon Press: Oxford (2006). ISBN: 978-0-19-921106-7.

Swinburne, TD and Dudarev, SL, *Phys. Rev. B* **92**, 134302 (2015). https://doi.org/10.1103/PhysRevB.92.134302.

Taylor, GI, *Proc. R. Soc. A* **145**, 388 (1934). https://doi.org/10.1098/rspa.1934.0106

Teodosiu, C, *Elastic models of crystal defects*, Springer-Verlag: Berlin (1982), sections 6.2 and 23.2. ISBN 0-387-11226-X.

Tewary, VK, *Adv. Phys.* **22**, 757–810 (1973). https://doi.org/10.1080/00018737300101389

Thomson, R, *Solid State Phys.* **39**, 2–129 (1986). https://doi.org/10.1016/S0081-1947(08)60368-9

Thomson, R, Hsieh, C, and Rana, V, *J. Appl. Phys.* **42**, 3154–60 (1971). https://doi.org/10.1063/1.1660699

Thomson, W (Lord Kelvin), Article 37 of Volume 1 of *Mathematical and Physical Papers* (1848), p.97. https://archive.org/details/mathematicaland01kelvgoog

Timoshenko, SP, *History of strength of materials*, Dover Publications: New York (1983), p.108. ISBN 0-486-61187-6.

Ting, TCT, *Anisotropic elasticity*, Oxford University Press: Oxford (1996), pp.159–61. ISBN: 0195074475.

Tittmann, BR and Bömmel, HE, *Phys. Rev.* **151**, 178–89 (1966). https://doi.org/10.1103/PhysRev.151.178

Todorov, TN, Dundas, D, Lü, J-T, Brandbyge M, and Hedegård, P, *Eur. J. Phys.* **35**, 065004 (2014). https://doi.org/10.1088/0143-0807/35/6/065004

Troitskii, OA, JETP Lett. **10**, 11–14 (1969). http://www.jetpletters.ac.ru/ps/1686/article_25672.shtml

Troitskii, OA, *Strength Mater.* **7**, 804–9 (1975). https://doi.org/10.1007/BF01522653.

Troitskii, OA and Likhtman, VI, *Dokl. Akad. Nauk SSSR* **148**, 332–4 (1963). In Russian.

Valladares, A, White, JA and Sutton, AP, *Phys. Rev. Lett.* **81**, 4903–6 (1998). https://doi.org/10.1103/PhysRevLett.81.4903

Verschueren, J, Gurrutxaga-Lerma, B, Balint, DS, Sutton, AP and Dini D, *Phys. Rev. Lett.* **121**, 145502 (2018). https://doi.org/10.1103/PhysRevLett.121.145502

Vitek, V, *Phil. Mag.* **18**, 773–86 (1968). https://doi.org/10.1080/14786436808227500

Vitek, V, *Prog. Mater. Sci.* **56**, 577–85 (2011). https://doi.org/10.1016/j.pmatsci.2011.01.002

Vitek, V and Paidar, V, *Dislocations in solids*, ed. FRN Nabarro, Vol. 14, Elsevier: Amsterdam (2008), pp.439–514. ISBN 9780444531667.

Volterra, V, *Annales scientifiques de l'École Normale Supérieure* **24**, 401–517 (1907). http://www.numdam.org/item?id=ASENS_1907_3_24__401_0

Wall, DP, Allen, JE and Molokov, S, *J. Appl. Phys.* **98**, 023304 (2005). https://doi.org/10.1063/1.1924871

Wallace, DC, *Thermodynamics of crystals*, John Wiley & Sons Inc.: New York (1972), section 2. ISBN 9780471918554.

Weertman, J, Lin, I-H and Thomson, R, *Acta Metall.* **31**, 473–82 (1983). https://doi.org/10.1016/0001-6160(83)90035-4

Westlake, DG, *ASM Trans. Q.* **62**, 1000–6 (1969).
Willis, JR, *J. Mech. Phys. Solids* **15**, 151–62 (1967). https://doi.org/10.1016/0022-5096(67)90029-4
Zener, CM, in *Fracturing of metals*, Symposium, American Society for Metals: Cleveland, OH (1948), p.3.

Author index

Subject index